AF577206

UCLA Symposia on Molecular and Cellular Biology, New Series

Series Editor, C. Fred Fox

RECENT TITLES

Volume 50
Interferons as Cell Growth Inhibitors and Antitumor Factors, Robert M. Friedman, Thomas Merigan, and T. Sreevalsan, *Editors*

Volume 51
Molecular Approaches to Developmental Biology, Richard A. Firtel and Eric H. Davidson, *Editors*

Volume 52
Transcriptional Control Mechanisms, Daryl Granner, Michael G. Rosenfeld, and Shing Chang, *Editors*

Volume 53
Progress in Bone Marrow Transplantation, Robert Peter Gale and Richard Champlin, *Editors*

Volume 54
Positive Strand RNA Viruses, Margo A. Brinton and Roland R. Rueckert, *Editors*

Volume 55
Amino Acids in Health and Disease: New Perspectives, Seymour Kaufman, *Editor*

Volume 56
Cellular and Molecular Biology of Tumors and Potential Clinical Applications, John Minna and W. Michael Kuehl, *Editors*

Volume 57
Proteases in Biological Control and Biotechnology, Dennis D. Cunningham and George L. Long, *Editors*

Volume 58
Growth Factors, Tumor Promoters, and Cancer Genes, Nancy H. Colburn, Harold L. Moses, and Eric J. Stanbridge, *Editors*

Volume 59
Chronic Lymphocytic Leukemia: Recent Progress and Future Direction, Robert Peter Gale and Kanti R. Rai, *Editors*

Volume 60
Molecular Paradigms for Eradicating Helminthic Parasites, Austin J. MacInnis, *Editor*

Volume 61
Recent Advances in Leukemia and Lymphoma, Robert Peter Gale and David W. Golde, *Editors*

Volume 62
Plant Gene Systems and Their Biology, Joe L. Key and Lee McIntosh, *Editors*

Volume 63
Plant Membranes: Structure, Function, Biogenesis, Christopher Leaver and Heven Sze, *Editors*

Volume 64
Bacteria–Host Cell Interactions, Marcus A. Horwitz, *Editor*

Volume 65
The Pharmacology and Toxicology of Proteins, John S. Holcenberg and Jeffrey L. Winkelhake, *Editors*

Volume 66
Molecular Biology of Invertebrate Development, John D. O'Connor, *Editor*

Volume 67
Mechanisms of Control of Gene Expression, Bryan Cullen, L. Patrick Gage, M.A.Q. Siddiqui, Anna Marie Skalka, and Herbert Weissbach, *Editors*

Volume 68
Protein Purification: Micro to Macro, Richard Burgess, *Editor*

Volume 69
Protein Structure, Folding, and Design 2, Dale L. Oxender, *Editor*

Volume 70
Hepadna Viruses, William Robinson, Katsuro Koike, and Hans Will, *Editors*

Volume 71
Human Retroviruses, Cancer, and AIDS: Approaches to Prevention and Therapy, Dani Bolognesi, *Editor*

Please contact the publisher for information about previous titles in this series.

UCLA Symposia Board

The Pharmacology and Toxicology of Proteins

The Pharmacology and Toxicology of Proteins

Proceedings of a Cetus-UCLA Symposium Held at
Lake Tahoe, California, February 21–27, 1987

Editors

John S. Holcenberg
Department of Hematology
Children's Hospital of Los Angeles
Los Angeles, California

Jeffrey L. Winkelhake
Department of Pharmacology
Cetus
Emeryville, California

Alan R. Liss, Inc. • New York

Address all Inquiries to the Publisher
Alan R. Liss, Inc., 41 East 11th Street, New York, NY 10003

Printed in the United States of America

Library of Congress Cataloging-in-Publication Data

CETUS-UCLA Symposium (1987 : Lake Tahoe, Calif.)
The pharmacology and toxicology of proteins.

(UCLA symposium on molecular and cellular biology ; new ser., v. 65)
Includes index.
1. Proteins—Therapeutic use—Testing—Congresses.
2. Proteins—Toxicology—Congresses. I. Winkelhake, Jeffrey L. II. Holcenberg, John S., 1935–
III. Cetus Corporation. IV. University of California, Los Angeles. V. Title. VI. Series. [DNLM: 1. Proteins —pharmacodynamics—congresses. W3 U17N new ser. v.65 / QU 55 C423p 1987]
RM666.P87C48 1987 615′.3 87-22654
ISBN 0-8451-2664-4

Contents

Contributors

Alfred A. Amkraut, ALZA Corporation, Palo Alto, CA 94303-0802 [131]

Joffre B. Baker, Department of Biochemistry, University of Kansas, Lawrence, KS 66045 [307]

N.U. Bang, Lilly Research Laboratories, Eli Lilly and Company, Indianapolis, IN 46202 [351]

John W. Baynes, Department of Chemistry, University of South Carolina, Columbia, SC 29208 [59]

Leslie Z. Benet, Department of Pharmacy, University of California, San Francisco, CA 94143-0446 [91]

Christopher D.V. Black, National Cancer Institute, National Institutes of Health, Bethesda, MD 20892 [75]

Emil Bogenmann, Division of Hematology-Oncology, Children's Hospital of Los Angeles, Los Angeles, CA 90027 [325]

Ernest C. Borden, Department of Human Oncology, University of Wisconsin Clinical Cancer Center, Madison, WI 53792 [273]

Jorge Carrasquillo, Clinical Center, National Institutes of Health, Bethesda, MD 20892 [75]

Dennis E. Chenoweth, RLT-02, Route 120 and Wilson Road, Round Lake, IL 60073 [255]

P.C. Comp, Oklahoma Medical Research Foundation, Oklahoma City, OK 73104 [351]

David G. Covell, National Cancer Institute, National Institutes of Health, Bethesda, MD 20892 [75]

Fitz-Roy E. Curry, Department of Human Physiology, University of Nebraska College of Medicine, Omaha, NE 68105-1065; present address: Department of Human Physiology, University of California, Davis, Davis, CA 95616 [23]

Tracy Deinhart, Department of Immunology, Cetus Corporation, Palo Alto, CA 94043 [255]

A.R. Dente, Department of Pharmacodynamics, Medical Research Division of American Cyanamid Company, Pearl River, NY 10965 [337]

Renee R. Eger, National Cancer Institute, National Institutes of Health, Bethesda, MD 20892 [75]

Paul H. Ehrlich, Monoclonal Antibody Department, Sandoz Research Institute, East Hanover, NJ 07936 [199]

The numbers in brackets are the opening page numbers of the contributors' articles.

S.C. Emerick, Indiana University School of Medicine, Indianapolis, IN 46223 **[351]**

C.T. Esmon, Oklahoma Medical Research Foundation, Oklahoma City, OK 73104 **[351]**

N.L. Esmon, Oklahoma Medical Research Foundation, Oklahoma City, OK 73104 **[351]**

John W. Fara, ALZA Corporation, Palo Alto, CA 94303-0802 **[131]**

Bruce D. Gitter, Department of Microbiology and Immunology, Duke University Medical Center, Durham, NC 27710; present address: Eli Lilly and Co., Indianapolis, IN 46202 **[221]**

Carol A. Gloff, Triton Biosciences Inc., Alameda, CA 94501 **[91]**

Murray Goodman, Department of Chemistry, University of California, San Diego, CA 92093 **[113]**

John W. Greiner, Laboratory of Tumor Immunology and Biology, National Cancer Institute, National Institutes of Health, Bethesda, MD 20892 **[287]**

Robert S. Gronke, Department of Biochemistry, University of Kansas, Lawrence, KS 66045 **[307]**

Elliott B. Grossbard, Department of Clinical Research, Genentech, Inc., South San Francisco, CA 94080 **[299]**

Fiorella Guadagni, Laboratory of Tumor Immunology and Biology, National Cancer Institute, National Institutes of Health, Bethesda, MD 20892 **[287]**

Elaine W. Hall, Department of Medicine, Division of Respiratory Medicine, Stanford University, Stanford, CA 94305 **[255]**

K. Elisabeth Harfeldt, Monoclonal Antibody Department, Sandoz Research Institute, East Hanover, NJ 07936 **[199]**

C.S. Harms, Lilly Research Laboratories, Eli Lilly and Company, Indianapolis, IN 46202 **[351]**

Maureane Hoffman, Department of Pathology, Duke University Medical Center, Durham, NC 27710 **[95]**

John S. Holcenberg, Department of Pediatrics, Division of Hematology-Oncology, University of Southern California, Children's Hospital of Los Angeles, Los Angeles, CA 90054 **[xvii, 369]**

V. Holford-Strevens, Department of Immunology, University of Manitoba, Winnipeg, Manitoba, Canada R3E 0W3 **[205]**

Oscar D. Holton III, National Cancer Institute, National Institutes of Health, Bethesda, MD 20892 **[75]**

Allen Honeyman, Department of Microbiology, University of Kansas, Lawrence, KS 66045 **[307]**

L.L. Houston, Department of Physiology and Biophysics, University of Nebraska College of Medicine, Omaha, NE 68105-1065; present address: Cetus Corporation, Emeryville, CA 94608 **[23]**

William J. Hubbard, Tumor Cell Biology/Biological Products, Biotherapeutics, Inc., Franklin, TN 37064 **[221]**

C.A. Huss, Lilly Research Laboratories, Eli Lilly and Company, Indianapolis, IN 46202 **[351]**

Akitoshi Ishizaka, Department of Medicine, Division of Respiratory Medicine, Stanford University, Stanford, CA 94305 **[255]**

C-J. C. Jackson, Department of Immunology, University of Manitoba, Winnipeg, Manitoba, Canada R3E 0W3 **[205]**

Dale E. Johnson, Division of Experimental Toxicology, International Research and Development Corporation, Mattawan, MI 49071 **[165]**

William L. Joyner, Department of Physiology and Biophysics, University of Nebraska College of Medicine, Omaha, NE 68105-1065 **[23]**

James C. Justice, Monoclonal Antibody Department, Sandoz Research Institute, East Hanover, NJ 07936 **[199]**

Andrew M. Keenan, Clinical Center, National Institutes of Health, Bethesda, MD 20892 **[75]**

Manzoor M. Khan, Departments of Medicine and Pharmacology, Stanford University School of Medicine, Stanford, CA 94305 **[113]**

Joanne R. Kopplin, Department of Toxicology, Cetus Corporation, Emeryville, CA 94608 **[161]**

Steven L. Kunkel, Department of Pathology, University of Michigan, School of Medicine, Ann Arbor, MI 48109 **[255]**

R.A. Lanc, Department of Pharmacodynamics, Medical Research Division of American Cyanamid Company, Pearl River, NY 10965 **[337]**

G. Lang, Department of Immunology, University of Manitoba, Winnipeg, Manitoba, Canada R3E 0W3 **[205]**

M. Lanzilotti, Department of Pharmacodynamics, Medical Research Division of American Cyanamid Company, Pearl River, NY 10965 **[337]**

James W. Larrick, Department of Immunology, Cetus Corporation, Palo Alto, CA 94303 **[255]**

Steven M. Larson, Clinical Center, National Institutes of Health, Bethesda, MD 20892 **[75]**

Walter E. Laug, Division of Hematology-Oncology, Children's Hospital of Los Angeles, Los Angeles, CA 90027 **[325]**

J.R. Lawter, Department of Pharmacodynamics, Medical Research Division of American Cyanamid Company, Pearl River, NY 10965 **[337]**

Anne Lewis, National Cancer Institute, National Institutes of Health, Bethesda, MD 20892 **[75]**

Hugh B. Lewis, School of Veterinary Medicine, Purdue University, West Lafayette, IN 47907 **[173]**

G.L. Long, Lilly Research Laboratories, Eli Lilly and Company, Indianapolis, IN 46202 **[351]**

Michael T. Lotze, National Cancer Institute, National Institutes of Health, Bethesda, MD 20892 **[75]**

P.K. Maiti, Department of Immunology, University of Manitoba, Winnipeg, Manitoba, Canada R3E 0W3 **[205]**

C.A. Marks, Lilly Research Laboratories, Eli Lilly and Company, Indianapolis, IN 46202 **[351]**

Frank Martin, Liposome Technology, Inc., Menlo Park, CA 94025 **[149]**

L.E. Mattler, Lilly Research Laboratories, Eli Lilly and Company, Indianapolis, IN 46202 **[351]**

Janet L. Maxwell, Department of Chemistry, University of South Carolina, Columbia, SC 29208 **[59]**

Michael McGrogan, Invitron Corporation, Redwood City, CA 94063 **[307]**

W.E. McWilliams, Department of Pharmacodynamics, Medical Research Division of American Cyanamid Company, Pearl River, NY 10965 **[337]**

Kenneth L. Melmon, Departments of Medicine and Pharmacology, Stanford University School of Medicine, Stanford, CA 94305 **[113]**

Geraldine P.G. Miller, Department of Medicine, Program in Infectious Diseases and Clinical Microbiology, University of Texas Medical School at Houston, Houston, TX 77030 **[187]**

Zeinab A. Moustafa, Monoclonal Antibody Department, Sandoz Research Institute, East Hanover, NJ 07936 **[199]**

James L. Mulshine, National Cancer Institute, National Institutes of Health, Bethesda, MD 20892 **[75]**

H. Murayama, Indiana University School of Medicine, Indianapolis, IN 46223 **[351]**

Laura J. Nell, Department of Medicine, Baylor College of Medicine, Houston, TX 77030 **[187]**

Kirstin C. Nichols, ALZA Corporation, Palo Alto, CA 94303-0802 **[131]**

Gabriela Nicolau, Department of Pharmacodynamics, Medical Research Division of American Cyanamid Company, Pearl River, NY 10965 **[337]**

Lars Östberg, Monoclonal Antibody Department, Sandoz Research Institute, East Hanover, NJ 07936 **[199]**

Robert J. Parker, National Cancer Institute, National Institutes of Health, Bethesda, MD 20892 **[75]**

Salvatore V. Pizzo, Department of Pathology, Duke University Medical Center, Durham, NC 27710 **[95]**

Matthew Pollack, Department of Medicine, Uniformed Services University School of Medicine, Bethesda, MD 20814 **[239]**

Thomas A. Raffin, Department of Medicine, Division of Respiratory Medicine, Stanford University, Stanford, CA 94305 **[255]**

Nigel P. Ray, ALZA Corporation, Palo Alto, CA 94303-0802 **[131]**

Joan H. Schiller, Department of Human Oncology, University of Wisconsin Clinical Cancer Center, Madison, WI 53792 **[273]**

Jeffrey Schlom, Laboratory of Tumor Immunology and Biology, National Cancer Institute, National Institutes of Health, Bethesda, MD 20892 **[287]**

Randy W. Scott, Invitron Corporation, Redwood City, CA 94063 **[307]**

A.H. Sehon, Department of Immunology, University of Manitoba, Winnipeg, Manitoba, Canada R3E 0W3 **[205]**

B.M. Silber, Department of Pharmacodynamics, Medical Research Division of American Cyanamid Company, Pearl River, NY 10965 **[337]**

Christian C. Simonsen, Invitron Corporation, Redwood City, CA 94063 **[307]**

Gregory T. Stelzer, Biotherapeutics, Inc., Franklin, TN 37064; present address: Center for Clinical Studies, International Clinical Laboratories, Nashville, TN 37202 **[221]**

Kenton E. Stephens, Department of Medicine, Division of Respiratory Medicine, Stanford University, Stanford, CA 94305 **[255]**

James P. Tam, Rockefeller University, New York, NY 10021 **[7]**

J.W. Thomas, Department of Medicine, Baylor College of Medicine, Houston, TX 77030 **[187]**

Suzanne R. Thorpe, Department of Chemistry, University of South Carolina, Columbia, SC 29208 **[59]**

Alfred P. Tonelli, Department of Pharmacodynamics, Medical Research Division of American Cyanamid Company, Pearl River, NY 10965 **[337]**

R. Reid Townsend, Department of Biology, Johns Hopkins University, Baltimore, MD 21218 **[41]**

Karen Toy, Department of Immunology, Cetus Corporation, Palo Alto, CA 94043 **[255]**

Jeff Wang, Department of Immunology, Cetus Corporation, Palo Alto, CA 94043 **[255]**

John N. Weinstein, National Cancer Institute, National Institutes of Health, Bethesda, MD 20892 **[75]**

I. Wilkinson, Department of Immunology, University of Manitoba, Winnipeg, Manitoba, Canada R3E 0W3 **[205]**

Jeffrey L. Winkelhake, Department of Pharmacology, Cetus Corporation, Emeryville, CA 94608 **[xvii, 3]**

A. Yacobi, Department of Pharmacodynamics, Medical Research Division of American Cyanamid Company, Pearl River, NY 10965 **[337]**

S.B. Yan, Lilly Research Laboratories, Eli Lilly and Company, Indianapolis, IN 46202 **[351]**

John R. Yannelli, Biotherapeutics, Inc., Franklin, TN 37064 **[221]**

Annie Yau-Young, Liposome Technology, Inc., Menlo Park, CA 94025 **[149]**

Preface

The order of topics in this book is the same as the order of presentations at the Cetus-UCLA Symposium, The Pharmacology and Toxicology of Proteins, which was held at Lake Tahoe, California, February 21–27, 1987. Each section has been preceded by a brief summary of the presentations including those of investigators who were unable to submit chapters.

This book begins with presentations of several examples of current protein and peptide drugs to illustrate the state of the art in defining the targets of these proteins and the use of techniques of genetic engineering to improve their biologic and pharmacologic properties. Next, the unique pharmacologic properties of proteins are presented with examples of receptor binding and transport so as to access membranes and the microvasculature. Three sections provide discussion of the pharmacokinetics of native, or chemically modified proteins and peptides, and controlled release dosage forms. Then, several chapters illustrate immune responses, the development of tolerance to these foreign materials and the toxicology of proteins.

Following this basic information, two sections are presented to describe the clinical pharmacology of proteins and peptides including monoclonal antibodies, interferons, interleukin-2, atriopeptin and protease inhibitors. Finally, the book concludes with two current challenges: 1) how to combine protein therapy with other treatments, and 2) to determine if exogenously administered enzymes can enter specific cells and intracellular compartments in sufficient amounts to treat genetic disorders.

We hope this book will provide the reader with a summary of the unique features and problems that proteins and peptides present as drugs, the current uses of these agents, and a glimpse of their future applications.

We gratefully acknowledge Cetus Corporation for the sponsorship of this meeting. Additional gifts were received from: Enzon, Inc.; Genentech, Inc.; ALZA Corporation; Riker Laboratories/3M; Biotherapeutics, Inc.; International Research and Development Corporation; Boehringer Mannheim GmbH; Bend Research, Inc.; Bristol-Myers Company, Pharmaceutical Research & Development Division; DNAX Research Institute; and The Squibb Institute.

John S. Holcenberg
Jeffrey L. Winkelhake

Preface

SECTION I: PROTEINS, PEPTIDES, AND THE PHARMACOLOGIST: STATE OF THE ART

This section begins with a preamble by J. Winkelhake that illustrates how specific proteins can act as pure signals to elicit specific responses and also help define complex homostatic systems. Next, D. Mark of Cetus Corporation discussed their current use of site specific mutagenesis to improve the properties of fibroblast interferon, tumor necrosis factor and interleukin-2 (IL-2) expressed in *E coli*. An example was the removal of one of three cysteine residues in interferon so that only the correct disulfide bridge will form upon reoxidation and folding of the protein. J. Tam of Rockfeller University presented structure activity relationships of transforming growth factor and epidermal growth factor. Various modifications of these structures were made by solid-phase peptide synthesis and tested for biologic and biochemical properties. All three disulfide bonds are required but some shortened analogues have activity. J. Farrar discussed the program he directs at Hoffman La Roche Company to define the receptor binding of IL-2 and develop a screen for testing natural and synthetic compounds that can specifically block this binding. IL-2 was found to interact with P55 (TAC) and P70 glycoproteins on T-cells. P55 has been overexpressed in CHO cells, the structure of this complex has been modeled and blocking antibodies developed. A synthetic organic compound of 300 daltons (RO22-0680) has been discovered which can specifically block IL-2 binding. This work illustrates that smaller molecules may be found that can mimic some of the actions of large proteins.

This session ended with a presentation by T. Deuel, Washington University, on platelet derived growth factor (PDGF) and its effects on normal and transformed cells. The human factor is a heterodimer while the v-sis and c-sis products and PDGF from certain tumors are homodimers. He raised the possibility that some of the effects of v-sis on infected cells may not require excretion into the media and binding to an extracellular domain of a receptor. Intracellular actions would preclude blockade by surface acting drugs.

The Pharmacology and Toxicology of Proteins, pages 3–5

PREAMBLE

Jeffrey L. Winkelhake
Department of Pharmacology
Cetus Corporation, Emeryville, CA 94608

Classical models for studying pharmacologic aspects of xenobiotics in animals often require significant modifications to be useful in answering pharmacologic questions about biological "drugs". There are two key driving forces behind the need for these modifications. First, is the need to move away from "drug discovery" to "model discovery"; that is, traditionally animal models for neoplastic, infectious or genetic diseases are established and thousands of compounds are screened for potential efficacy prior to clinical development. With the new biotechnology-based drugs, time, cost and uncertainties about their real modes of action provide a much more limited number of "products" (in the dozens). Coupled with this is a keen desire to show that these novel protein drugs actually work in ways conceived by test tube experiments or in the minds of the developers. Secondly, many of the new protein/peptide drugs call upon the host's biology as an intricate component of the drug's ultimate action - thus host cellular responses are modifed and the products electing these change require different types of animal models than do biochemical modifiers.

With this caveat, the challenges and complexities encountered in the therapeutic use of protein/peptide drugs will be discussed as the basic format for this meeting. What we hope will emerge are "first principles" of protein pharmacology, and since drug delivery always relates to potential therapeutic efficacy, key issues we will cover at this first meeting on the Pharmacology and Toxicology of Proteins relate to pharmacokinetics and the problems of protein drug delivery.

The concept that the new wave of proteotherapeutics are generally designed to "turn on/off" rather than poison systems is not new. In the cancer field, this idea was last tried without a lot of successes with BCG therapy in the 1970s. There is a key difference, however, between such an "immunostimulatory" drugs and the pure, genetically engineered proteins in testing today. The salient features of that difference can be exemplified by analogy with the electrical engineer's "black box".

If we consider the body as a black box with sophisticated circuitry and (biological) systems, there is a way that the engineer can determine the contents and operations of the electronic analog without opening it; namely, by in-putting a signal and observing the response(s) from the box. In all likelihood, there will be more than one output (for our purpose, several toxic effects and several beneficial effects (see Fig. 1)). If the engineer had in-put white noise, he would probably see every signal the box could make. But if the original input was a pure signal, a limited output would ensue. Next, the engineer changes or modulates the signal intensity, schedule, and may sequence it with another pure signal. By making systematic changes, in no time at all the engineer can tell exactly what the black box is designed to do and he can orchestrate the box's function. By analogy, BCG is white noise and synthetic peptides/recombinant proteins are pure signals. BCG (white noise immunotherapy) did not allow us to either figure out how the black box works or to orchestrate it's systems to our (therapeutic) advantage. The salient feature of the new proteinacious therapeutics then is that they allow us to begin to systematically perturb homeostatic mechanisms such as the immune system - hopefully to our advantage.

THE BLACK BOX RESPONSE MODIFIER ANALOGY

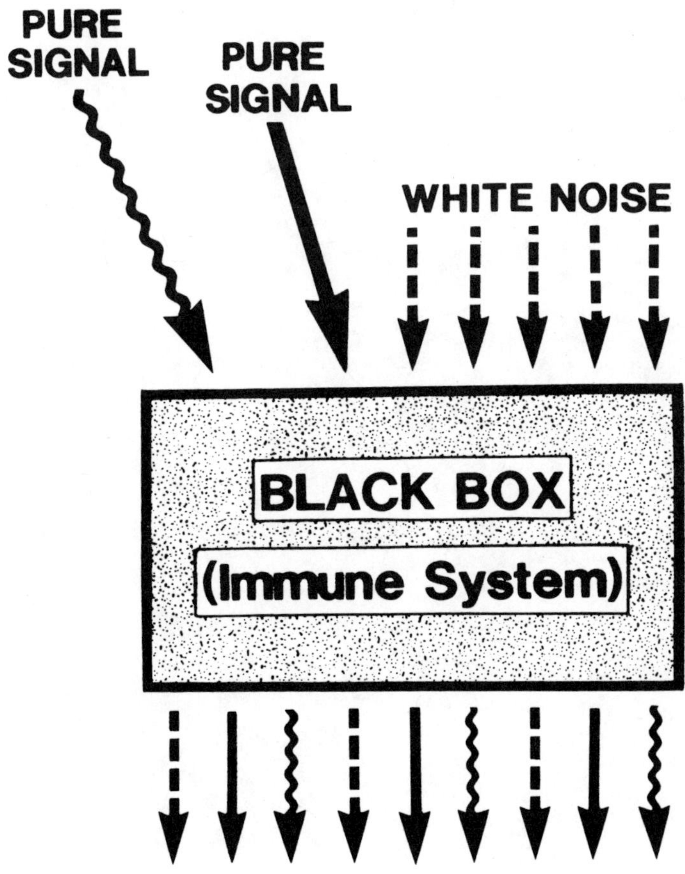

Fig. 1

The Pharmacology and Toxicology of Proteins, pages 7–20

PEPTIDE SYNTHESIS: STRUCTURE-ACTIVITY RELATIONSHIPS OF TRANSFORMING GROWTH FACTOR TYPE ALPHA[1]

James P. Tam

The Rockefeller University, 1230 York Ave., New York, N.Y.10021

ABSTRACT Transforming growth factor type α (TGFα) and epidermal growth factor (EGF) are small mitogenic proteins that share considerable similarities in sequence homology and structural motif. While EGF is strongly associated with normal physiological functions, TGFα has been found to be closely associated with oncofetal states. To understand the relationships between the chemical structures of these growth factors and their biological properties, we used solid-phase peptide synthesis to produce these growth factors, their structural variants, and shortened analogs. Our results showed that full biological activity of these growth factors required a tertiary structure maintained by the three disulfide pairs. Shortened analogs or extended structures without the disulfide-restraint, greatly reduced biological activity. However, several shortened analogs showed partial agonist activity and may provide us with useful leads to the design of therapeutic useful agents.

INTRODUCTION

Transforming growth factor type α (TGFα, 1-4)[2] and epidermal growth factor

[1]This work was supported by PHS Grant number CA36544, awarded by the National Cancer Institute, and DHHS.

[2] Abbreviations follow the tentative rules of the IUPAC-IUB commission on Biochemical Nomenclature, published in J. Biol. Chem. (1972) 247,979-982. Others: Boc, tertbutyloxycarbonyl, Bzl, benzyl, DCC, dicyclohexylcarbodi-imide, DMF, N,N-dimethylformamide, DMS, dimethylsulfide EGF, epidermal growth factor, HF, hydrogen fluoride HOAc, acetic acid, HPLC, high-performance liquid chromatography, HOBt, 1-hydoxybenzotriazole, TGFα, transforming growth factor type α, TFA, trifluoroacetic acid, TFMSA, trifluoromethanesulfonic acid.

TABLE 1
TGFα-EGF AND SEQUENCE HOMOLOGS IN PROTEINS

Sequence	Occurrence/Origin	Ref
EGF	normal tissues	
	mouse	5
	humnan	6
TGFα	transformed cells and tissues	
	rat, mouse	2,3
	human	4
DNA-virus gene product		
	19 KD protein, Vaccinia	42
	Shope fibroma virus	44
	Myxoma virus	45
Blood clotting factor		
	Factor IX and X	42,43
	Protein C, S and Z	46,47
	plasminogen activator	42
Extracellular matrix		
	laminin	42
Embryonal protein		
	Lin-12 (nematode)	48
	notch locus (Drosophilia)	49
Protein precursor		
	EGF precursor	50
	LDL receptor	51

(EGF, 5-6) are mitogenic protein growth factors of about 50 residues. Because they are also the smallest known protein growth factors, they are particularly suitable for the synthetic approach to study the relationships between the chemical structures and their biological activities.

Structurally, the TGFα-EGF family is distinguished by the placement of their six cysteinyl residues that give a general consensus formula of Cys X_6 Cys X_5 Cys X_{10} Cys X Cys X_8 Cys (where X denotes other amino acids). The first four cysteinyl residues form a fused ring of first and second loop (A- and B-loop) and the last two cysteinyl residues form the third loop (the C-loop, Fig. 1). However, from sequence comparisons, homologous sequences of the EGF-TGFα family are also present as domains in a diverse group of proteins including blood coagulation factors, complements, tissue plasminogen activiator, DNA-virus gene products, and extracellular matrix. (Table I). Both EGF and TGF apparently interact with a single class of cell surface receptor *in vitro* (7) in eliciting similar biological responses such as stimulation of tyrosine-specific protein kinase and induction of anchorage-independent growth of indicator cells in soft agar. However, they are clearly different in origin, biosynthesis, and occurrence.

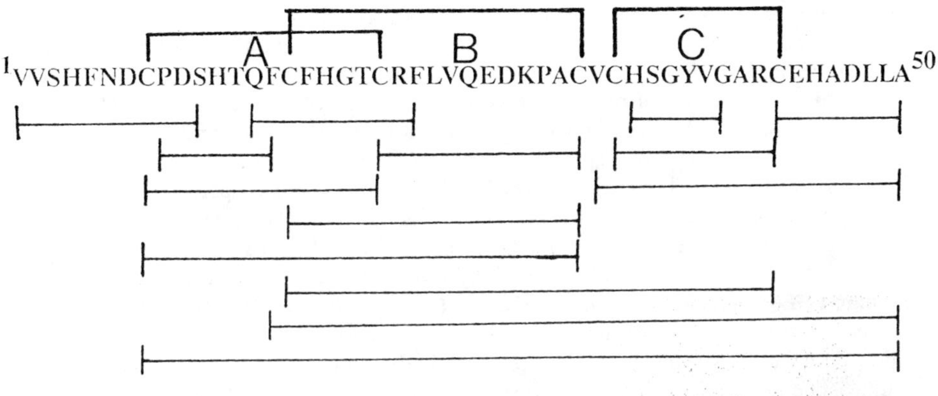

FIGURE 1. Structure of human TGFα. Synthetic fragments for the structure-activity study are shown in lines.

Whereas EGF is derived from a precursor protein containing more than 1,200 amino acids, TGFα is from a precursor protein of 160 amino acids (4). Furthermore, cleavage of EGF from its precusor is likely by a protease with the specificity of trypsin, the cleavage of TGFα from its precusor may require a unique protease with the specificity to cleave between alanine and valinylvaline. More significantly, EGF is a mitogen found in normal physiological states, but the occurrence of TGFα has been known to be closely associated with oncofetal states (8-10). Because of this important difference, the structure-activity study of TGFα may provide useful leads for the design of new therapeutic agents for the disruption of tumor.

RESULTS AND DISCUSSION

Peptide Synthesis

Recent improvements in many aspects of the solid-phase methodologies have generally facilitated the synthesis of a complex peptide such as TGFα (11-12). The synthesis of murine and human EGF, rat and human TGFα has been achieved (13-18). In general, it is important to attend to every detail of each synthesis and devised plans to avoid predictable side reactions. For example, aggregation of peptide chain is a potential but a serious problem in the synthesis of a long and complex peptide, which leads to incompletion of coupling. To ensure that the coupling will go to completion, the coupling is, at times, performed at elevated temperature (19). Furthermore, it is equally important to adopt the mildest possible means to remove the protecting groups at the completion of each synthesis so that the integrity of the peptide chain can be preserved. A key improvement in the deprotection method is the introduction of the low-high HF deprotection procedure that allows an efficient and mild condition for the removal of all protecting groups (20-21). The low-high HF procedure consists of two steps in a one-pot reaction. Furthermore under the low-HF condition, the N^1-formyl of Trp and sulfoxides of protected cysteine and methionine are smoothly converted to their deprotected forms in one single operation. Thus, the low-high HF procedure combines three conventional steps (HF-cleavage, base treatment of Trp(For) and aqueous thiol reduction of Met(O)) into a one-step reaction. Using this procedure, the syntheses of fully biologically active murine EGF, rat and human TGFα were achieved in good yields (13-16). Because the low-high HF procedure provids an efficient method for the deprotection of cysteinyl-containing peptides, it was used for the synthesis of human TGFα analogs for this structure-activity study. The synthetic strategy of TGFα is briefly discussed to emphasize recent improvements.

(i) Resins. The improved, trifluoroacetic-acid stable p-phenylacetamidomethyl (Pam) resin (22) was used for the preparation of TGFα and its analogs. The Pam-resin offers superior performance than the usual chloromethyl resin. It reduces considerable side reactions and losses of peptide chains during the synthesis. Another resin that has the acid-stable property of Pam-resin is the benzhydrylamine resin that provides a peptide with an α-carboxamide. However, benzhydrylamine resin usually requires very strong acid such as HF for its cleavage from the resin support. A more versatile benzhydrylamine resin that is recently introduced is the p-acyloxybenzhydrylamine resin (Aba) resin (22-23). The Aba-resin is one of the multidetachable resins that allow flexibility in their modes of detachment from the resin support. It can be used in the gradative deprotection strategy (22) which allows the use the trifluoromethanesulfonic acid-trifluoroacetic acid-dimethylsulfide cleavage method in the reaction vessel and avoids the use of HF.

(ii) Protecting group strategy. A combination of tertbutoxycarbonyl for the N^{α}- and benzyl alcohol derivatives for the side chain protecting group strategy was used in most syntheses. The Boc-benzyl strategy is convenient to use and the starting protected amino acids are readily available in high quality from commercial source. To minimize base-catalyzed imide formation during the repetitive base neutralization steps, cyclohexyl protecting group for the side chain protection of Asp is also used (25).

(iii) Coupling strategy. A mixed-mode strategy was used, i.e. symmetrical anhydride in CH_2Cl_2 for the first coupling and dicyclohexylcarbodiimide/1-hydroxybenzotriazole in dimethylformamide for the second coupling. This coupling procedure is suitable for most amino acids except Boc-Asn, Boc-Gln, Boc-Arg(Tos), and Boc-Gly. For Boc-Asn and Boc-Gln, the DCC/HOBt method was used to prevent dehydration of the side chain amide during the activation step. For Boc-Arg(Tos) and Boc-Gly only DCC in CH_2Cl_2 will be used to prevent respectively, lactam formation and dipeptide rearrangement. For the coupling to N^{α}-amino glutamic acid or glutamine residue, activated amino acids (symmetrical anhydride form for most amino acids) in dimethylformamide was used to prevent pyroglutamic acid formation (26).

A major problem in the coupling step for the assembly of a complex peptide is the occurrence of one or several steps during the chain assembly to go into completion. The mechanism of such a slow coupling rate remains unknown but could be explained by the physical aggregation of a peptide chain. A method to overcome this difficulty has been found by coupling of an activated amino acid at elevated temperature in a dipolar aprotic solvent such as N-methylpyrrolidinone to force the coupling reaction to completion (19). To monitor the progress of the coupling reaction, a quantitative ninhydrin monitoring method was employed (27). This monitoring method is sensitive to detect about 0.05% of the uncoupled N^{α}-amino group.

(iv) Low-high HF deprotection. One of the most difficult and persistent obstacle for the synthesis of a long and complex peptide has been the inability to eliminate modification products due to alkylation side reactions occurring during the strong acid cleavage step (20). Under the conventional HF deprotection condition (28), the strong acidity of HF promotes and stabilizes carbocations which tend to alkylate amino acids such as Trp, Tyr, Cys, and Met to form modification products. It also leads to dehydration of the side chain carboxylic acid of glutamic and aspartic acid. The low-high HF deprotection procedure (20-21) was designed to minimize these modification products and would provide a mild and effective deprotection procedure for the synthesis of a growth factor (14-16).

The low-high HF cleavage involves operationally two distinct steps but for all practical purposes, is always carried out as a one-pot reaction. The low HF step contains a 1 to 1 molar complex of HF and dimethylsulfide (DMS) and in volume ratio a mixture of about 25%HF-75%DMS. Usually, aromatic scavengers such as p-cresol and thiocresol are added to effect better swelling of the resin and to serve as scavengers. Because the acidity of the low HF condition is drastically lower than the conventional HF condition, many of the side reactions encountered in the high acidity of the anhydrous HF cleavage reaction is eliminated. Furthermore, the low HF condition removes protecting groups by a S_N2 mechanism whereby the benzyl moieties of the side chains are removed as the stable sulfonium salts and thus avoids the alkylation side reactions of Trp, Tyr, Cys, and Met. Probably the most important contribution of the low-high HF deprotection in the synthesis of Met or Cys-rich peptide such as the growth factor is its ability to reduce methionine sulfoxide or cysteinyl sulfoxide to its respective sulfide.

Two amino acid residues: Cys and Arg, pose the greatest problems in the present scheme. This is in part due to the lack of satisfactory protecting groups for these residues that can be removed under mild low-HF condition. A compromise is a two step cleavage reaction. All oxygen-linked benzyl protecting groups are removed under the low-HF step and the resulting sulfur-linked protecting groups of Cys and Arg are removed under highly acidic conditions, S_N1, high-HF step. However, since most of the benzyl-based carbocation source that causes modification products has been removed under the low-HF condition, the subsequent high-HF treatment is found to be less destructive to the peptide chain.

Refolding, renaturation, and purification of TGFα

The family of EGF-TGFα growth factor is both heat and acid-stable and has been calculated to be one of the most energetically stable-proteins (29). Indeed both EGF and TGFα only starts to unfold in extreme denaturing conditions of 7 M urea or 6 M guanidine hydrochloride (Gn.HCl). More importantly, the complete denatured and reduced TGFα or EGF could be restored with full biological activity by removal of the thiol and denaturant and subsequent re-oxidation by air.

Based on these observations, a purification and refolding scheme was designed for the purification and refolding of synthetic TGF First, the crude synthetic TGFα after the low-high HF cleavage step was directly extracted from the resin into a strongly denaturating and reducing solution containing 8 M urea (or 7.2 M guanidine hydrochloride) and 0.2 M dithiothreitol at pH 8.0. Under such a condition, the synthetic product such as TGFα that contains multiple cysteine would be denatured and dissociated from the aromatic scavengers such as p-cresol and p-thiocresol to facilitate dissolution and extraction. Furthermore, the extracted solution was then extensively dialyzed in the same condition to exclude the small peptides and organic molecules.

Secondly, the concentration of the denaturant was sequentially lowered to allow refolding and at the concentration 2-3 M urea (or 2 M Gn.HCl) a mixture of reduced and oxidized glutathione was added to form the disulfide linkages of the synthetic TGFα. The refolding and oxidation under a mild denaturating condition by the mixed disulfide method allows the folding and refolding to arrive at the most thermodynamic stable, presumably the properly refolded, TGFα molecule. In fact, this rationale was found to be correct and TGFα was obtained in a highly yield. Thus, our refolding strategy departs from the conventional method that uses extensive chromatography to purify the reduced synthetic proteins under denaturing conditions prior to the refolding the molecule in nondenaturing condition.

Finally, our refolding strategy also allowed us to simplify the purification procedure. The refolded TGFα was purified in one-step by the C_{18}-reverse phase chromatography. The correctly folded TGFα is expected to be most compact and will elute early in the chromatography. Indeed, this was our finding and very little difficulties were encountered in purifying the correctly folded TGFα from the impurities.

Characterization of TGFα and other bioactive TGFα-like compounds

After purification of the synthetic products by C_{18} or C_8-reverse-phase HPLC and showing that it was homogenous in the reverse phase system, it must also be shown by additional criteria that a single molecular species has been obtained and demonstrated that it is the desired target molecule. The requirements for the proof of homogeneity and identity of solid-phase synthetic products with the desired target peptide, resembled rather closely those that have been laid down for the characterization of proteins. At least four independent analytical test should show uniformity. A good amino acid analysis, though not discriminating, is a necessary requirement for homogeneity of a peptide. Thus, amino acid analysis by 6 N HCl and enzymatic hydrolysis such as pronase, was performed on all synthetic products. Since the integrity of cysteine is important for this synthesis, quantitative measurements of cysteine will be performed independently by two procedures. The free sulfhydryl groups were determined by 5,5'-dithiobis-(2-nitrobenzoic acid) in 8 M urea, pH 8.0 sodium phosphate buffer and by performic acid oxidation to cysteic acid which was quantitated by amino acid analysis.

The availabilities of several high performances of ion-exchange coloumns also make it possible to use these high resolving columns as analytical tools. Since ion exchange chromatography operates mainly on the basis of charge, it will complement well with the products that are obtained mainly through C_{18} reverse-phase HPLC. Furthermore, we had also examined the enzymatic digestion fragments of the synthetic products (by thermolysin at 45° C) using the high performance ion-exchange chromatography (also C_{18} reverse phase HPLC) to determine the disulfide pairing as well as to verify the predicted homogeneity of the product.

A powerful method to show the proof of identity has been from the fission-fragment mass spectrometry (31-32). Synthetic TGFα (both human and rat sequences) has been analyzed by Cf-252 fission ionization mass spectrometry and the molecular ion $(M+H)^{+}$ can be clearly identified. It was found to correspond closely to the calculated mean value m/z molecular weight. The resolution of the fission instrument at this mass ion range of about 6,000 was about a 2 unit mass. Furthermore, strong peaks corresponding to the $(M+2H)^{++}$ and $(M+3H)^{+++}$ can also be identified in the range of 3,000 and 2,000 mass unit (in the case of TGFα). Under such circumstances, the calculated and theoretical dimeric form of TGFα $(2M+3H)^{+++}$ would be at 3,700 mass unit. These analyses provided both strong evidence that the synthetic material was monomeric and unambiguous proof for the identity of the synthetic growth factor.

Biological Activities of Synthetic TGFα

Synthetic TGFα were found to be in general as active as EGF in stimulation of anchorage-independent growth of normal rat kindney fibrolasts in culture, inhibition of histamine-stimulated gastric acid secretion, and stimulation of plasminogen activator production (33-38). The primary *in vitro* assays for TGFα have been the mitogenic and radioreceptor assay for the competition of ^{125}I-EGF for receptor-binding in formalin-fixed A431 cells. Since a good-correlation between mitogenicity and the radioreceptor assay were, the latter assay was used for screening of biological activities in the present study.

Tertiary Structure and Activity

What role can the tertiary structure of a compact but small protein be expected to play in biological activity? Since biological activities are highly sensitive to the unique tertiary folding of a protein, it is expected that the synthetic molecules other than that of the correctly "folded" one will be inactive. However, since the binding region of a protein to its receptor is usually confined to a small cluster of determinants, the incorrectly folded molecules may still contain the necessary determinants for activity. It appeared that TGFα was ideally suited to test these questions.

Three forms of rat TGFα were used: the correctly folded TGFα, a S-glutathione conjugated TGFα (GS-STGF and a TGFα dimer whose disulfide arrangements were not established. These three forms were easily distinguished from each other in the gel permeation chromatography and C_{18}reversed-phase HPLC. In P-10 chromatography, the dimer and TGFα were eluted off as defined peaks at the lysozyme and insulin marker respectively, and the GS-STGFα eluted between these two forms. The elution profile in C_{18} reversed-phase HPLC using CH_3CN as the mobile phase reflected the compactness of these compounds with TGFα being the first to be eluted at about 26% CH_3CN, followed by GS-STGFα at 33% CH_3CN and finally the dimer at about 40% CH_3CN.

These different forms were tested for their ability to bind and activate the receptor-tyrosine kinase. TGFα which exhibited similar binding activity as EGF was found to activate the receptor-protein kinase to phosphorylate exogenous substrates such as angiotensin. When compared with TGFα, the dimeric TGFα showed considerable binding activity (1% of TGFα) to the EGF-receptor, but was essentially devoided of any activity to activate the receptor-tyrosine protein kinase prepared from the A431 cell membrane vesicles. Similarly, the extended form of GS-STGFα also showed weak binding activity (0.1% of TGFα), but unlike the dimeric form, the GS-STGFα was an inhibitor of the receptor-kinase at μmolar concentrations. A plausible explanation is that the GS-STGFα which is rich in carboxylates (each glutathione adduct will provide an addition net carboxyl charge) may serve as a competitive substrate to produce the observed apparent inhibition.

Here, we can conclude that the determinants of TGFα for the binding to its receptor is likely to be limited to a small cluster(s) of amino acids since the extended or the dimeric form of TGFα has a limited ability to bind to the receptor. However, to activate the receptor-protein kinase for biological activities and signal transduction, a correct tertiary structure is necessary.

Structure-Activity Study of human TGFα

Structurally, the TGFα-EGF family is distinguished by the placement of their six cysteinyl residues that form the three disulfide loops. The first four cysteinyl residues form a fused ring of the first and second loop (A- and B-loop) and the last two cysteinyl residues form the third loop (the C-loop, Fig. 1). For the structure-activity studies, overlapping linear peptides containing 7 to 17 residues covering the whole molecule of TGFα (residues 1 to 50, TGFα-peptides:[1-9], [9-16], [16-21], [8-21], [21-32], [16-32], [34-43] and [34-50]) were synthesized and tested for their radioreceptor binding activities in A431 cells. None of these linear analogs were found to show significant radioreceptor competing activity at concentrations lower than 1 mM.

TABLE 2
STRUCTURE-ACTIVITY STUDY OF HUMAN TGFα

TGFα	Disulfide loop	Binding A431 IC_{50} (μM)[a]
H-(1-50)-OH	TGFα	0.01
H-(8-50)-OH	[ABC-loop]	0.03
H-(16-50)-OH	[BC-loop]	300
H-(34-50)-OH	[C-loop]	>10,000
H-(8-43)-OH	[ABC-loop]	0.1
H-(8-32)-OH	[AB-loop]	300
H-(16-43)-OH	[BC-loop]	400
H-(8-21)-NH_2	[A-loop]	4,000
H-(16-32)-NH_2	[B-loop]	>10,000
H-(34-43)-NH_2	[C-loop]	>10,000

[a] IC_{50} inhibition of binding of 3 nM I^{125}-EGF to A431 cells.

Since the linear synthetic fragments of TGFα were found to be inactive, our attention was then turned to frangments that contain disulfide linkages. These include synthetic fragments: (1) with only one disulfide linkages such as loop A (residues are represented in parenthesis, 8-21), B (16-32) and C (34-43); (2) with two disulfide linkages: loop AB (8-32) and BC (16-43); and (3) with three disulfide linkages: loop ABC (8-50). Synthetic peptides of loops B or C showed no significant competing activities in the EGF-radioreceptor assay even in high concentrations (10^5 fold > TGFα). However, synthetic peptides the A-loop was found to be more promising and showed weak competing activity with I^{125}-EGF at submicromolar level (Table 2). Peptides containing two disulfide linkages such as loop AB and BC showed strong activities at range of 50-400 μM and at an effective concentration of of 0.01% of TGFα. The shortened TGFα analog containing all three disulfide loops, TGFα-(8-50), was found to be 10 to 30% as active as TGFα. Taken together, these results suggest the the spatial arrangements of different groups defined by the tricyclic structure of TGFα are responsible for its potency but a fragment containing loop A may have the necessary determinants for biological activities.

Previous structure-activity studies by others on EGF and rat TGFα have shown two regions that exhibit low but distinct activities: portion of B-loop in murine EGF (residue 22-31) and C-loop of rat TGFα (39-40). These corresponding fragments in human TGFα were found to be inactive in both the receptor binding and the mitogen assays at a concentration of 1 mM. However, it is necessary to point out that the EGF and TGFα B-loop fragment share little common sequence homology.

TABLE 3
ANTIGENICITY OF hTGFα ANTIBODIES

Loop	Half-maximal Response (nM)[a]	
	Pab	Mab
TGFα	6	80
[ABC-loop]	8	80
[BC-loop]	300	95
[AB-loop]	1,000	120
[B-loop]	3,000	100
[A-loop]	>500,000	>10,000
[C-loop]	>500,000	>10,000

[a] solid-phase immunoabsorbent assay, Pab, polyclonal antibody, Mab, monoclonal antibody

Antigenic Site of TGFα

The development of antibodies against TGFα would provide an possibility to obtain information concerning the tertiary structure via the internal imaging of the antigen-antibody interactions. Of the three disulfide loops of TGFα and EGF (A, B and C, Fig. 1), B-loop represents the longest of the three with 17 residues. Previous work by others have shown that the antigenic site of polyclonal antibody raised against EGF is confined to a portion of B-loop (39). Recently, we have also raised polyclonal and monoclonal antibodies against human and rat TGFα. Using synthetic peptide fragments covering the entire molecule of TGFα to map the antigenic site, we found that the major immunodominant epitope of all TGFα-specific antibodies was confined to B-loop. It has been demonstrated that the segmental motility of a protein correlates with the antigenicity (40). Thus, it is not surprising that immunodominant epitope lies within that the longest loop of TGFα (B-loop) that has the greatest segmental mobility. Thus, our data show that B-loop is the most exposed and flexible loop of the TGFα-EGF family. Interestingly, B-loop is also found to be the least homologous and longest of all three loops in other members of the TGFα-EGF family, which includes domain portions of a diverse group of proteins. Thus, B-loop provides a more suitable candidate for the development of a specific antiserum using a synthetic peptide than the region containing the C-loop as previously suggested (39). Antibody specific for the B-loop will be most desirable and will greatly minimize the possibility of cross-reaction to EGF. Indeed, our results confirmed that the polyclonal and monclonal antibodies specific to the B-loop did not show cross-reactivity with EGF even at a high concentration.

Conclusion

Potent biological activity of human TGFα requires multiple disulfide restraint but weak receptor binding activity is observed from the A-loop alone. The major immunodominant epitope, however, is contained in the B-loop. Further refinements of the A-loop may lead to increase potency.

REFERENCES

1. DeLarco JE, Todaro, GJ (1978). Growth factors from murine sarcoma virus-transformed cells. Proc Natl Acad Sci 75:4001.
2. Marquardt H, Hunkapiller MW, Hood LE, Twardzik DR, De Larco JR, Todaro GJ (1983). Transforming growth factors produced by retrovirus-transformed rodent fibroblast and human melanoma cells: Amino acid sequence homology with epidermal growth factor. Proc Natl Acad Sci 80:4684.
3. Massague J (1983). Epidermal growth factor-like transforming growth factor I. Isolation, chemical characterization and potentiation by other transforming growth factors from feline sarcoma virus-transformed rat cells. J Biol Chem 258:13606.
4. Derynck R, Roberts A B, Winkler M E, Chen EY, Goeddel DV (1984). Human transforming growth factor-α precursor structure and expression in E. coli. Cell 38:297.
5. Savage CR Jr, Inagami J, Cohen S (1972). The primary structure of epidermal growth factor. J Biol Chem 247:7612.
6. Gregory H (1975). Isolation and structure of urogastrone and its relationship to epidermal growth factor. Nature 257:325.
7. Todaro GJ, Fryling C, De Larco JE (1980). Transforming growth factors produced by certain human tumor cells poly peptides that interact with epidermal growth factor receptors. Proc Natl Acad Sci 77:5258.
8. Ozanne B, Fulton J, Kaplan PL (1980).Kirsten murine sarcoma virus transformed cells lines and a spontaneously transformed rat cell line produce transforming factors. J Cell Physiol 301:163.
9. Moses HL, Branum EL, Proper JA, Robinson RA (1981). Transforming growth factor production by chemically transformed cells. Cancer Res 41:2842.
10. Twardzik DR, Ranchalis JE, Todaro GJ (1982). Mouse embryonic transforming growth factors related to those isolated tumor cells. Cancer Res 42:590.
11. Merrifield RB (1963). Solid phase peptide synthesis I. The synthesis of a tetrapeptide. J Am Chem Soc 85:2149.
12. Merrifield RB (1986). Solid-phase synthesis. Science 232:341.
13. Tam JP, Marquardt H, Rosberger DF, Wong TW, Todaro GJ (1984). Synthesis of biologically active rat transforming growth factor I. Nature (London) 309:376.

14. Tam JP (1987). Synthesis of biologically active transforming growth factor alpha. Int J Peptide Protein Res 29:421.

15. Tam JP, Shiekh MA, Salomon DS, Osowski L (1986). Efficient synthesis of human type α transforming growth factor: Its physical and biological characterization. Proc Nat Acad Sci 83:8082.

16. Heath WF, Merrifield RB (1976). A synthetic approach to structure-function relationships in the murine epidermal growth factor molecule. Proc Natl Acad Sci USA 83:6367.

17. Hagiwara D, Neya M, Miyazaki Y, Hemmi, K, Hashimoto M (1984). Total synthesis of urogastrone (human epidermal growth factor, h-EGF). J Chem Soc Chem Commun:1676.

18. Akaji K, Fujii N, Yajima H, Hayashi K, Mizuta K, Aono M, Moriga M (1985). Studies on peptides. CXXVII.Synthesis of a tripentacontapeptide with epidermal growth factor activity. Chem Pharm Bull 33:184.

19. Tam JP (1985). Enhancement of coupling efficiency in solid phase peptide synthesis by elevated temperature. In Deber CM, Kopple KD, Hruby VJ (eds): "Proceeding Am Peptide Symposium, 9th". Ill: Pierce Chem Co, p 423.

20. Tam JP, Heath WF, Merrifield RB (1983). S_N2 deprotection of synthetic peptides with a low concentration of HF in dimethyl sulfide: Evidence and application in peptide synthesis. J Am Chem Soc 105:6442.

21. Tam, JP, Heath, WF, Merrifield, RB (1986). Mechanism for the Removal of Benzyl Protecting Groups in Synthetic Peptides by Trifluoromethanesulfonic Acid-Trifluoroacetic acid-Dimethylsulfide. J Am Chem Soc 108:5242.

22. Mitchell AR, Erickson BW, Ryabtsev MN, Hodges RS, Merrifield RB (1976). tert-Butoxycarbonylaminoacyl-4-(oxymethyl)- phenylacetamidomethyl-Resin, a more acid-resistant support for solid phase peptide synthesis. J Am Chem Soc 98:7357.

23. Tam JP (1985). A gradative deprotection strategy for the solid-phase synthesis of peptide amides using p-(acyloxy) benzhydrylamine resin and the S_N2 deprotection method. J Org Chem 50:5291.

24. Tam JP, Tjoeng FS, Merrifield RB (1980). Design and synthesis of multidetachable resin supports for solid-phase peptide synthesis. J Am Chem Soc 102:6117.

25. Tam JP, Wong T-W, Riemen MW, Tjoeng F-S, Merrifield RB (1979). Cyclohexyl ester as a new protecting group for aspartyl peptides to minimize aspartimide formation in acidic and basic treatments. Tetrahedron lett 42:4033.

26. DiMarchi RD, Tam JP, Kent SBH, Merrifield RB (1982). Weak acid-catalyzed pyrrolidone carboxylic acid formation from glutamine during solid phase peptide synthesis. Int J Peptide Protein Res 19:88.

27. Sarin VK, Kent SBH, Tam JP, Merrifield RB (1981). Quantitative monitoring of solid-phase peptide synthesis by the ninhydrin reaction. Analytical Biochemistry 117:147.

28. Sakakibara S (1971). Hydrogen flouride in peptide synthesis. In Weinstein N (ed): "Chemistry and Biochemistry of Amino Acids, Peptides, and Proteins" vol.1, p51. Dekker, New York, 1971.

29. Holladay LA, Savage CR, Cohen S, Puett D (1976). Conformation and unfolding termodynamics of epidermal growth factor and derivaties. Biochemistry 15:2624.

30. Ahmed AK, Schaffer SW, Wetlaufer DB (1975). Nonenzymic Reactivation of Reduced Bovine Pancreatic Ribonuclease by air oxidation and by glutathione oxidoreduction buffers. J Biol Chem 250:8477.

31. Chait BT, Gisin BF, Field FH (1982). Fission fragment ionization mass spectrometry of alamethicin I. J Am Chem Soc 104:5157.

32. Macfarlane, RD, Torgerson DF (1976). Californium-252 plasma desorption mass spectroscopy. Science 191:920.

33. Rhodes JA, Tam JP, Finke U, Saundersm M, Bernanke J, Silen W, Murphy, RA (1986). Transforming growth factor alpha inhibits secretion of gastric acid. Proc Natl Acad Sci USA 83:3844.

34. Nakhla AM, Tam JP. (1985) Transforming growth factor is a potent stimulator of testicular orithnine decarboxylase in immature mouse. Biochem Biophys Research Comm 132:1180.

35. Tam JP (1985). Physiological effects of transforming growth factor in newborn mouse. Science 229:673.

36. Tam JP (1985). Physiological effects of transforming growth factor. In Moody TW (ed): Vth International Washington Spring Symposium "Neural and Endocrine Peptides and Receptors", New York: Plenum Publishing Co., p.407

37. Ibbotson KJ, D'Souza SM, Ng KW, Osborne CK, Niall M, Mundy GR (1983). Tumor-derived growth factor increases bone resorption in a tumor associated with humoral hypercalcemia of malignancy. Science 221:1292.

38. Tashijian AH, Voelkel EF, Lazzaro M, Singer FR, Roberts AB, Derynck R, Winkler ME, Levine E (1985). α and β human transforming growth factors stimulate prostaglandin production and bone resorption in cultured mouse calvaria. Proc Natl Acad Sci USA 42:4535.

39. Komoriya A, Hortsch M, Meyers C, Smith M, Kanety H, Schlessing J (1984). Biologically active synthetic fragments of epidermal growth factor:localization of a mjor-binding region. Proc Natl Acad Sci USA. 81:1351.

40. Nestor JJ, Newman SR, DeLustro BM, Todaro GJ, Schreiber, AB (1985). A synthetic fragment of rat transforming growth factor α with receptor binding and antigenic properties. Biochem Biophys Res Commun. 129:226.

41. Westhof E, Altschuh D, Moras D, Bloomer AC, Mondragon A, Klug A, Van Regenmortel MHV (1984). Correlation between segmental mobility and the location of antigenic determinants in proteins. Nature (London) 311:123.

42. Bloomquist MC, Hunt LT, Barker WC (1984). Vaccinia virus 19-kilodalton protein:relationship to several mammalian proteins, including two growth

factors. Proc Natl Acad Sci USA 81:7363.

43. Choo KH, Gould KG, Rees DJG, Brownlee GG (1982). Molecular cloning of the gene for human anti-haemophilic factor IX. Nature 299:178

44. Chang W, Upton C, Hu SL, Purchio AF, McFadden G (1987). The genome of shope firbroma virus, a tumorigenic poxvirus, contains a growth factor gene with sequence similarity to those encoding epidermal growth factor and transforming growth factor alpha. Mol and Cell Biol 7:535.

45. Upton C, Macen JL, McFadden G (1987). Mapping and sequencing of a gene from myxoma virus that is related to those encoding epidermal growth factor and transforming growth factor alpha. J Virology 61:1271.

46. Lundwall A, Dackowski W, Cohen E, Shaffer M, Mahr, Dahlback B, Stenflo J, Wydro R (1986). Isolation and sequence of the cDNA for human protein S, a regulator of blood coagulation. Proc Natl Acad Sci USA 83:6716.

47. Foster D, Davie EW (1984). Characterization of a cDNA coding for human protein C. Proc Natl Acad Sci USA 81:4766.

48. Greenwald I (1985). Lin-12, a nematode homeotic gene, is homologous to a set of mammalian proteins that includes epdiermal growth factor. Cell 43:583.

49. Wharton KA, Johansen KM, Xu T, Artavanis-Tsakonas S (1985). Nucleotide sequence from the neurogenic locus notch implies a gene product that shared homology with proteins containing EGF-like repeats. Cell 43:567.

50. Scott J, Urdea M, Quiroga M, Sanchez-Pescador R, Fong N, Selby M, Rutter WJ, Bell GI (1983). Structure of a mouse submaxillary messenger RNA encoding epidermal growth factor and seven related proteins. Science 221:236.

51. Russell DW,Schneider WJ,Yamamoto T,Luskey KL,Brown MS and Goldstein JL (1984). Domain map of the LDL receptor: sequence homology with the epidermal growth factor precursor. Cell 37:577.

SECTION II: BIOCHEMICAL PHARMACOLOGY: RECEPTORS, TRAFFICKING, AND DEGRADATION

The session began with a discussion of the signal sequences and mechanisms of membrane localization of specific domains of proteins by V. Lingappa, University of California, San Francisco. This translocation can be assayed by protease protection. The process appears to require ATP for energy and intact ribosomes plus mRNA. One can engineer changes in insertion of proteins by moving signal sequences for transfer and stop of transfer. Next W. Joyner, University of Nebraska, illustrated the hydrodynamic properties of the flux of macromolecules across vascular beds. This work demonstrated how difficult it is to achieved appreciable concentrations of large proteins in interstitial spaces. R. Townsend, Johns Hopkins University discussed the complexity of glycoprotein binding of mammalian lectins. The spatial configuration of sugar moieties is critical for tight surface binding. Thus, production of glycoproteins with the desired binding affinities may require careful selection of a cell culture line that will synthesize the correct oligosaccharide chains.

J. Barnes, University of South Carolina, presented a method to label proteins with biologically inert radioactive or fluorescent tags so that one can localize the sites of uptake and degradation of native and modified proteins. This technique should aid in the testing of chemical or genetic modifications designed to target proteins to specific tissue.

The Pharmacology and Toxicology of Proteins, pages 23–40

PROTEIN PASSAGE ACROSS MICROVESSELS AND ITS MODULATION[1]

William L. Joyner, Fitz-Roy E. Curry[2] and L.L. Houston[3]

Department of Physiology and Biophysics
University of Nebraska College of Medicine
Omaha, Nebraska 68105-1065

ABSTRACT Transvascular exchange of macromolecules across the microvascular barrier proceeds via porous and non-porous pathways. The major mechanisms determining the characteristics of transvascular exchange through porous pathways are diffusive and convective transport. Diffusive transport is a dissipative process that depends upon the difference in protein concentration and the permeability-surface area properties of the membrane. Convective transport is dominated by the balance of hydrostatic and osmotic gradients, the protein concentration and the hydraulic and reflection coefficients of the restrictive barrier; thus it is a process of ultrafiltration or solvent drug. Porous pathways may be related to intra-cellular junctions, whereas non-porous pathways may include vesicular mechanisms and some type(s) of carrier-mediated transport. All of these mechanisms can be modulated by physiological, pharmacological and pathological factors in a time- and spatial-dependent manner. Transport characteristics and their modulation can be different in different vascular beds, e.g. newly

[1]These studies in part were supported by NIH Grants: HL17902 and HL28607 and Cetus Corporation.
[2]Present address: Department of Human Physiology, University of California, Davis, Davis, California 95616
[3]Present address: Cetus Corporation, Emerysville, California 94608

developing vessels and vessels exposed to metabolic products. Alterations in the chemical composition of a macromolecule can modify drastically the permeability properties of the molecule. Thus, to facilitate the delivery of various solutes to specific sites, an understanding of the permeability properties of the microvessels in relation to that particular solute is mandatory.

DYNAMICS OF THE TRANSVASCULAR EXCHANGE PROCESS FOR MACROMOLECULES

General Aspects of the Exchange Process.

Transvascular exchange of large proteins across the vascular wall is a complex process under normal as well as stimulated states; the process involves the physical principles governing mass transport by diffusive and convective mechanisms (1,2). Anatomically the process also encompasses the broad regions of the endothelial cell and its surrounding matrix, as well as the immediate structures associated with the ablumenal surface of the vascular wall, e.g. the basement membrane, pericyte and its basement membrane and any anchoring or attachment proteins enmeshed in the network (3). The complexity further resides in considering the interactions of the biological systems with each of these cellular elements and their structural components during exposure to varying environments. Modulation of the exchange process involves the interaction of the formed elements in the circulating blood bathing the endothelial cells and the endothelial cell, itself, and its surrounding structures (4,5). The cell-cell interactions, the humoral environment surrounding the vascular wall and the metabolic state of the individual cells determine the basal levels of protein fluxes and their potential modulation by various mediators which may alter this complex balance. These aspects are depicted schematically in Figure 1. The basic

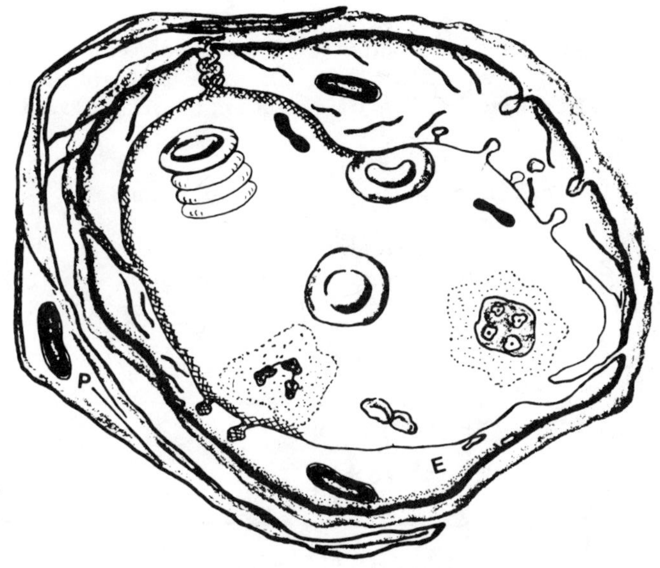

FIGURE 1. Schematic representation of structural elements of the microvascular wall involved in maintaining a barrier for the passage of macromolecules from blood to tissue, the potential pathways available for the exchange of large molecules and possible cell-cell interactions that could modulate the barrier function of the microvascular wall.

components represented in the figure are: the endothelial cell (E) with junctions, one filled with a fiber matrix and one exposed to the basement membrane, the basement membrane and the pericyte (P). All of these structures may be involved in determining the restrictive barrier for the movement of proteins across the vascular wall (6). Specific pathways are represented by vesicles, coated and uncoated, transvascular channels, and intercellular junctions

(3,7). All of these pathways may be involved in the dynamic process of protein flux. The presence of red blood cells, white blood cells and platelets along the endothelial cell surfaces of the vascular wall certainly determines the state of protein flux and the further interaction between these cells. Also, the cells and matrix of the vascular wall provide the environment for the dynamic modulation of protein transfer across the vascular wall.

Physiological and Anatomical Correlates.

The thermodynamic principles governing the process of protein fluxes across the microvascular wall through porous pathways have been reviewed recently (8,9).

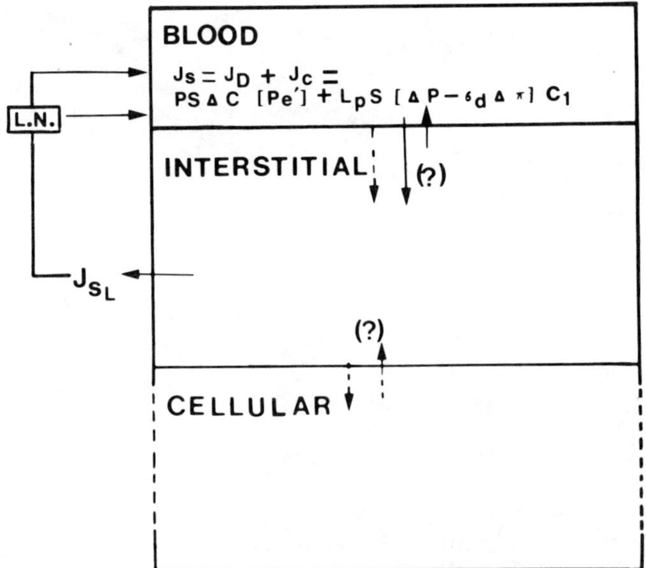

FIGURE 2. Graphic design representing the basic compartments for storage and transport of large molecules, e.g. proteins, and one mathematic representation for describing the transport of proteins across the microvascular wall and their return via the lymphatic system to the vascular compartment.

Basically, the total flux of proteins from the vascular to the extravascular compartment depends upon the summation of the diffusive and convective components, Figure 2. These processes depend upon electrochemical potentials for solute and water flow across membranes and the membrane coefficients for the respective processes. The diffusive component (J_D) depends upon the concentration of protein (ΔC) across the microvascular wall, the permeability of the vascular wall to that particular protein (P) and the surface area (S) available to the protein. Further, the magnitude of the diffusive component depends upon the coupling of the convective to the diffusive process as expressed by the Péclet (Pe') number (8). The convective flux (J_C) is a function of the transvascular movement of water (J_V) as described by the hydrostatic (ΔP) and osmotic pressure ($\Delta \Pi$) differences across the microvascular wall, the hydraulic conductivity (L_p), the surface area (S) and the reflection coefficient (σ) for the specific osmotic solute across the microvascular wall. Thus, bulk flow and/or ultrafiltration of protein can be described by the total water flow across the microvascular wall and the average protein concentration ($\bar{C}$). The thermodynamics of transvascular exchange have been utilized to predict various models than might reasonably explain the physiological data (2,8,9). These models have been described as the "pore" and/or fiber-matrix theory (9,10,11) and these theories have been useful in predicting the physical and anatomical features of the exchange pathways and their potential alterations with inflammatory mediators (11,12). The movement of proteins through the interstitium, their possible binding and/or uptake with interstitial cells and thus their flux through the interstitial compartment is still poorly understood (13). The dynamics of lymphatic return of protein to the vasculature and its alteration in edematous states has been reviewed (13,14).

Modulation of the Exchange Process.

The modulation of transcapillary exchange involves a complex array of humoral pathways, intracellular events and extracellular interactions. Many agents have been described as pharmacological mediators of inflammation and/or modulators of microvascular permeability to proteins (15). The classical agent scrutinized as a permeability-altering substance is histamine (16). Substances, such as histamine, have both a time-dependent and concentration-dependent effect on the transvascular exchange of macromolecules (17,18). Time-dependent effects would indicate that histamine produces a transient increase in protein flux across microvessels; this process also is dose-related (17). Increased flux of macromolecules may be due to the transient openings of large gaps between endothelial cells and their formation is thought to result from the contractions of endothelial cells (19). Thus, the increased transvascular exchange of macromolecules could be coupled to an increased diffusive and/or convective component. The requirement of calcium in this response of increased transvascular exchange appears to link the phenomenon to a contractile mechanism (20). This increased escape of large proteins after an inflammatory mediator such as histamine has been described as occurring in post-capillary venules only (17,21). Thus, the phenomenon is both time- and spatial-dependent. Although, the paradigm seems complete and well understood, a number of confounding and coexisting problems exist. Why does the process occur only in post-capillary venules; is the modulation through widening of endothelial junctions; do the surrounding structures provide any restrictivity and/or selectivity and if so, can they be modulated; what is the role of intracellular messengers, transducers and organelles; and do the extravascular cells contribute to the modulation of protein flux? Thus, some important questions remain to be investigated and the basic mechanisms as presented remain questionable.

TECHNOLOGICAL APPLICATIONS TO THE INVESTIGATION OF MACROMOLECULAR TRANSPORT

Techniques, in General.

The resolution of the phenomenon of transvascular exchange of proteins and its modulation may depend upon the techniques and/or methods utilized to investigate the process. Various methods have been directed towards the study of protein exchange in microvessels (22,23). Basically, they range from whole animal techniques to the isolation and/or cannulation of microvessels and now include the culture of endothelial cells (5). Using these techniques, the physiological, pharmacological, biochemical, anatomical and pathological aspects of the exchange process have been described (1,5,12,15). To delineate the problems associated with investing the modulation of the exchange process for proteins, two techniques utilizing direct, in vivo, microscopy will be described further.

In Vivo Microscopy for Extravasation of Fluorescein-labeled Solutes.

Earlier, Majnö and coworkers (16,21,24) using electron microscopy described the effect of histamine-type mediators on the vascular permeability of the microvessels in the cremaster muscle of the rat. They described the formation of endothelial openings/gaps occurring in postcapillary venules (24). Later, these results were corroborated and extended using direct, intravital microscopy of the microvessels in the cheek pouch of the hamster (17,20,25,26). After exposing the cheek pouch for intravital microscopy, these investigators gave a large molecular-weight, fluorescein-labeled dextran intravenously and the microvessels of the cheek pouch were observed by fluorescent microscopy. After a control period the cheek pouch then could be suffused with a histamine-type mediator and the time-dependent changes recorded. Basically, histamine and other mediators of inflammation induced a localized extravasation of the fluorescein-labeled dextran from the microvessels of the cheek pouch; these areas were observed as "leaky" sites and counted as a function of time and membrane area. The formation of these "leaky" sites was transient and they were observed only in postcapillary venules (17,25). Numerous studies have emerged subsequently utilizing this technique to study the pharmacology and pathology of inflammatory mediators (15,19). An advancement in the technique was produced by collecting the suffusate from the cheek pouch over discrete intervals of time and measuring the concentration of the extravasated dextran in the fluid (27). Simultaneous measurements of the suffusate flow and the fluorescein-dextran concentration allowed for the measurement of the clearance of the dextran in normal and stimulated states. The microvascular pattern of fluorescein-dextran extravasation from microvessels in a normal cheek pouch is depicted in Figure 3. These profiles were obtained

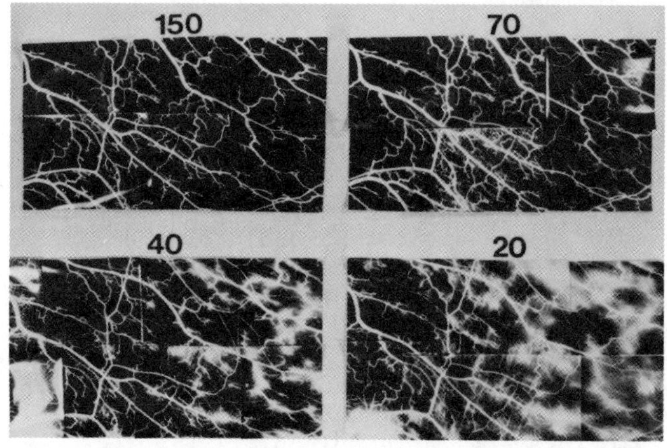

FIGURE 3. The photographs were obtained from microvessels in the cheek pouch of one hamster 30-45 minutes after specific fluorescein-labeled dextrans were given intravenously. The larger molecular weight dextran was given first and the other fractions were administered at one hour intervals in order of decreasing molecular weight.

from a single hamster after the serial administration of fluorescein dextrans (27). An inverse relation for "leaky" sites versus the molecular size of the fluorescein-dextran fraction was observed. In the same series of experiments the number of "leaky" sites and measurements of the clearance for these fluorescein dextrans were recorded from the microvessels of the cheek pouch in non-stimulated states and then after graded administrations of histamine. These results are shown in Figure 4. For each dextran fraction the number

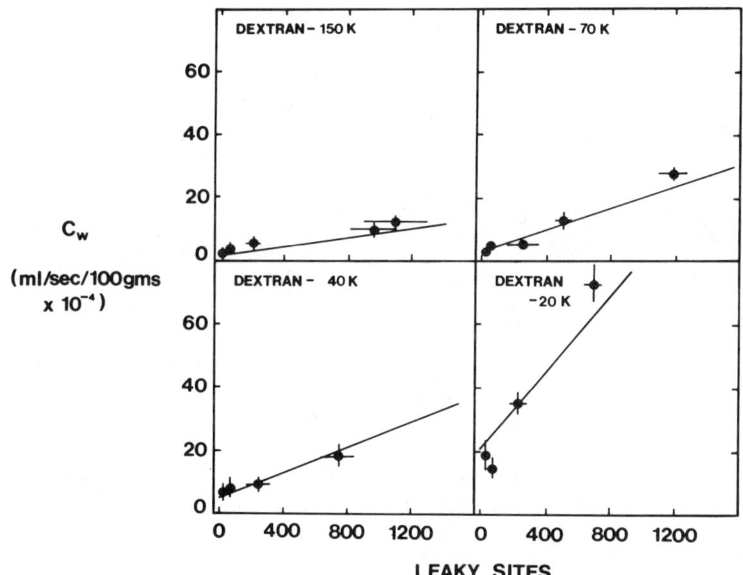

FIGURE 4. Plot of the clearance for each fraction of dextran in relation to the number of "leaky" sites that were measured simultaneously in cheek pouches from eight hamsters before and after stimulation with histamine.

of "leaky" sites was linearly related to the clearance of the fluorescein dextran and the clearance of the fluorescein dextran per "leaky" sites was inversely related to the molecular size and/or radius of the fluorescein dextran. The results would indicate that histamine augments the formation of "leaky" sites in the vascular wall and that these "leaky" sites are related directly to the extravasation of the dextrans. However, the technique does not allow the quantitative determination of dextran fluxes into diffusive and convective components and further, the basic characteristics of the vascular wall cannot be described in terms of membrane coefficients; i.e. the effects of histamine cannot be related to changes in membrane permeability (P) and/or surface area (S). This aspect is exemplified by observing the clearance of each dextran fraction when there were no observable "leaky" sites (Figure 4). It can be seen that the clearance is not zero and that this zero intercept decreases as the molecular weight of the dextran increases; thus, some other mechanisms and/or pathways not accounted for with this technique may be involved in the overall process. Selective control of the microvascular system is not sufficient with this technique as described to date; consequently, the environment within the microvessels cannot be adequately controlled and/or measured. For example, it is readily apparent that the interaction of white cells with the vascular wall modifies the normal state of the restrictive barrier to protein flux (28) and the subsequent modification may change the response of the vessel wall to histamine-type mediators. This type of interaction will be difficult to study using the presently described techniques. Of course, the advantages of these techniques reside in the ease of preparation and recording, the use of a wide spectrum of microvessels, the measurement of time-dependent responses, and the access to pharmacological manipulation for dose-response testing of agonists and antagonists.

In Vivo Photometry of Isolated, Perfused Microvessels.

Recently a new technique has been developed for measuring the flux of large proteins across the walls of single microvessels which had been cannulated and perfused with a rhodamine-labeled solute (29,30,31). A schematic representing the technique is shown in Figure 5. A small, post-capillary venule was selected from the mesentery of a frog for cannulation. One

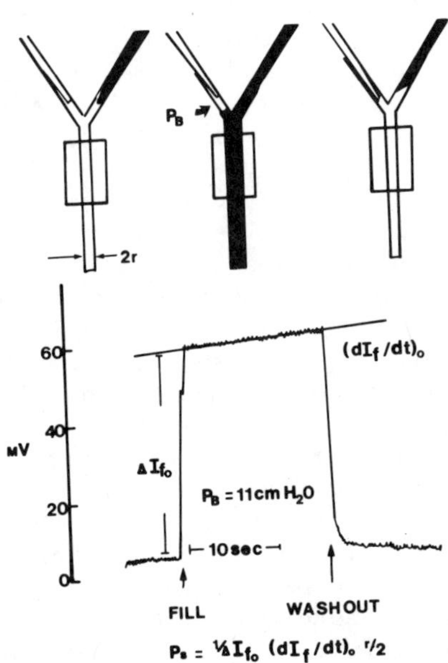

FIGURE 5. A diagrammatic representation of a microvessel with its bifurcation cannulated with a pipette containing non-labeled albumin, clear, and one containing a rhodamine-labeled albumin, dark. The photomultiplier window is depicted as a box superimposed over the vessel and the lower portion contains a tracing of the output from this photomultiplier before, during and after the vessel is perfused with the rhodamine-labeled albumin.

branch of the microvessel was cannulated with a micropipette containing frog Ringer's and unlabeled albumin (1%) while observing the vessel by direct, in vivo microscopy; then, the second branch of the microvessel was cannulated with a pipette containing frog Ringer's plus rhodamine (TRITC)-labeled albumin (1%). Initially, perfusion was maintained via the clear, non-labeled albumin solution. The intensity over one portion of the vessel was monitored and recorded continuously through a rectangular window of a photomultiplier system. The output (mV) from the photomultiplier is shown in the lower portion of the figure. Perfusion was switched rapidly to the rhodamine-labeled pipette and the microvessel filled briskly as indicated by the step-increase in intensity recorded over the vessel. While the fluorescent dye front was balanced in the other branch of the vessel there was a continual increase in intensity recorded from the photomultiplier. The relation of the slope of this slower increase in intensity to both the step-increase in intensity and the surface to volume ratio (r/2) of the microvessel was used to calculate the total flux of the albumin (P_S); these measurements and the calculation are described at the bottom of the figure. The pressure required to balance the florescent dye was recorded as the pressure within the post-capillary venule (P_B). The perfusion of the microvessel was switched back to the clear pipette and the rhodamine-labeled albumin was cleared from the microvessel, as indicated by the return of the intensity back to the original levels. To determine the relation of capillary pressure to solute flux, this procedure could be repeated at different capillary pressures; to investigate a different solute one and/or both of the pipettes could be removed and the microvessels recannulated with a new test solute. Also, a test agent could be applied to the microvessel by an intra- or an extra-vascular route and the dynamics of the potential changes in solute flux monitored rapidly for a given interval. Therefore, unlike other techniques this technique can be utilized to follow rapid transients in the process of transvascular exchange of large molecules; it can be repeated and of course, a combination of these alternatives can be selected for investigation as a specific protocol demands. One such

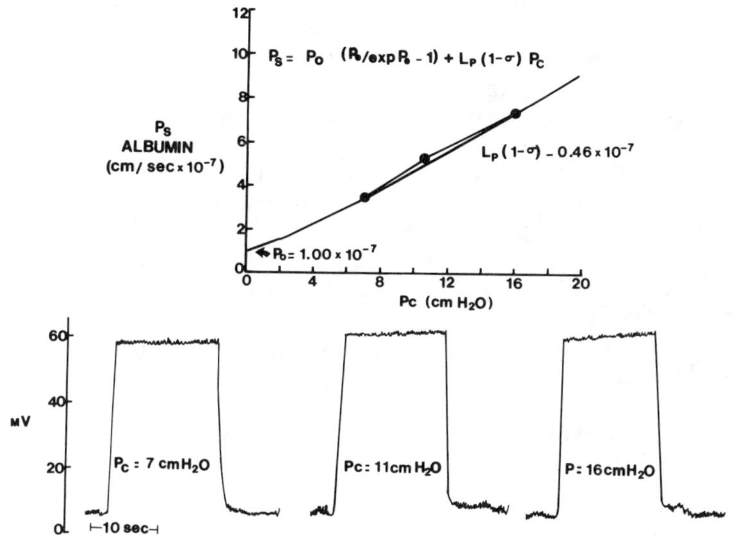

FIGURE 6. Relation of capillary pressure (P_c) to the transvascular flux of albumin (P_s) in a single, doubly-cannulated and perfused microvessel in the mesentery of the frog.

protocol is shown in Figure 6 where the capillary pressure was altered from 7-16 cm H_2O and the solute flux of albumin measured and calculated as described previously. These calculated fluxes for albumin (P_s) are plotted in the upper portion of the figure as a function of capillary pressure and using a nonlinear, mass balance equation describing the membrane coefficients, a line of best fit was ascribed for these points (30,31). From these manipulations the diffusive (P_o) and convective [$L_p(1-\sigma)$] components were calculated for this normal, non-stimulated post-capillary venule. The transport of albumin was coupled to the hydrostatic gradient and thus the convective component contributed 98% of the total flux at a capillary pressure of 10 cm H_2O. This system has been utilized to investigate the effects of permeability-altering agents on the alteration of transvascular exchange for macromolecules and for describing potential characteristics of the membrane coefficients and their modulation with agents that mediate inflammation and alter calcium fluxes

(30,31,32). The advantages of this system reside in: (1) the rigid control that can be imposed on the microvessel to be studied, (2) the ability to quantitate the transport characteristics and/or membrane coefficients by varying specific driving forces for convective and/or diffusive transport, (3) the capability of quantifying capillary pressure, (4) the ability to rapidly and selectively fix microvessels for morphometric study. One aspect of this specific control indicates that these studies can be conducted in microvessels perfused with cell-free perfusates; thus, the role of circulating blood cells can be investigated quite readily. The major disadvantage is that for any given experiment only a few microvessels can be examined; consequently, the role of blood flow regulation on the exchange process is difficult to investigate.

Selective Permeation of Cytotoxic Proteins.

One potential of this technique lies in the ability to describe the selective properties of the micro-vascular wall that might be related to the delivery of specific proteins to the cells residing in the inter-stitium. One set of such experiments is shown in Table 1 for cytotoxic agent(s) prepared from the castor bean and recombinant genes expressed in E. coli.
Each protein was tested in combination with albumin, rhodamine labeled and these preliminary results indicate that for its molecular weight the entire ricin molecule is slightly more permeable than albumin. It is curious that the native ricin-A chain, which is glycosylated, is 3-4 times more permeable than the recombinant form, which is not glycosylated. Ricin binds to cell surface receptors that contasin galactose and it is possible that these values increased in the presence of

TABLE 1.
MEASUREMENTS OF PERMEABILITY COEFFICIENTS FOR SPECIFIC PROTEINS IN NON-STIMULATED POST-CAPILLARY VENULES.

Protein	Mol. Wt.	P_o	$Lp(1-\sigma)$
Albumin	69,000	5.99 ±2.69 (10)	0.97 ±0.35
Ricin-D	65,000	11.40 ±1.41 (3)	2.45 ±0.86
Native Ricin - A Chain	32,000	160.00 ±17.00 (4)	10.00 ±4.00
Recombinant Ricin A Chain	30,000	45.00 ±5.00 (3)	4.43 ±0.64

P_o = permeability (cm/sec x 10^{-7}); $Lp(1-\sigma)$ = hydraulic coupling (cm/sec/cm H_2O x 10^{-7}) All values are means ± SEM; ( ) = number of microvessels.

galactose. These values are more striking when compared to the molecule, α-lactalbumin (mol. wt. 14,176): in similar studies (19) the permeability was 19.6 cm/sec x 10^{-7} and the hydraulic coupling was 2.36 cm/sec/cm H_2O x 10^{-7}). These proteins from ricin are extremely permeable or cytotoxic to the vascular wall and also they may be coupled to water flow across the vascular wall. These preliminary studies indicate the potential for utilizing similar techniques to further investigate aspects of genetic modification of proteins on their subsequent passage across the vascular wall in normal, stimulated and pathological states.

ACKNOWLEDGEMENTS

The authors would like to acknowledge the expert technical assistance of J.C. Lenz and S. Aloni for the preparation of FITC proteins. Also, Drs. T.C. Wang, W.G. Mayhan, T.O. Myers and C.A. Eccleston-Joyner contributed significantly.

REFERENCES

1. Renkin EM (1985). Capillary transport of macromolecules: Pores and other endothelial pathways. J Appl Physiol 58:315.
2. Landis EM, Pappenheimer JR (1963). Exchange of substances through capillary walls. In Dow P, Hamilton WF (eds): "Handbook of Physiology", Section 2, "Circulation", Vol II, Washington DC: Am Physiol Soc, p 961.
3. Simionescu N (1983). Cellular aspects of transcapillary exchange. Physiol Rev 63:1536.
4. Crone C (1986). Modulation of solute permeability in microvascular endothelium. Fed Proc 45:77.
5. Shepro D, D'Amore PA (1984). Physiology and biochemistry of vascular wall endothelium. In Renkin EM, Michel CC (eds): "Handbook of Physiology", Section 2, "The Cardiovascular System", Vol IV, "The Microcirculation", Part I, Bethesda, MD: Am Physiol Soc, p 103.
6. Renkin EM (1978). Multiple pathways of capillary permeability. Circ Res 41:735.
7. Bundgaard M, Frøkjaer-Jensen J (1980). Functional aspects of the ultrastructure of terminal blood vessels: A quantitative study on consecutive segments of the frog mesenteric microvasculature. J Ultrastruct Res 73:9.
8. Curry FE (1984). Mechanics and thermodynamics of transcapillary exchange. In Renkin EM, Michel CC (eds): "Handbook of Physiology", Section 2, "The Cardiovascular System", Vol IV, "The Microcirculation", Part 1, Bethesda, MD: Am Physiol Soc, p 309.

9. Taylor AE, Granger DN (1984). Exchange of macromolecules across the microcirculation. In Renkin EM, Michel CC (eds): "Handbook of Physiology", Section 2, "The Cardiovascular System", Vol IV, "The Microcirculation", Part 1, Bethesda, MD: Am Physiol Soc, p 467.
10. Curry FE (1986). Determinants of capillary permeability: A review of mechanisms based on single capillary studies in the frog. Circ Res 59:367.
11. Curry FE, Michel CC (1980). A fiber matrix theory of capillary permeability. Microvasc Res 20:96.
12. Simionescu M, Simionescu N (1984). Ultrastructure of the microvascular wall: Functional correlations. In Renkin EM, Michel CC (eds): "Handbook of Physiology", Section 2, "The Cardiovascular System", Vol IV, "The Microcirculation", Part 1, Bethesda, MD: Am Physiol Soc, p 41.
13. Bert JL, Pearce RH (1984). The interstitium and microvascular exchange. In Renkin EM, Michel CC (eds): "Handbook of Physiology", Section 2, "The Cardiovascular System", Vol IV, "The Microcirculation", Part 1, Bethesda, MD: Am Physiol Soc, p 103.
14. Granger HJ (1979). Role of the interstitial matrix and lymphatic pump in regulation of transcapillary fluid balance. Microvasc Res 18:209.
15. Grega GJ, Adamski SW, Dobbins DE (1986). Physiological and pharmacological evidence for the regulation of permeability. Fed Proc 45:96.
16. Majnö G, Palade GE (1961). Studies on inflammation. I. The effect of histamine and serotonin on vascular permeability: An electron microscopic study. J Biophys Biochem Cytol 11:571.
17. Svensjö E, Arfors K-E (1979). Dimensions of postcapillary venules sensitive to bradykinin and histamine-induced leakage of macromolecules. Uppsala J Med Sci 84:47.
18. Curry FE, Joyner WL (1986). The effects of histamine, 48/80 and A23187 on albumin permeability in frog venular capillaries. Fed Proc 45:1159.
19. Grega GJ, Adamski SW, Svensjö E (1985). Is there evidence for venular large junctional gap formation in inflammation. Microcirc Endothelium and Lymphatics 2:211.

20. Mayhan WG, Joyner WL (1984). The effect of altering the external calcium concentration and a calcium channel blocker, verapamil, on microvascular leaky sites and dextran clearance in the hamster cheek pouch. Microvasc Res 28:159.
21. Majnö G, Palade GE, Schoefl GI (1961). Studies on inflammation. II. The site of action of histamine and serotonin along the vascular tree: a topographic study. J Biophys Biochem Cytol 11:607.
22. Curry FE, Huxley VH, Sarelius IH (1983). Techniques in the microcirculation: Measurement of permeability, pressure and flow. In Linden RJ (ed): "Techniques in the Life Sciences", Section P3/I, "Techniques in Cardiovascular Physiology", Part 1, Vol P309. New York: Elsevier, p 1.
23. Renkin EM, Kramer GC (1986). Measurement of vascular and microvascular permeability to large molecules. In Baker CH, Nastuk WL (eds): "Microcirculatory Technology", Academic Press, p 471.
24. Majnö G, Shea SM, Leventhal M (1969). Endothelial contraction induced by histamine type mediators: an electron microscopic study. J Cell Biol 42:647.
25. Hulström D, Svensjö E (1977). Simultaneous fluorescence and electron microscopical detection of bradykinin induced macromolecular leakage. Bibl Anat 15:466.
26. Svensjö E, Arfors K-E, Arturson G, Rutili G (1978). The hamster cheek pouch preparation as a model for studies of macromolecular permeability of the microvasculature. Uppsala J Med Sci 83:71.
27. Joyner WL, Wang TC (1983). Venular leaky sites and clearance of various dextrans after histamine. Microvasc Res 25:240.
28. Lewis RE, Granger HJ (1986). Neutrophil-dependent mediation of microvascular permeability. Fed Proc 45:109.
29. Huxley VH, Curry FE, Adamson RH (1987). Quantitative fluorescence microscopy on single capillaries: α-lactalbumin transport. Am J Physiol 252:H188.
30. Joyner WL, Curry FE (1986). Measurements of albumin permeability coefficients in single capillaries of hamster mesentery. Fed Proc 45:583.
31. Curry FE, Joyner WL (1987). Modulation of capillary permeability: Methods and measurements in individual perfused mammalian and frog microvessels. In Ryan U (ed): "Endothelial Cells", CRC Press (In Press).
32. Jain RK (1985). Transport of macromolecules in tumor microcirculation. Biotech Prog 1(2):81.

The Pharmacology and Toxicology of Proteins, pages 41–57

CARBOHYDRATE MEDIATED CELLULAR UPTAKE: LEVELS OF STRUCTURAL ORGANIZATION OF THE LIGAND[1]

R. Reid Townsend

Department of Biology and the McCollum-Pratt Institute, Johns Hopkins University, Baltimore, Maryland 21218

ABSTRACT The binding of glycoproteins to mammalian lectins with high affinity is a result of the spatial arrangement of determinant monosaccharides in oligosaccharides at specific loci on the peptide chain. The hepatic Gal/GalNAc lectin is discussed as a paradigm of binding and endocytosis mediated by mammalian lectins. The "rigid" structure of potent oligosaccharides optimally positions the terminal galactose molecules (Gal) which can result in 1000-fold increases in affinity by the addition of a single Gal to the structure. Glycoproteins, with multiple oligosaccharide chains, compared to glycopeptides and glycosides, define one-third fewer cell surface sites, dissociate much more slowly from the cell surface, and cannot saturate all cell surface Gal-combining sites. A model consisting of a lattice of ordered Gal-combining sites on the cell surface is proposed to explain binding data using glycoproteins, glycosides and glycopeptides. Endocytosis of glycoproteins results in degradation of most of the ligand that is bound to the cell surface. A large proportion (60%) of glycopeptides and glycosides escape from the cell after internalization. Kinetic modeling of endocytosis suggests that the exocytosis of glycopeptides and glycosides is due both to their faster dissociation rates and their different modes of surface binding.

[1]This work was supported by National Science Foundation Grant DCB8509638 and National Institutes of Health Research Grant AM31376.

INTRODUCTION

Since the discovery of the hepatic Gal/GalNAc lectin or asialoglycoprotein receptor (1), an ever-increasing number of mammalian lectins has been recognized. Table I summarizes some features of these carbohydrate binding proteins. Despite their diversity in molecular weight, monosaccharide specificity, and cell type of origin, the lectins are characterized by their ability to discriminate glycoproteins according to their carbohydrate structures. Since certain carbohydrate structures may be shared among a set of glycoproteins, lectins can sort glycoproteins by their oligosaccharide structures. The functional significance of this capacity is evident from the ability of the Man-6-PO_4 receptor to identify lysosomal enzymes endowed with Man-6-PO_4 and direct them to lysosomes (2). Similarly, the Gal/GalNAc lectin can bind to galactose of plasma glycoproteins, independent of their function, and deliver them to lysosomes in hepatic parenchymal cells (3). Although monosaccharide specificity is commonly used to classify lectins, two observations indicate that the binding of lectins to glycoproteins is more complicated. First, glycoproteins bind with an affinity of $K_d \simeq$ nM while monosaccharides bind at $K_d \simeq$ mM and secondly, glycoproteins, presumably the natural ligands for lectins, possess multiple terminal monosaccharides grouped as oligosaccharides each located at specific glycosylation sites. The well studied N-linked oligosaccharide chains are such branched structures and have multiple penultimate monosaccharides, frequently of the same type (13). For example, the Gal/GalNAc lectin, macrophage Man receptor, and the Man-6-PO_4 receptor recognize terminal galactoses (3), mannoses (7), and mannose-6-phosphates (6), respectively, in N-linked oligosaccharides. It can be postulated that the binding of multiple monosaccharides produces the affinities ($K_d \simeq$ nM) usually observed for glycoproteins.

The relationship of ligand structure and binding can be viewed at four levels. DIAGRAM 1 shows different aspects of carbohydrate structure using ligands recognized by the hepatic Gal/GalNAc lectin. Most lectins recognize a limited number of monosaccharides (Table 1). Binding may be confined to one or two monosaccharides or a hierarchy of sugar specificity can be observed as in the case of the macrophage Man receptor. The hepatic Gal/GalNAc lectin binds to either galactose or N-acetylgalactosamine (DIAGRAM 1)(14). The second level can be observed from the binding of glyco-

TABLE 1
MAMMALIAN LECTINS

Lectin (Receptor)[a]	M_r	Monosaccharide Specificity	Cell Type
Fucose (4) Lectin	88,000 77,000	Gal=Fuc	Kupffer cells (5)
Gal/GalNAc Lectin (3)	42,000 53,000 58.000	GalNAc>Gal	Hepatic parenchymal cells
Man-6-PO_4 Receptor (6)	215,000	Man-6-PO_4	Fibroblast, macrophages and others
Mannose Receptor (7)	175,000(8)	Fuc=Man> GlcNAc=Glc	Macrophages (8)
Plasma Man/GlcNAc Lectin (9)	32,000	Man, GlcNAc ManNAc	Secreted by hepatic parenchymal cells(10)
Leucocyte Lectin (11)	45,000	Gal	Lymphocytes
Lung Lectin (12)	29,000 18,000 14,500	Lactose	Lung, liver, and other tissues

[a]Mammalian carbohydrate binding proteins of non-immune origin have been designated, in the literature, as either lectins or receptors. For this and previous reports, we have termed the asialoglycoprotein receptor or Gal/GalNAc lectin as a lectin when binding specifics were emphasized and as a receptor when its role in endocytosis was discussed.

DIAGRAM 1

I

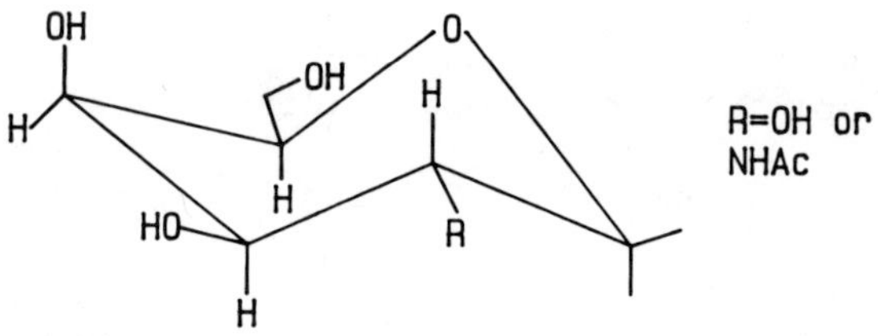

OH OH O H H H HO R H

R=OH or NHAc

II

Galβ(1,4)GlcNAcβ(1,2)Manα(1,6)
\
ManOH
/
Galβ(1,4)GlcNAcβ(1,2)Manα(1,3)
/
Galβ(1,4)GlcNAcβ(1,4)

III

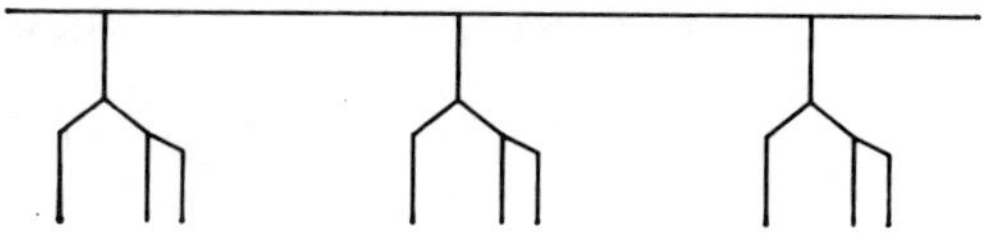

IV

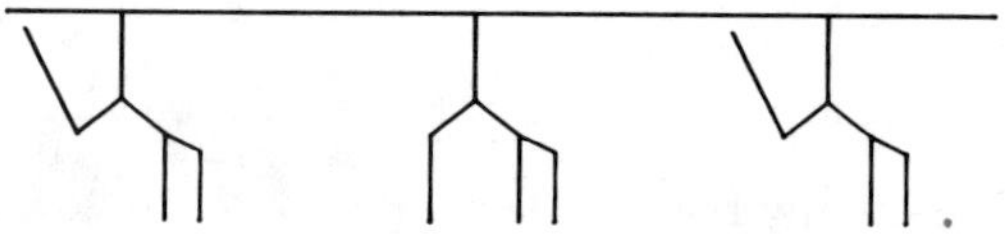

peptides or oligosaccharides which contain clusters of the determinant monosaccharides. The affinity of oligosaccharides is usually higher than that found for the monosaccharides and, in some cases, oligosaccharides can possess affinities comparable to glycoproteins (15,16). In N-linked oligosaccharides of plasma glycoproteins, 2,3, or 4 terminal Gal are often found which are referred to as bi-, tri- and tetraantennary structures (13). For example, the second structure shown in DIAGRAM 1 is a triantennary oligosaccharide that binds to the hepatic Gal/GalNAc lectin with $K_d \simeq$ nM (16). Glycoproteins can utilize multiple oligosaccharides (DIAGRAM 1, Structure III) for lectin binding. Additional complexity is usually present since there can exist a large diversity of oligosaccharide chains and each glycosylation site may possess a plethora of structures (13). Finally, the peptide chain may interact with the carbohydrate and change the "availability" of monosaccharides for the binding sites, possibly by altering the conformation of the oligosaccharide. This level of structural complexity in lectin-glycoprotein interaction has been proposed for the binding of some ligands to the Man-6-PO_4 receptor (17). Evidence now exists for the interaction of lactosamine arms with the peptide as shown in DIAGRAM 1, Structure IV (18). The remainder of this review will focus on carbohydrates of the above complexities that have been used to study binding to the hepatic Gal/GalNAc lectin.

RESULTS AND DISCUSSION

Binding of Monosaccharides.

The Gal/GalNAc lectin recognizes two monosaccharides, Gal and GalNAc (DIAGRAM I, Structure I). Studies using modified monosaccharides and neoglycoproteins possessing different sugars have been used to determine which groups of the pyranose ring are involved in binding. For Gal (19,20), the 2-OH, 3-OH, 4-OH, 6-CH_2- and the aglycon are important for optimal binding. Replacement of the 6-OH with either positively charged or a bulky hydrophobic groups does not significantly alter affinity whereas a negatively charged group is not tolerated. The anomeric oxygen or sulfur does not participate significantly in binding.

The 2-deoxy-2-acetamido group enhances binding such that GalNAc has a 10-50 fold greater affinity than Gal (14,21). The N-substituents can be as large as a benzamido or phtalimido group and as polar as a trifluoroacetate without affecting affinity by more than 10-fold. Methylation of the 3-OH or 4-OH and 6-OH results in a 155 to 350-fold decrease in affinity whereas methylation of the 6-OH has little effect on binding. Replacement of the 4-OH position with a fluorine results in poor binding which suggests that the 4-OH is a proton donor in hydrogen bond formation.

Binding of Oligosaccharides and Cluster Glycosides.

Soon after the discovery of the Gal/GalNAc lectin it was recognized that monosaccharides did not possess as high an affinity as glycoproteins. Asialoorosomucoid (ASOR), a glycoprotein frequently used to study this lectin, had an affinity of $K_d \simeq$ 1nM while the monosaccharides were in the range of 1mM - 50μM. It was also shown that, on the average,

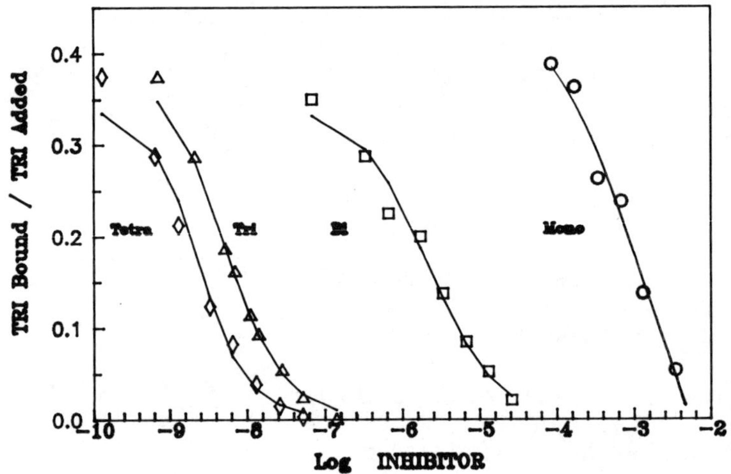

FIGURE 1. Inhibition of triantennary glycopeptide binding to rabbit hepatocytes. Unlabeled ligands (DIAGRAM 2) and ^{125}I-TRI were incubated with freshly isolated hepatocytes at the indicated concentrations in medium (16). After a 2h the bound labeled ligand was determined as previously described (16).

two Gal molecules were required for clearance of infused partially desialylated glycoproteins (22). The binding of neoglycoproteins with different levels of sugar substitution demonstrated that affinity increased 1000-fold from 3 - 6 Gal coupled to bovine serum albumin (BSA) molecule to 15 - 20 Gal/BSA (23). Studies on the binding of bovine fetuin glycopeptides to rat hepatocytes indicated that three Gal could produce an affinity of $K_d \simeq$ 1nM (15). However, the availability of defined branched synthetic mimetic oligosaccharides and the resulting binding studies has provided the clearest picture of hepatic lectin binding specificity for N-linked oligosaccharides (16).

Figure 1 shows the binding potency of mono-, bi-, tri-, and tetra-branched oligosaccharides whose structures are given in DIAGRAM 2. Clustering two Gal molecules together as a biantennary structure increases affinity 1000-fold to $K_d \simeq 1$ μM. A one-million fold affinity difference is observed between Galβ(1,4)βGlcNAc(1,2)Man (mono-), which constitutes one branch, and the three lactosamine-branched oligosaccharide, triantennary (TRI). Positioning the third lactosamine in β(1,6) linkage to the α(1,6) linked Man (DIAGRAM 2, TRI' structure) results in a 50-100 decrease in affinity relative to TRI. A significant increase in affinity is not observed with a tetraantennary structure compared to TRI and, thus, we consider TRI as a minimal structure required for binding with $K_d \simeq$ 1nM.

Comparisons of the binding of other oligosaccharides reveal that 1) the number of Gal residues per oligosaccharide, 2) the linkage positions of the branches to the trimannosyl core, 3) the number of monosaccharides in each branch and 4) the linkage position of the Gal to the GlcNAc (24) markedly alter binding affinity. Determination of the three-dimensional structure of these oligosaccharides, based on HSEA calculations (25,26), provides insight into the spatial arrangement of Gal's required for optimal binding. Correlation of the inter-Gal distances (DIAGRAM 2) of minimal energy conformers with binding affinity revealed that a major structural feature for high affinity binding is the distance between the third and fourth branches, which

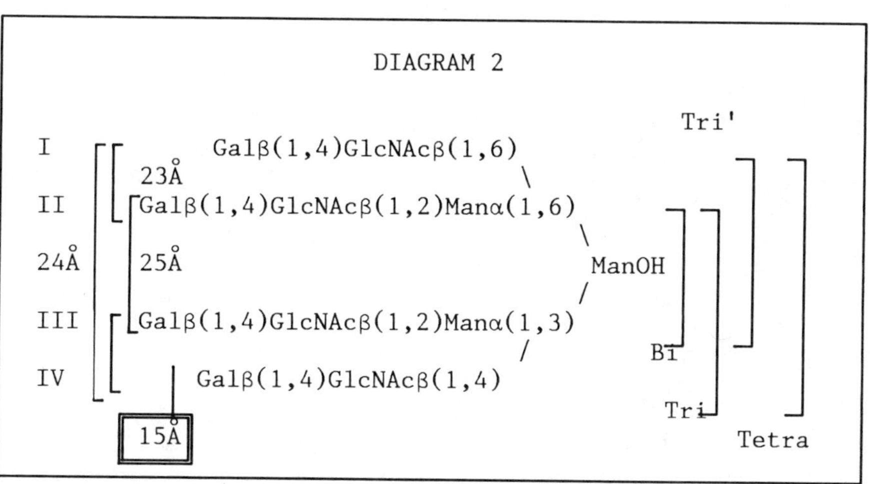

DIAGRAM 3

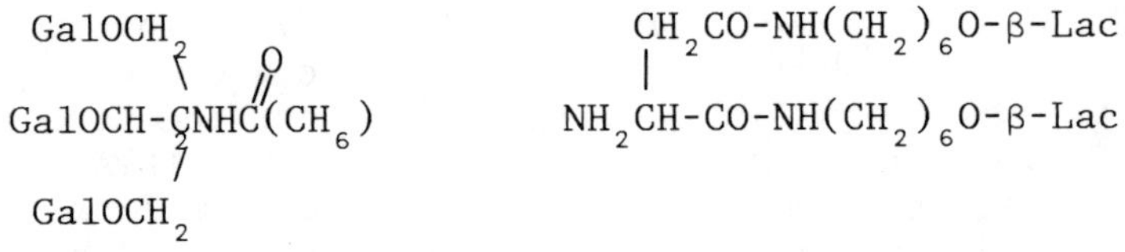

```
GalOCH2
       \   O
        \  //
GalOCH-CNHC(CH6)
        /
       /
GalOCH2

Tris-galactoside
```

```
     CH2CO-NH(CH2)6O-β-Lac
     |
NH2CH-CO-NH(CH2)6O-β-Lac

Asp(LacAH)2
```

was found to be significantly shorter (15 Å). We have recently found that widening of this inter-Gal distance by 3-4 Å by linkage of the fourth arm Gal β(1,3) instead of β(1,4) yields a 50-100 fold decrease in affinity (24).

The finding that the Gal residues had to be extended by more than 15 Å and that the only apparent function of the non-terminal sugars was to govern the spatial arrangement of Gal residues prompted the synthesis of branched sugar structures shown in DIAGRAM 3 (27). The maximal inter-Gal distance for a tris-galactoside is 8 Å and it has an affinity of only 8 mM. Glycosides with affinities comparable to the above synthetic oligosaccharides were synthesized by extending the Gal as aminohexyl-lactosyl arms or aminohexyl tris galactosides (27). This approach results in Gal distances of up to ≃ 30 Å (27).

Binding of Glycoproteins.

Ligand structure-affinity correlates of glycoproteins are presently uncertain. Glycoprotein ligands which bind with high affinity and contain a single oligosaccharide structure at each glycosylation site are not available, since methods to purify glycoproteins according to carbohydrate structure and efficiently analyze oligosaccharides at each site are just evolving. The magnitude of glycoprotein diversity is illustrated in DIAGRAM 4 with a glycoprotein with three glycosylation sites and each site contains either a bi-, tri, or tetraantennary oligosaccharide. Nine of the 27 possibilities are shown in which a biantennary oligosaccharide is at the first site. Thus, current binding studies employing a single glycoprotein, according to peptide structure, are probably measuring the

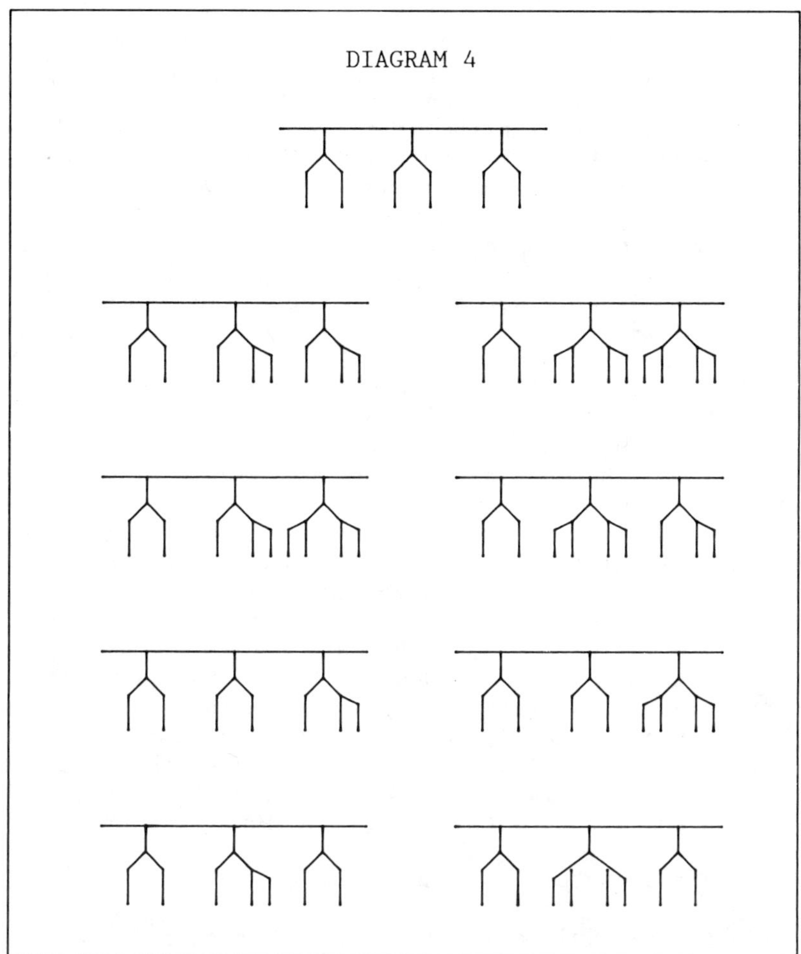

average binding of multiple structures. It becomes imperative to develop capabilities to purify glycoproteins with single structures at each glycosylation site. In this regard, the number of different structures at each glycosylation site appears to be limited (28,29).

With the lack of 'pure' glycoprotein ligands, we have been studying glycoprotein binding in parallel with defined synthetic glycosides (30). We have found the following differences in the binding of glycoproteins and glycosides. 1) There are 3-4 times more binding sites for glycosides than for glycoproteins (ASOR).2) Although certain glycosides are in the same affinity range as glycoproteins, the dissociation rates for glycosides are much more rapid. 3) Glycoproteins cannot saturate all the hepatocyte cell surface Gal-combining binding sites. 4) The isolated lectin binds glycoproteins with the same affinity as the lectin on the cell surface but the affinity of glycopeptides and glycosides are reduced as much as 800 fold (27).

We have explained these observations by proposing that the Gal-combining sites are clustered on the cell surface in adjacent arrays close enough to accommodate the multiple oligosaccharides of glycoproteins as shown in DIAGRAM 5. Equivalent Gal-combining sites on the cell surface are depicted as ordered groups of three ellipses, inferred from the tight binding of the triantennary structure in DIAGRAM 2 in which the Gal residues are separated by 15, 25, and 22 Å. Two triantennary structures from a glycoprotein are depicted occupying adjacent triads of Gal-combining sites. This observation is consistent with above observations 1 and 2. Three additional sites are occupied by three Gal residues of a synthetic glycoside similar in structure to those shown in DIAGRAM 3 (Di-tris-lac) which possesses 6 Gal but has engaged only three. The Gal-combining sites occupied by Di-tris-lac are inaccessible to the Gal's of other glycoproteins. This concept is supported by the inability of glycoproteins to saturate all the Gal-combining sites on the cell surface(30). Finally, the structural organization of the lectin on the cell surface which produces the dramatic synergism shown in Fig. 1 is lost upon isolation presumably due the disruption of subunit organization by detergents.

Carbohydrate Structure and Endocytosis Events.

The hepatic Gal/GalNAc lectin, viewed as a receptor, is an integral membrane protein which mediates the efficient delivery of glycoproteins to lysosomes (3). A complete description of the receptor-mediated endocytosis of asialoglycoproteins must also address the effect of carbohydrate structure on events subsequent to binding. We have found that ligands with different carbohydrate structures are released from hepatocytes to different extents after internalization (31). Ligands which can bind receptor by multiple oligosaccharide chains (e.g. ASOR) recycle back to the cell surface to a lesser extent than glycopeptides and glycosides. Glycoproteins remain bound to receptor and are not lost from the cell upon resurfacing. A greater proportion of triantennary glycopeptides (31) and glycosides recycle back to the hepatocyte surface unbound to

DIAGRAM 5

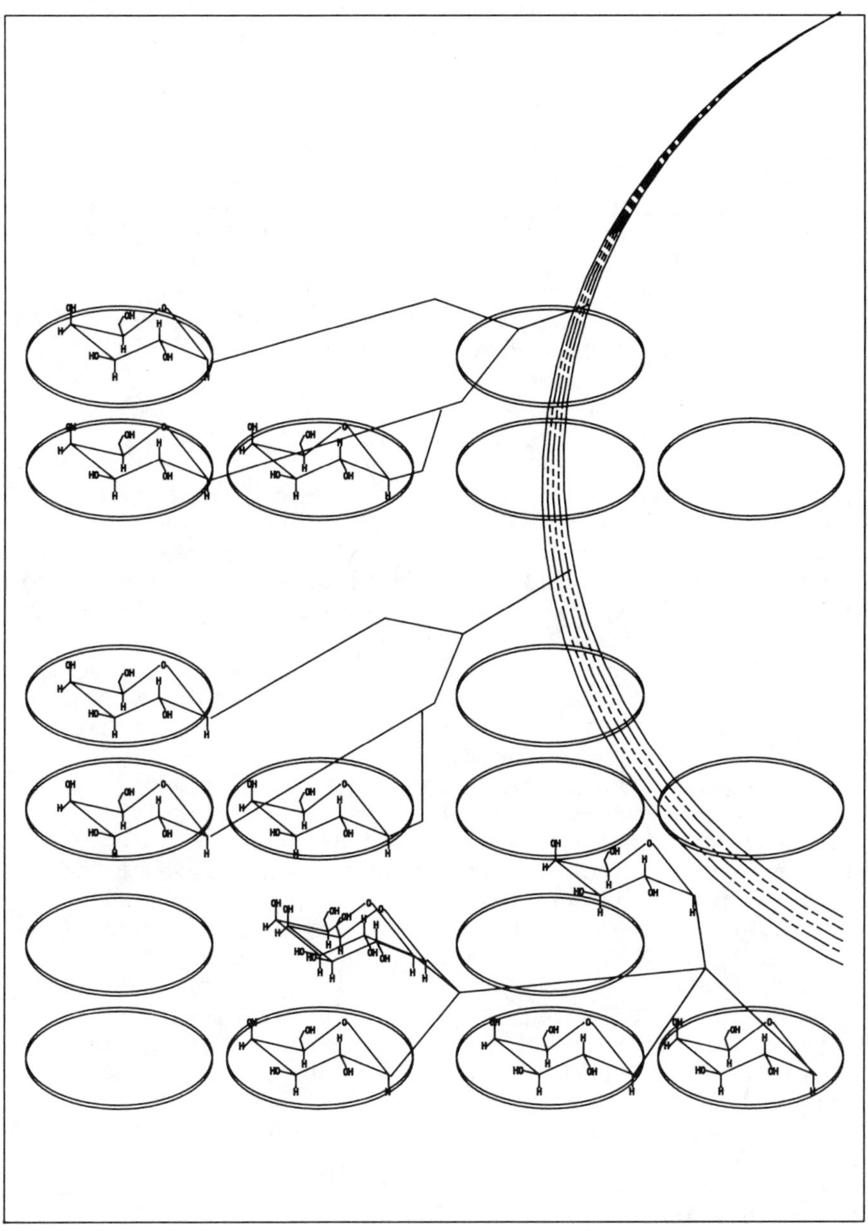

receptor and also escape from the cell. Extensive studies using ASOR have revealed evidence for two distinct populations of surface receptors (reviewed in Ref.32). It has been demonstrated that receptor which resurfaces with ASOR is delivered to an intracellular pool in which ligand dissociates more slowly relative to another intracellular pool (32).

Using a compound which contains two tris-galactosides linked to aspartic acid (DIAGRAM 3), Di-tris-lac, we have recently found that 1) the kinetics of exocytosis are identical for both ASOR and Di-tris-lac, 2) Di-tris-lac, unlike ASOR, is released from the cells after internalization, in the absence of binding inhibitors and 3) the maximum amount of internalized ligand that is released from cells is 50% for ASOR and 80% for Di-tris-lac. We have concluded that the kinetics of ligand movement back to the cell surface (exocytosis) is independent of carbohdyrate structure. Whether a ligand is released from the cell upon recycling of receptor-ligand complex is related to ligand dissociation rate and therefore, to carbohdyrate structure. We have found by kinetic modeling that the faster dissociation rate of Di-tris-lac will not explain the increased proportion (50 versus 80%) of Di-tri-lac recycling back to the cell surface using the model shown in Figure 2. We have not been able to alter this proportion by temperature modulation, length of uptake time, dose of ligand, or co-internalization with other ligands (ASOR) (R. Townsend and T. Chai, unpublished results). This partition of 80% of ligand bound for exocytosis and 20% destined for degradation appears to be a unique attribute of the ligand. In the case of ASOR the proportion is equivalent. We conjecture that the above discussed different mode of binding of glycosides and glycoproteins is related to this partitioning of ligand along two different pathways.

Figure 2 shows a working model for endo/exocytosis consisting of two inter-related receptor populations. We propose that both ligands take the same pathways but due to their dissociation rates and mode of binding manifest different dynamics. First if we follow ASOR through the pathways, ASOR binds to the two population of surface receptors in equal proportions (Steps 1 and 5). ASOR's dissociation rate is very slow and release within the cell (Step 4) does not occur prior to Steps 7 and 8. The (RL_i) complexes moves back to the cell surface and changes state (R') such that upon reinternalization delivery occurs to lysosome by either of the previously proposed "slow" and

$$
\begin{array}{ccccccc}
 & & & R'_s + L & & & \\
 & & & \downarrow\uparrow 5 & & & \\
 & & & (R'L)_s & \overset{6}{\rightarrow} (R'L)_i & \overset{8}{\rightarrow} & (R \text{ or } R')_s + L_d \\
 & & & \uparrow 3 & \downarrow 7 & & \\
R_s + L & \overset{1}{\leftrightarrow} & (RL)_s & \overset{2}{\rightarrow} (RL)_i & (R \text{ or } R')_s + L_d & & \\
 & & & \downarrow 4 & & & \\
 & & & (R \text{ or } R')_s + L & & &
\end{array}
$$

FIGURE 2. Mechanism of endo/exocytosis mediated by the hepatic Gal/GalNAc lectin. The designations are as follows. $(R)_s$ and $(R')_s$ represent two states of surface receptor with $(RL)_s$ and $(R'L)_s$ corresponding surface receptor-ligand complexes. The internal complexes are represented by $(RL)_i$ and $(R'L)_i$. Ligand destined for lysosomal degradation and extracellular unbound ligand are L_d and L, respectively. Steps 1 and 5 are reversible cell surface binding. Exocytosis steps are 3 and 4. Internalization is shown as steps 2 and 6. Fast and slow receptor ligand dissociation are given as steps 7 and 8.

"fast" dissociating pathways (32) (Steps 7 and 8) Receptor is recycled in either the R_s or R'_s state. Di-tris-lac binds with 80% to R_s and 20% to R'_s. Since this ligand rapidly exchanges with receptor, it recycles free from receptor in both Steps 4 and 5. This model explains in kinetic terms, the differences in exocytosis of ASOR and Di-tris-lac provided the proportions of the two cell surface receptor populations occupied by each ligand is different and the faster dissociation rate for Di-tris-lac is included. How carbohydrate structure can "signal" partitioning to either exocytosis or lysosomal degradation merits further study.

The function of exocytosis is not clear; it is possible that exocytosis is important to control ligand flux into the cell or that exocytosis is essential to receptor dynamics. Exocytosis may be a second control point, after binding, to insure that ligand is "ready" for degradation or it may be a mechanism to link the plasma milieu of carbohydrate structures to endocytosis events subsequent to binding. Exocytosis may be part of an interconversion of receptor states as I have proposed in the working model in Figure 2.

Clearly, a better understanding of the role of carbohydrate structure in mammalian lectin function will require improved definition of glycoprotein structure and their mulivalent binding to the *in situ* lectins.

ACKNOWLEDGMENTS

The generous gift of Di-tris-lac from Dr. R. T. Lee and the support from Dr. Y.C. Lee are gratefully acknowledged.

REFERENCES

1. Morell AG, Irvine, RA, Sternlieb I, Scheinberg HI and Ashwell G (1968). Physical and chemical studies on ceruloplasmin: V. Metabolic studies on sialic-free ceruloplasmin in vivo. J. Biol. Chem. 243:155.

2. Goldberg D, Gabel C, and Kornfeld S (1984) Processing of lysosomal enzyme oligosaccharide units. In Dingle JT, Dean RT and Sly W (eds): "Lysosomes in Biology and Pathology, Vol 7," New York: Elsevier, p 45.

3. Ashwell G and Harford J (1982). Carbohydrate-specific receptors of the liver. Ann. Rev. Biochemistry 51:531.

4. Lehrman MA, Haltiwanger RS and Hill RL (1986) The binding of fucose-containing glycoproteins by hepatic lectins: The binding specificity of the rat liver fucose lectin. J. Biol. Chem. 261:7426.

5. Haltiwanger RS, Lehrman MA, Eckhardt AE and Hill, RL (1986) The distribution and localization of the fucose-binding lectin in rat tissues and the identification of a high affinity form of the mannose/N-acetylglucosamine-binding lectin in rat liver. J. Biol. Chem. 261:7433.

6. Creek KE and Sly WS (1984) The role of the phosphomannosyl receptor in the transport of acid hydrolases to lysosomes. In Dingle JT, Dean RT and Sly W (eds): "Lysosomes in Biology and Pathology, Vol 7," New York: Elsevier, p 63.

7. Shepherd VL and Stahl PD (1984) Macrophage receptors for lysosomal enzymes. In Dingle JT, Dean RT and Sly W (eds): "Lysosomes in Biology and Pathology, Vol 7," New York: Elsevier, p 83.

8. Wileman TE, Lennartz MR and Stahl PD (1986) Identification of the macrophage mannose receptor as a 175-kDa membrane protein. Proc. Natl. Acad. Sci. (USA) 83:2501.

9. Mizuno Y, Kozutsumi Y, Kawasaki T and Yamashina I (1981) Isolation and characterization of a mannan-binding protein from rat liver. J. Biol. Chem. 256:4247.

10. Maynard Y and Baenziger JU (1982) Characterization of a mannose and N-acetylglucosamine-specific lectin present in rat hepatocytes. J. Biol. Chem. 257:3788.

11. Bezouska K, Táborský O, Kubrycht J, Pospísil and Kocourek J (1985) Carbohydrate-structure-dependent recognition of desialylated serum glycoproteins in the liver and leukocytes. Biochem. J. 227:345

12. Cerra RF, Gitt MA and Barondes SH (1985) Three soluble rat β-galactoside-binding lectins. J. Biol. Chem. 260:10474.

13. Montreuil J (1980) Primary structure of glycoprotein glycans: Basis for the molecular biology of glycoproteins Adv. in Carbohydr. Chem. and Biochemistry 37:157.

14. Sarkar M, Liao, J, Kabat EA, Tanabe T and Ashwell G (1979) The binding site of rabbit hepatic lectin. J. Biol. Chem. 254:3170

15. Baenziger JU and Fiete D (1980) Galactose and N-acetylgalactosamine-specific endocytosis of glycopeptides by isolated rat hepatocytes. Cell 22:611.

16. Lee YC, Townsend RR, Hardy MR, Lönngren J, Arnarp J, Haraldsson M and Lönn H (1983) Binding of synthetic oligosaccharides to the hepatic Gal/GalNAc lectin: Dependence on fine structural features. J. Biol. Chem. 258:199.

17. Freeze HH (1985) Interaction of Dictyostelium discoideum lysosomal enzymes with the mammalian phosphomannosyl receptor: The importance of oligosaccharides which contain phosphodiesters. J. Biol. Chem. 260:8857.

18 Savvidou G, Klein M, Grey AA, Dorrington KJ and Carver JP (1984) Possible role for peptide-oligosaccharide interactions in differential oligosaccharide processing at aspraragine-107 of the light chain and asparagine-297 of the heavy chain in a monoclonal $IgG_1\kappa$. Biochemistry 23:3736.

19. Lee RT and Lee YC (1980) Preparation and some biochemical properties of neoglycoproteins produced by reductive amination of thioglycosides containing an ω-aldehydoglycon. Biochemistry 19:156.

20.Lee RT, Myers RW and Lee YC (1982) Further studies on the binding characteristics of rabbit liver galactose/N-acetylgalactosamine-specific lectin. Biochemistry 24:6292.

21. Wong TC, Townsend RR and Lee YC (1987) Synthesis of D-galactosamine derivatives and binding studies using isolated rat hepatocytes. Carbohydr. Res. in press.

22. Ashwell G and Morell AG (1974) The role of surface carbohydrate on the hepatic recognition and transport of circulating glycoproteins. Adv. Enzymol 41:99.

23. Krantz, MJ, Holtzman, NA, Stowell, CP and Lee YC (1976) Attachment of thioglycosides to proteins: Enhancement of liver membrane binding. Biochemistry 15:3963.

24. Townsend RR, Hardy MR, Wong TC and Lee YC (1986) Binding of N-linked bovine fetuin glycopeptides to isolated rabbit hepatocytes: Gal/GalNAc hepatic lectin discrimination between Galβ(1,4)GlcNAc and Galβ(1,3)GlcNAc in a triantennary structure. Biochemistry 25:5716.

25. Bock K, Arnarp J and Lönngren J (1982) The preferred conformation of oligosaccharides derived from the complex-type carbohydrate portions of glycoproteins. Eur. J. Biochemistry 129:171.

26. Lee YC, Townsend RR, Hardy MR, Bock K and Lönngren J (1984) Binding of synthetic clustered ligands to the lectin on isolated rabbit hepatocytes. In Lo T-B, Liu T-Y and Li C-H (eds): "Biochemical and Biophysical Studies of Proteins and Nucleic Acids," Amsterdam: Elsevier, p 349.

27. Lee RT, Lin P and Lee YC (1984) New synthetic cluster ligands for galactose/N-acetylgalactosamine-specific lectin of mammalian liver. Biochemistry 23:4255.

28. Mellis SJ and Baenziger JU (1983) Structures of the oligosaccharides present at the three asparagine-linked glycosylation sites of human IgD. J. Biol. Chem. 258:11546.

29. Anderson DR, Atkinson PH and Grimes WJ (1985) Major carbohydrate structures of five glycosylation sites on murine IgM determined by high resolution ^{1}H-NMR spectroscopy. Arch. Biochem. Biophys. 243:605.

30. Hardy MR, Townsend RR, Parkhurst SM and Lee YC (1985) Different modes of ligand binding to the hepatic galactose/N-acetylgalactosamine lectin on the surface of rabbit hepatocytes. Biochemistry 24:22.

31. Townsend RR, Wall DA, Hubbard AL and Lee YC (1984) Rapid release of galactose-terminated ligands after endocytosis by hepatic parenchymal cells: Evidence for a role in carbohydrate structure in the release of internalized ligand from receptor. Proc. Natl. Acad. Sci. (USA) 81:466

32. Weigel PH (1985) Receptor recycling and ligand processing mediated by hepatic galactosyl receptor: A two-pathway system. In Parent JB and Olden K (eds): "Vertebrate Lectins," New York: Van Nostrand Reinhold, p 65.

The Pharmacology and Toxicology of Proteins, pages 59–72

RESIDUALIZING LABELS FOR IDENTIFYING CELLS[1] ACTIVE IN PROTEIN UPTAKE AND CATABOLISM

Janet L. Maxwell, John W. Baynes
and Suzanne R. Thorpe

Department of Chemistry and School of Medicine,
University of South Carolina, Columbia, SC 29208

ABSTRACT Residualizing labels are biologically inert radioactive or fluorescent tags for proteins which are retained in lysosomes following cellular uptake and catabolism of the carrier protein. The label, dilactitol-^{125}I-tyramine has been used for autoradiographic identification of skin fibroblasts as a major site of albumin catabolism in rats. N,N-dilactitol-N'-fluoresceinyl-ethylenediamine has been used to isolate albumin-degrading fibroblasts from cell cultures by fluorescence activated cell sorting. Residualizing labels should also be useful for assessing the efficiency of targeting of antibodies, immunotoxins and other protein pharmaceuticals in vivo.

INTRODUCTION

One of the major challenges of modern pharmacology is the development of methods for selective targeting of toxic pharmaceutical agents to affected tissues. Current strategies for targeted drug delivery include immunological techniques which employ antibodies against surface antigens on target cells, and receptor-based techniques in which the drugs are coupled to protein ligands for receptor-mediated endocytic systems in target cells. The immunological approach makes use of tissue-specific mono- or polyclonal antibodies which may rely on either endogenous complement

[1]This work was supported by National Institutes of Health Research Grant AM-25373.

activity or covalently coupled drugs or toxins (1-4) for target cell destruction. Coupling of drugs or radiopharmaceutical agents to hormones (2,4,5) or protein ligands, such as asialofetuin (ASF)[2] (2,4), also promotes the delivery of these compounds to tissues expressing the appropriate receptor.

Radiochemical measurements and imaging are frequently used to evaluate the targeting efficiency of drug delivery systems in vivo. Unfortunately, some of the more convenient labels, such as radioactive iodine (*I) coupled directly to tyrosine residues in protein, are lost from tissues following endocytosis and degradation of the labeled antibody or carrier protein. The loss of radioactivity from target tissues can lead to an underestimate of the actual efficiency of the delivery system, as well as an increase in background radioactivity in all tissues. This problem is more serious when the kinetics of targeting are slow, such as might be observed when a tumor is poorly vascularized or when a low capacity receptor system is temporarily saturated by an excess of drug-ligand conjugate. In these cases it may be difficult to evaluate the efficiency of the targeting process if the kinetics of degradation of the carrier protein and release of degradation products from the cell are comparable to the kinetics of the delivery system.

Ideally, the radioactive label used in these pharmacokinetic studies would have a long residence time in the target tissue, regardless of the metabolic fate of the labeled protein. For studies on the sites of catabolism of plasma proteins which have slow rates of uptake in tissues, i.e., long circulating half-lives, we have developed a series of radioactive labels, which we call "Residualizing" labels (R-labels). These labels are retained, i.e., residualize, in tissues following degradation of the carrier protein. They have been used in model studies on enzyme therapy to identify the tissue and cellular sites of uptake and catabolism of lysosomal enzymes (6) and should also be useful for identifying the sites of uptake and assessing the cumulative, long-term delivery of drug-protein conju-

[2] Abbreviations used: ASF, asialofetuin; DLF, N,N-dilactitol-N'-fluoresceinyl-ethylenediamine; DLT, dilactitol-tyramine; FL, fluorescein isothiocyanate; *I, radioactive iodine; InTn, inulin-tyramine; R-label, Residualizing label; RSA, rat serum albumin.

gates to target cells. We will describe here the development of R-labels and their application in our studies on the sites of catabolism of the plasma protein, albumin, and will also outline potential applications of these labels in studies on the targeting of protein pharmaceutical agents.

RATIONALE FOR DEVELOPMENT OF RESIDUALIZING LABELS

The scheme shown in Figure 1 describes a typical sequence of events involved in the turnover of a circulating protein, whether an endogenous protein such as albumin or an injected drug-protein conjugate. In either case the overall kinetic process is complex, involving either fluid-phase or receptor-mediated endocytosis, followed by lysosomal degradation and then loss of the degradation products from the cell. In our studies on albumin catabolism, we set out first to identify the tissue and cellular sites of degradation of the protein, and then to characterize mechanisms regulating the protein's catabolism at these sites.

For a short-lived protein, such as ASF (half-life less than 3 min at low dose in rats), it is relatively easy to identify its site of degradation in vivo. This protein is recognized by the hepatocyte receptor for galactose terminal glycoproteins (7) and is removed from the circulation at a rate significantly greater than its rate of degradation in lysosomes and release of degradation products from liver (Fig. 1: $k_1 >> k_2$ and k_3). The catabolism of ASF is readily monitored by labeling its tyrosine residues with *I, using reagents, such as chloramine T, lactoperoxidase or Iodogen. Following injection of *I-ASF into the circulation, radioactivity accumulates rapidly in liver (90% of injected dose within 15 min), permitting the identification of hepatocytes as the site of degradation of ASF. However, the degradation products, iodine and iodotyrosines, are rapidly released from liver with a half-life of about 30 min, and only traces of radioactivity are detectable in liver after a few hours.

Because of the relatively rapid kinetics of exocytosis of degradation products, direct labeling of proteins with *I is not useful for identifying the sites of catabolism of longer-lived, slowly catabolized proteins such as albumin or immunoglobulins. In these cases the proteins are delivered to cells involved in their catabolism at a rate which is significantly slower than rates of lysosomal degradation and exocytosis (Fig. 1: $k_1 << k_2$ and k_3). Thus, degradation

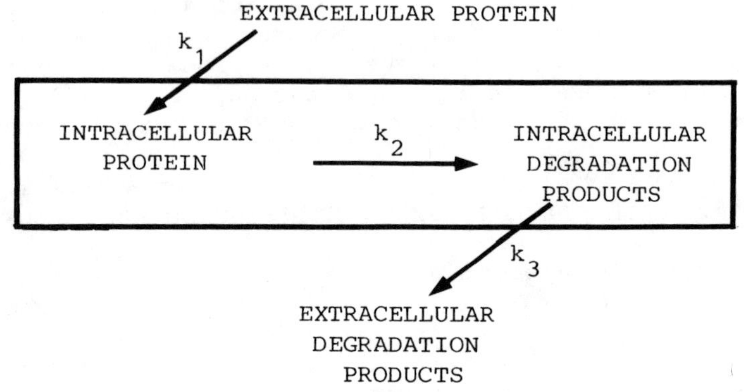

Labeled Probe	Kinetic Features	Degradation Products
*I-Asialofetuin	$k_1>>k_2,k_3$	Temporary Accumulation
*I-Albumin	$k_1<<k_2,k_3$	No Accumulation
*R-Protein	$k_3<k_1,k_2$	Long-term Residualization

FIGURE 1. Kinetic model for catabolism of plasma proteins and disposition of degradation products. The diagram at the top shows a simplified model of compartments and kinetic processes involved in degradation of extracellular or circulating proteins. The tabular data at the bottom describe observations expected on injection of proteins labeled with conventional or R-labels.

products do not accumulate in the cells involved in the proteins' catabolism. To circumvent this problem we have developed the R-labels which, as represented on the bottom line in Figure 1, are radioactive or fluorescent labels for protein which are retained in lysosomes following degradation of the carrier protein (Fig.1: $k_3<k_1$ and k_2). These types of labels have proven useful in our laboratory for

identifying the tissue and cellular sites of degradation of albumin (8,9) and IgG (10) in the rat, and in studies on lipoprotein metabolism in other mammals and in cell culture by Daugherty et al. (11) and Steinberg, Pittman et al.(12).

PROPERTIES OF RESIDUALIZING LABELS

Figure 2 shows several compounds which have been used as R-labels in studies on plasma protein catabolism. The original prototypes were ^{3}H- or ^{14}C-labeled compounds, such as [^{3}H]raffinose (13) or [^{14}C]sucrose (12), but the present generation of compounds contains higher specific activity iodine isotopes (12,14) or fluorescent tags which allow for a range of detection techniques such as autoradiography, fluorescence microscopy and non-invasive tomographic imaging. The R-labels are characteristically hydrophilic in nature, i.e., are extracted into the aqueous phase during solvent partitioning, and have molecular weights in excess of 500 amu to limit their diffusion through the lysosomal membrane after degradation of the carrier protein. Each of the compounds is first tested to confirm that its coupling to protein does not affect the kinetics of the protein's clearance from the circulation. The absence of an effect on the fractional catabolic rate of a long-lived protein, such as albumin, is interpreted as evidence that the labeling procedure does not affect normal mechanisms and sites of catabolism of the protein. The usefulness of R-labels to identify sites of protein uptake and catabolism relies on their entrapment in lysosomes, both because of their large size and hydrophilicity and their resistance to degradation by lysosomal enzymes. The resistance to enzymatic degradation results from either the absence of an appropriate lysosomal hydrolase or steric inaccessibility to the enzyme. Thus, when *I-DLT is coupled to protein, as shown in Figure 3, the galactose residue is inaccessible to lysosomal β-galactosidase. Similarly, because of the absence of lysosomal fructofuranosidase in mammals, the inulin component of *I-InTn is

---

FIGURE 2 (opposite page). Structure of typical residualizing labels. *I-DLT and DLF are coupled to proteins by activation of the terminal galactose residues, as described in Figure 3. *I-InTn is coupled to protein through the tyramine residue using cyanuric chloride (12,14).

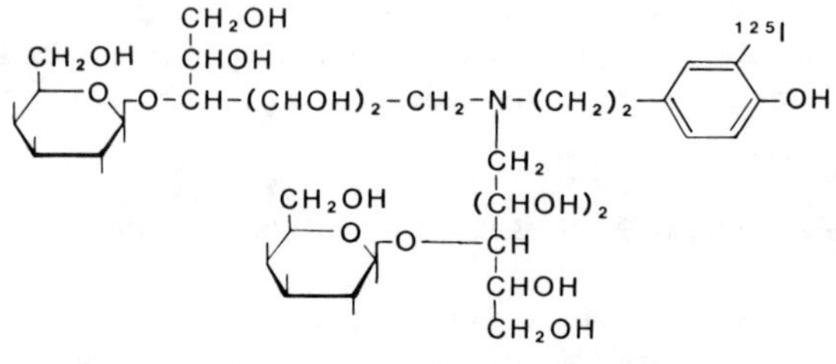

DILACTITOL-¹²⁵I-TYRAMINE (*I-DLT)

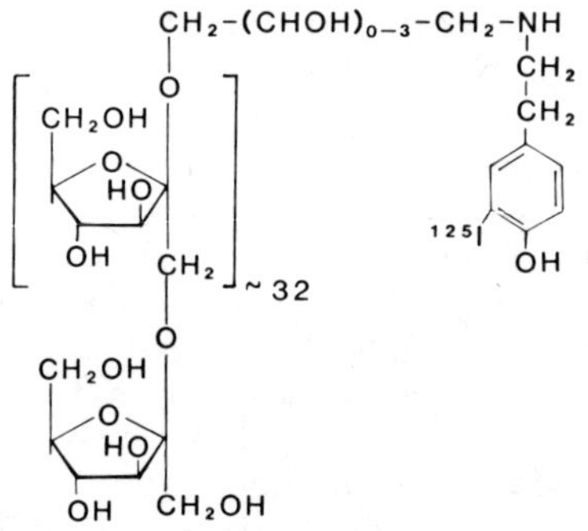

INULIN-¹²⁵I-TYRAMINE (*I-InTn)

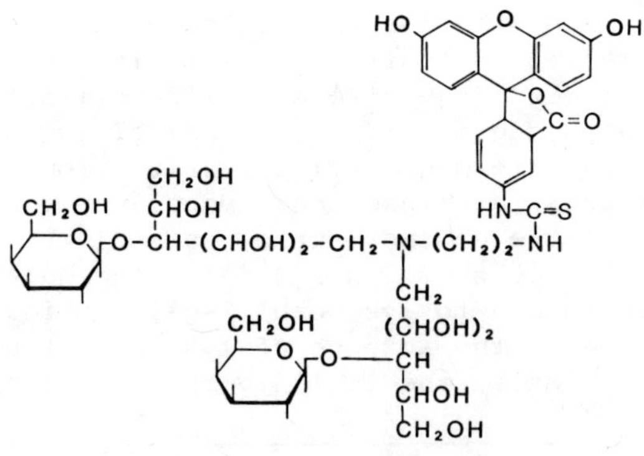

N,N′-DILACTITOL-N′-FLUORESCEINYL-
ETHYLENEDIAMINE (DLF)

Fig. 2.

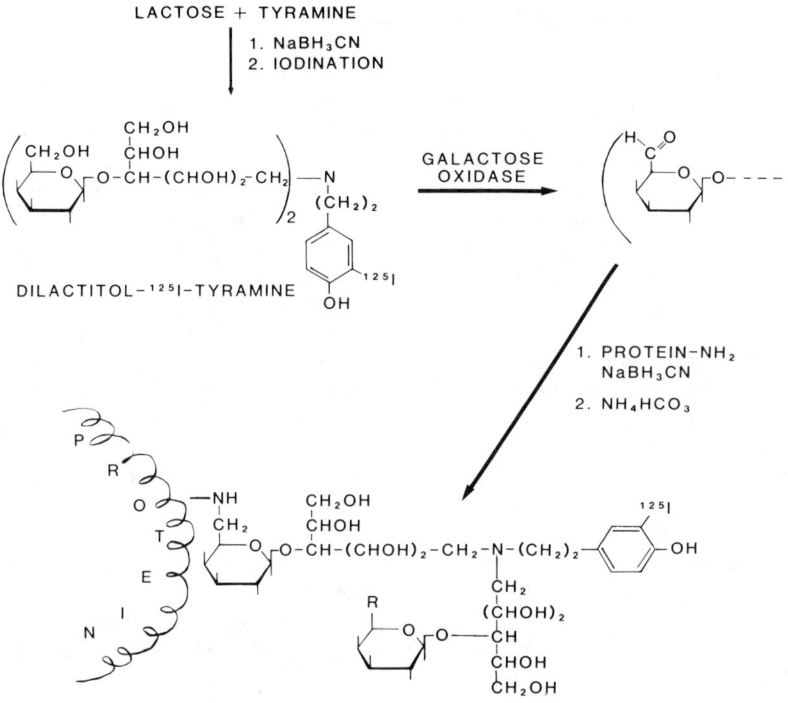

FIGURE 3. Chemistry of coupling of *I-DLT to protein (14). *I-DLT is first activated by oxidizing the C-6 residue of galactose to an aldehyde using galactose oxidase. The product is then coupled to amino groups on protein by reductive amination with $NaBH_3CN$. The reactions are carried out sequentially at pH 7.5, typically within a 2 hour period. Coupling efficiencies range from 20-50%, depending on the protein being labeled. Despite the potential for oxidizing both galactose residues in DLT, no crosslinking of any protein has been observed.

also stable in lysosomes. The various labels are readily detected in cells, both in cell culture and in vivo, where they are recovered as low molecular weight, acid soluble or dialyzable degradation products in the lysosomal subcellular fraction.

EVALUATION OF RESIDUALIZING LABELS

ASF is frequently used as a test protein for evaluating R-labels in vivo. Essentially all of the labeled protein can be delivered quickly to hepatocytes where it is almost completely degraded to acid soluble products in lysosomes within an hour. Thus, by measuring the rate of loss of the radioactive label from liver, one can obtain a measure of the efficiency of residualization of the label in lysosomes. We have determined that, for the radioactive labels shown in Figure 2, the degradation products, once released from liver, do not redistribute into other tissues, but are excreted nearly quantitatively in urine and feces. Thus, as shown in Figure 4, it is convenient to measure the rate of loss of radioactivity from the whole body as an indicator of retention of the radioactive label in hepatocyte lysosomes. For fluorescent labels, such as DLF, it is necessary to measure residual fluorescence in livers from a series of animals sacrificed at various times after injection. The data in Figure 4 show that none of the labels designed thus far are completely efficient with respect to residualization. However, half-lives for retention of these labels range from 2 - 5 days in rat liver, compared to less than 1 hour for conventional labels such as *I and fluorescein isothiocyanate (FL). InTn, which is larger and more hydrophilic than DLT, is retained more efficiently in liver and peripheral tissues, and should prove useful in studies on longer-lived proteins, such as the IgG immunoglobulins. While more efficient labels are being sought, the present labels are useful since their rate of leakage and/or exocytosis from cells is generally equal to or slower than the rate of catabolism of most plasma proteins in the rat.

APPLICATIONS OF RESIDUALIZING LABELS IN STUDIES ON PLASMA PROTEIN CATABOLISM IN VITRO AND IN VIVO

The [^{3}H]raffinose label was used originally to identify skin and muscle as the primary tissues sites of catabolism of rat serum albumin (RSA) in vivo (8). We then used *I-DLT-RSA and autoradiography to identify fibroblasts as the cells involved in catabolism of RSA in these tissues (9). Preliminary studies with *I-DLT-IgM in the rat indicate that liver has a major role in IgM turnover. *I-DLT has also been used to identify the liver and adrenal glands

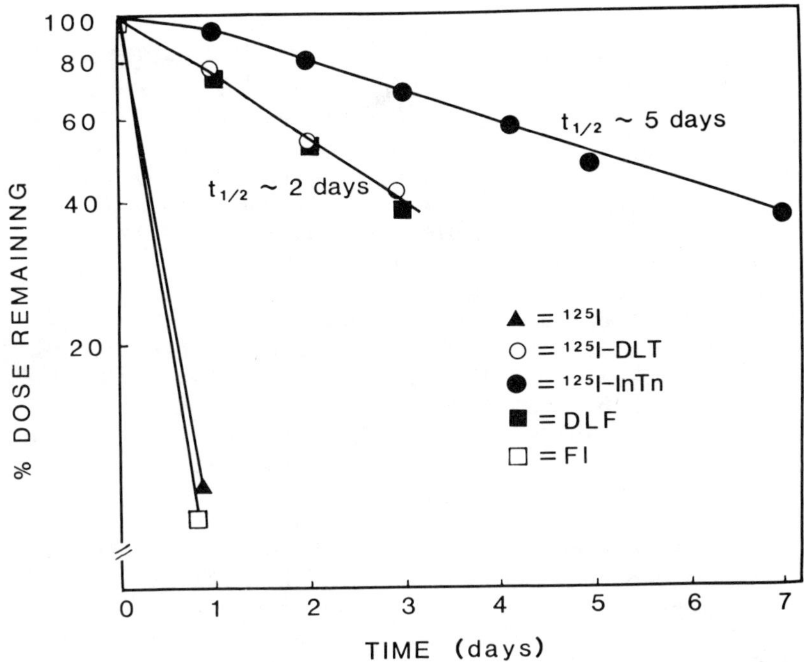

FIGURE 4. Kinetics of loss of various labels from liver, as measured by whole-body counting and fluorescence measurements. Rats were injected with *I-, *I-DLT-, *I-InTn-, DLF- or FL-labeled ASF. Measurements of whole-body radioactivity were begun at 30 min (defined as 100% dose) using a well-type gamma counter (14). For the DLF- and FL-ASF experiments, animals were sacrificed at each time point for measurement of fluorescence in a detergent-solubilized supernatant fraction of a liver homogenate. Studies conducted in parallel with these experiments, indicated that more than 90% of hepatic radioactivity and fluorescence was in the form of low molecular weight degradation products by 1 hour, and that at all times more than 90% of recovered radioactivity was localized in liver. The rapid loss of radioactivity and fluorescence from animals injected with *I-ASF and FL-ASF, respectively, illustrates the poor retention and rapid exocytosis of degradation products from conventionally labeled proteins.

as important sites of catabolism of very low and low density lipoproteins (β-VLDL and LDL) in the rabbit (11). R-labels have also proven useful in studies on lipoprotein (11) and albumin catabolism in fibroblasts in cell culture. The accumulation of labeled degradation products in cells eliminates possible artifacts resulting from catabolism of the protein at the cell surface or spontaneous loss of label from the protein in the incubation medium.

In our studies on the catabolism of *I-DLT-RSA in the rat we observed that only about 10% of fibroblasts in skin were labeled with autoradiographic grains (9). This suggested the possibility that either a subset of fibroblasts was active in albumin catabolism or that fibroblasts were active in albumin catabolism only at specific stages in their cell cycle. We developed techniques for isolation of the intact labeled cells in good yield from rat skin (9), and established that the radioactivity inside cells was acid-soluble and accounted for more than 80% of the estimated degradation products in the skin sample. The labeled cells were concentrated on the low density side of the total cell population in a Metrizamide density gradient, however we were unable to obtain good separation of labeled from unlabeled cells by density gradient fractionation.

We reasoned that fractionation of catabolically active from inactive cells might be achieved by fluorescence activated cell sorting using fluorescent R-labels. Thus far, we have synthesized DLF as the prototype of fluorescent R-labels. Fluorescence from DLF-ASF was observed to residualize efficiently in hepatic lysosomes (Fig. 4). In preliminary studies we have also shown that modification of RSA with less than 1 mol DLF/mol RSA does not affect the kinetics of catabolism of the protein, and that accumulation of the fluorescent label in skin and muscle and in isolated skin cells can be detected and quantitated at 4 days (2 half-lives) after injection of DLF-RSA. These in vivo experiments require the injection of larger amounts of protein than needed with the radioactive labels, i.e., about 5 mg DLF-RSA vs. 100 ug *I-DLT-RSA per 100 g body weight. Although this quantity of protein does not appreciably alter the total body or plasma mass of albumin, it could cause problems in studies on other proteins which are present at lower concentrations in blood. This problem can be overcome by the synthesis of labels which fluoresce with greater quantum efficiency and at wavelengths at which there is less endogenous tissue background.

We have begun to use DLF-RSA in studies on the

catabolism of albumin by rat fibroblasts in tissue culture. Preliminary experiments have established that fluorescence from DLF-RSA accumulates in these cells at about the same rate as radioactivity from *I-DLT-RSA, and that the cellular fluorescence is readily measured by flow cytometry. As shown in Figure 5A and 5C, fibroblasts are labeled with nearly equal efficiency by FL-RSA and DLF-RSA after a 24 hour incubation. However, after a second 24 hour incubation in the absence of labeled protein, i.e., a wash-out or chase experiment, the cells exposed to FL-RSA have lost the majority of their fluorescence (Fig 5B), while the fluorescence in cells labeled with DLF-RSA is apparently unchanged (Fig. 5D). The fluorescence associated with the cells exposed to FL-RSA (Fig. 5A) is about 80% in the form of intact protein, while that from cells exposed to DLF-RSA (Fig. 5C) is more than 80% in the form of low molecular weight degradation products. Measurements of fluorescence in detergent extracts of the cells indicated that, despite apparently comparable fluorescence by flow cytometry (Fig. 5A vs. 5C), the cells exposed to DLF-RSA contained 3-4 times as much fluorescence as those exposed to FL-RSA. Similarly, although it is not apparent from the fluorescence profiles in Figure 5C vs. 5D, about 40% of the label from DLF-RSA is lost during the 24 hr wash-out period. The discrepancies probably result from variable quenching of fluorescence in the cells, e.g., in the acid milieu of the lysosome. We are now in the process of isolating the populations of least and most fluorescent cells from the cell culture experiments in order to characterize the cells most active in catabolism of RSA.

PHARMACOLOGICAL APPLICATIONS OF R-LABELS

Thus far, R-labels have been applied almost exclusively in studies on the uptake and catabolism of plasma proteins or enzymes in whole animals and in cell culture. However, the technology is clearly applicable to studies on the distribution of protein pharmaceuticals and drugs and toxins targeted to cells via conjugates with immunoglobulins or with ligands for receptor-mediated endocytic systems. Either the pharmaceutical agent itself or the protein carrier may be labeled, or both simultaneously with different isotopes. R-labels should also be especially useful for evaluating effects of conjugation reactions on

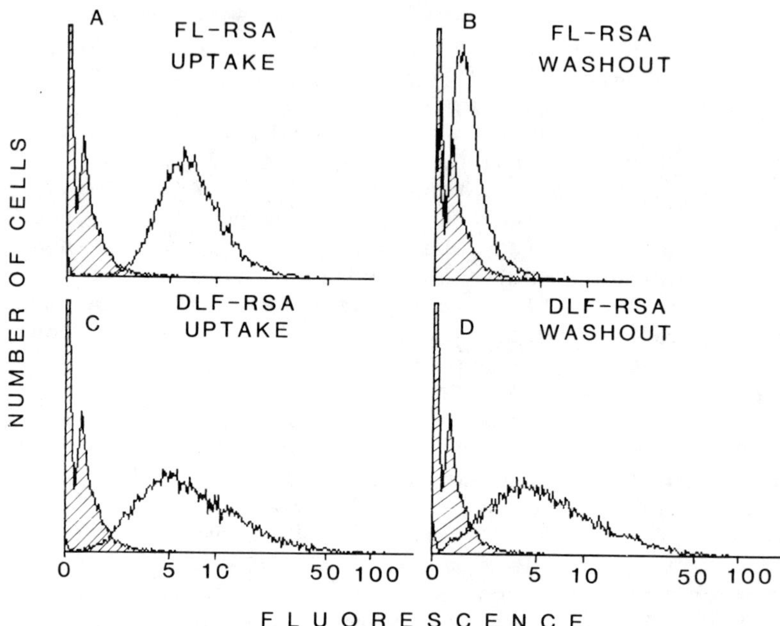

FIGURE 5. Uptake of FL- and DLF-RSA and release of degradation product by rat fibroblasts in cell culture. Confluent rat fibroblasts were incubated with similar amounts of either FL- or DLF-RSA in Ham's F-10 medium containing 10% fetal calf serum. After 24 hours one half of the plates were rinsed, and the cells reincubated for an additional 24 hours (wash-out) in fresh medium in the absence of fluorescent protein. Fluorescence analyses were carried out on a Coulter-Epics Flow Cytometer. (A) Cells incubated with FL-RSA for 24 hours, compared to control cells. (B) Reincubation of cells in Frame A in absence of FL-RSA, showing loss of majority of fluorescence during wash-out period. (C) Cells incubated with DLF-RSA as in Frame A. (D) Reincubation of cells in Frame C in absence of DLF-RSA, showing residualization of fluorescent degradation products from DLF-RSA.

the metabolic handling of proteins. Using the R-labels it would be possible to determine whether release of radioactivity from target tissue results from sluffing-off of immunoglobulin or receptor-ligand conjugates from the surface of target cells or from loss of radioactivity following uptake and catabolism of the protein.

While the current group of R-labels rely on an amino group for conjugation to protein, we are developing R-labels which can also be reacted with sulfhydryl and hydroxyl groups on proteins and carbohydrates. In addition, although assays for measuring the distribution of R-labels in whole animals have relied, thus far, on invasive techniques, usually an autopsy of the animal, we are now designing and evaluating R-labels which should be useful for non-invasive imaging, such as scintigraphy and positron emission tomography. These labels should provide high resolution imaging of the kinetics of transport and final disposition of proteins in intact animals. In general, the R-labels are conveniently prepared and are coupled to proteins under gentle conditions. Their long-term residualization in tissues should make them useful in a wide range of studies on the transport, distribution and targeting of protein pharmaceutical agents.

REFERENCES

1. Scheinberg DA, Strand M (1983). Kinetic and catabolic considerations of monoclonal antibody targeting in erythroleukemic mice. Cancer Res 43:265.
2. Gregoriadis G (1981). Targeting of drugs: implications in medicine. Lancet ii:241.
3. Vitetta ES, Uhr JW (1985). Immunotoxins: Redirecting Nature's Poisons. Cell 41:653.
4. Sikora K, Smedley H, Thorpe P (1984). Tumour imaging and drug targeting. Brit Med Bull 40:233.
5. Lindner HR, Kohen F, Amsterdam A (1980). An approach to site-directed chemotherapy of hormone-sensitive cancer. In Iacobelli S, King RJB, Lindner HR, Lippman ME (eds): "Hormones and Cancer," Raven Press, NY, p 541.
6. Morrone S, Pentchev PG, Baynes JW, Thorpe SR (1981). Studies in vivo of the tissue uptake, cellular distribution and catabolic turnover of exogenous gluco-glucocerebrosidase in the rat. Biochem J 194:733.

7. Neufeld EF, Ashwell G (1980). In Lennarz WJ (ed): "The Biochemistry of Glycoproteins and Proteoglycans," New York, Plenum Publishing Corp., p 241.
8. Baynes JW, Thorpe SR (1981). Identification of the sites of albumin catabolism in the rat. Arch Biochem Biophys 206:372.
9. Strobel JL, Cady CG, Borg TK, Terracio T, Baynes JW, Thorpe SR (1986). Identification of fibroblasts as a major site of albumin catabolism in peripheral tissues. J Biol Chem 261:7989.
10. Henderson LA, Baynes JW, Thorpe SR (1982). Identification of the sites of IgG catabolism in the rat. Arch. Biochem Biophys 215:1.
11. Daugherty A, Thorpe SR, Lange LG, Sobel BE, Schonfeld G (1985). Loci of catabolism of β-very low density lipoprotein in vivo delineated with a residualizing label, ^{125}I-dilactitol-tyramine. J Biol Chem 260:14564.
12. Pittman RC, Taylor CA Jr. (1986). Methods for assessment of tissue sites of lipoprotein degradation. Meth Enzymol 129:612.
13. Van Zile J, Henderson LA, Baynes JW, Thorpe SR (1979). [^{3}H]Raffinose, a novel radioactive label for determining organ sites of catabolism of proteins in the circulation. J Biol Chem 254:3547.
14. Strobel JL, Baynes JW, Thorpe SR (1985). ^{125}I-Glycoconjugate labels for identifying sites of protein catabolism in vivo: effect of structure and chemistry of coupling protein on label entrapment in cells after protein degradation. Arch Biochem Biophys 240:635.

SECTION III: PHARMACOKINETICS

L. Benet, UC San Francisco, presented the general principles. Pharmacokinetics was defined as what the body does to the drug while pharmacodynamics is what the drug does to the body. Clearance was shown to define dosing rate, half-life to define dosing interval, bioavailability to define dose adjustment and volume of distribution to define the maximum and minimal plasma concentrations. A detailed discussion followed about incomplete bioavailability which may be due to problems with dosage form (irregular release, poor solubilization), poor membrane permeability, and site effects (first pass metabolism, enterohepatic circulation, poor perfusion at the site of action or inadequate lymphatic drainage). Finally he showed that one must use the full pharmacokinetic equations including terminal half-lives to predict the total exposure of the body to a drug and its organ toxicity.

J. Young, Cetus Corporation, showed how these principles apply to their recombinant interleukin-2. In the rat, IL-2 has half lives of 5 minutes, 0.5 hours and 4 hours with the third phase accounting for only 3% of the total exposure (area under the curve). Total body clearance was linear with dose. Intramuscular, intraperitoneal and subcutaneous doses produced similar curves. Bioavailability is about 25% with major metabolism in the kidney. He pointed out that the biodynamic effects on T cell growth are quite prolonged relative to the short plasma half-lives.

Next J. Weinstein, National Cancer Institute, showed that some of the bioavailability problems after intravenous administration of monoclonal antibodies may be circumvented by subcutaneous injections into areas drained by lymphatics. He gave examples of this approach to image and treat draining lymph nodes. Administration into the lung parenchyma was shown to deliver antibodies to the draining nodes. Dr. Benet then summarized the discussion of these papers.

A. Abuchowski, Enzon Corporation, demonstrated how chemical modification of certain enzymes with polyethylene glycol (PEG) can increase their circulating half-lives, stability under physiologic conditions and greatly decrease their immunogenicity. Examples were given of adducts with asparaginase, uricase, superoxide dismutase and adenosine deaminase (ADA). Prolonged treatment with the PEG-ADA has improved the immune status of 2 patients with combined immune difficiency syndrome.

S. Pizzo, Duke University, described chemical modifications of alpha-2 macroglobulin which probe its activity as a general protease inhibitor, its receptor binding and its ability to suppress macrophage functions. K. Melmon, Stanford University, discussed a different role for peptides: the modification of the pharmacologic properties of bioactive amines.

Controlled release dosage forms were first discussed by A. Amkraut, Alza Corporation. Various osmotic pumps are available to deliver continuous or preprogrammed intermittent doses. Techniques to target administration to specific tissues were presented. Next, F. Martin of Liposome Technology, discussed the uses of these biodegradable lipid vesicles to deliver peptides. He presented their approaches to optimize lipid composition, particle size, drug loading, storage and other properties needed to create a useful formulation.

K. Smith, Bend Research, described bioerodible water-soluble polyester polymers that can trap up to 20 wt% of peptides and proteins. With drugs of molecular weight greater than 3000 daltons release was controlled by bioerosion. Release was effected by diffusion and erosion for smaller molecules. These polymers could block tumor capillaries and locally release bioactive materials.

The Pharmacology and Toxicology of Proteins, pages 75–89

DELIVERY OF MONOCLONAL ANTIBODIES TO LYMPH NODES VIA THE LYMPHATICS

John N. Weinstein[1], Christopher D.V. Black[1], Oscar D. Holton, III[1], David G. Covell[1], Robert J. Parker[1], James L. Mulshine[2], Michael T. Lotze[1], Jorge Carrasquillo[3], Renee R. Eger[1], Anne Lewis[1], Steven M. Larson[3], and Andrew M. Keenan[3]

1. National Cancer Institute (NCI), National Institutes of Health, Bethesda, MD 20892, U.S.A.

2. NCI-Navy Medical Oncology Branch, Naval Hospital, Bethesda

3. Clinical Center, National Institutes of Health

ABSTRACT Subcutaneous injection provides an efficient way to deliver monoclonal antibodies via the lymphatics to normal cell targets, to lymphoma cells, and to early metastases in lymph nodes. Possible applications of the technique include "immunolymphoscintigraphy", "immunolymphotherapy", and regional "immunomodulation". One barrier to successful use of the lymphatic route (and other routes as well) is the "percolation" problem: bindable antibodies may not easily penetrate solid tumors. There is also a second type of access problem: Some lymph node groups are not accessible from subcutaneous injection sites. Particularly important are the lymph node groups that drain the lung and

---

[1] R.R.E. received support from a gift by Hybritech, Inc. to the Cancer Immunotherapy Fund managed by the Foundation for Advanced Education in the Sciences.

constitute sites for metastasis of lung cancers. One possible solution is the injection of antibody through a fiberoptic bronchoscope into lung parenchyma. Preliminary studies in dogs demonstrate efficient, selective delivery by this approach. Analogous endoscopic injection techniques could be useful in other settings such as that of carcinoma of the colon and rectum.

Preliminary findings with respect to immunolymphoscintigraphy in melanoma have not been encouraging, but those using T101 antibody in patients with T-cell lymphoma have yielded strongly positive results. Uptake in the regional lymph nodes exceeded 20 percent of injected dose, and most of that uptake appeared to be antigen-specific. However, the antibody bound to normal T-cells, hence the uptake was not necessarily tumor-specific.

INTRODUCTION

For scintigraphic imaging and for other in vivo applications, monoclonal antibodies have generally been administered intravenously to gain access to all tissues served by the systemic circulation. However, regional delivery should be considered when the location of the target antigen is anatomically restricted. If the target is a malignant cell or an immune subset within lymph nodes, then delivery via the lymphatic system can be quite efficient (1-4).

Diagnostic assessment of lymph node status is important in the staging of most malignancies, including melanoma, Hodgkin's disease, non-Hodgkin's lymphoma, and carcinomas of the breast, testis, prostate, colon, lung, and cervix. Techniques in current practice for noninvasive detection of tumor are relatively insensitive, hence the interest in antibodies. Physical examination, x-ray, computerized tomography, gallium scanning, and magnetic resonance imaging are all unreliable for detecting tumors much less than about 1 cm in diameter (5). Lymphangiography with radiologic contrast material is a cumbersome procedure, and the only site of injection in routine clinical practice is the foot.

Lymphoscintigraphy with ^{99m}technetium colloid has been used for detection of internal mammary metastases of breast cancer (6). Radiolabeled colloid is injected under the costal margin, and its uptake by macrophages in the regional lymph nodes monitored scintigraphically. A positive reading is recorded only if the lymph node chain is grossly disrupted.

The problem is that each of these approaches is essentially nonspecific. In principle, imaging with antibodies directed against tumor cells could be thousands of times more sensitive (7). Regional delivery via the lymphatic system represents an attempt to overcome some of the pharmacologic inefficiencies associated with intravenous administration of antibodies.

The term "lymphatic delivery" will be taken to mean either direct injection into cannulated lymphatic vessels or injection into an interstitial space for uptake by the lymphatic capillaries.

LYMPHATIC PHYSIOLOGY

The lymphatic system is responsible for scavenging proteins that have leaked out of the circulation, filtering them through lymph nodes, and returning them to the bloodstream. This function provides the basis for lymphatic delivery of antibodies and is therefore worth reviewing here briefly.

The lymphatic system is shown schematically in Figure 1. If a low molecular weight, water soluble substance is injected subcutaneously, almost all of it passes directly into blood capillaries. Macromolecules, on the other hand, are prevented by the basement membrane and endothelium of blood capillaries from passing readily into those vessels. Terminal lymph capillaries, to the contrary, are specialized for uptake of macromolecules, particles, and cells. They lack a formed basement membrane (8) and have clefts in their endothelial walls (9). The clefts are thought to open and close in response to hydrostatic pressure differences and local muscle activity. Most of the IgG molecules mobilized from a site of subcutaneous injection appear to pass into the lymphatics rather than the blood capillaries.

Once formed, lymph flows into larger and larger vessels under the impetus of hydrostatic pressure, extrinsic skeletal muscle action, and rhythmic contraction

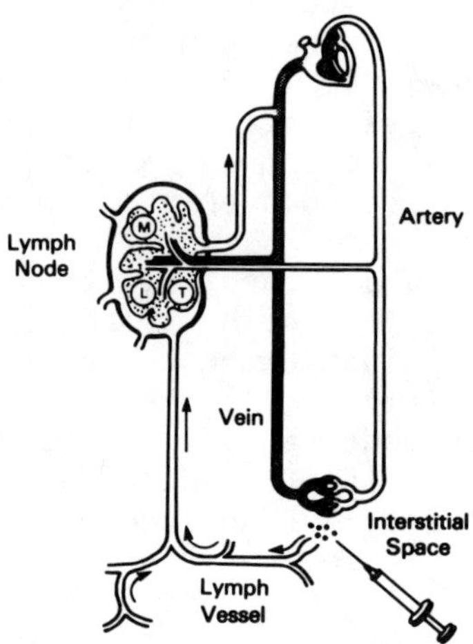

FIGURE 1. Schematic view of subcutaneous antibody administration. Immunoglobulin molecules pass through endothelial clefts into a terminal lymphatic vessel and flow with the lymph toward regional lymph nodes. In the first node encountered, or in more distant nodes of the chain, antibody may bind to normal cells such as the macrophages (M) and lymphocytes (L), or to tumor cells (T). Antibody not cleared from the lymph flow finally passes into the bloodstream, largely via the thoracic duct. Modified from Ref. 1.

of smooth muscle in the walls of some vessels. Retrograde flow is prevented by a system of unidirectional valves. Upon reaching a lymph node, lymph flows into the subcapsular space, then through sinusoids in the cortex and medulla. Antibodies in the lymph may bind specifically or nonspecifically to the various cell types or stromal elements in the node. Immunoglobulin not bound in the first node passes to others in the chain. If still not removed from the lymph, it passes into the bloodstream, in large part via the thoracic duct and right lymph duct. The lymphatics thus constitute a single-pass system, not a true "circulation".

MONOCLONAL ANTIBODIES DIRECTED AGAINST NORMAL LYMPH NODE CELLS

Initial studies of lymphatic delivery using monoclonal antibodies were performed with antibodies directed against normal lymph node cells, rather than against tumor (1,4). Those studies were intended to define the pharmacology in a quantitatively reproducible system and to provide the background for later clinical studies of lymphoma and immunomodulation.

Figure 2 shows scintigraphic images of two mice injected in the hind footpads with ^{125}I-labeled antibody

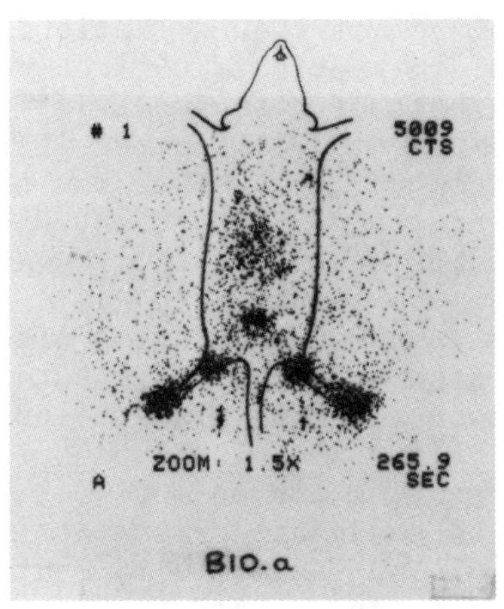

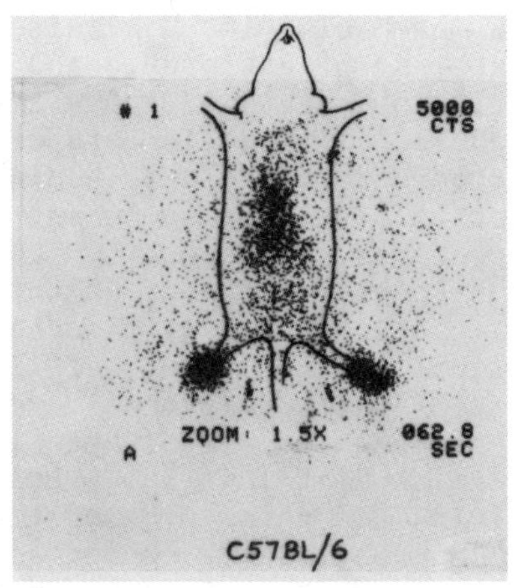

FIGURE 2. Scintigraphic images of H-2K^k-positive (left) and H-2K^k-negative (right) mice 2 hours after injection in both hind footpads with ^{125}I-labeled anti-H-2K^k IgG$_{2a}$ (approx. 0.5 μCi). The most prominent foci seen are the injection sites. Popliteal nodes (behind the knees) and lumbar nodes (in the lower abdomen) are also prominent in the antigen-positive mouse. Body background is more pronounced in the negative animal because more antibody has spilled over into the systemic circulation. Modified from ref. 4.

directed against the class I major histocompatibility determinant H-$2K^k$. Two hours after injection the popliteal and lumbar lymph nodes were well visualized in the K^k-positive mouse but not in the K^k-negative one. Each popliteal node weighed, on average, 0.8 mg. Hence, good images were obtained from about a million cells. In concentration units, the uptake in popliteal nodes exceeded 10,000 percent of dose per gram.

The following briefly summarizes our findings after injection of IgG into the hind footpads of mice (10): (i) Both specific and nonspecific IgG's (and their fragments) pass efficiently to the popliteal and lumbar nodes, principally during the first 2 to 3 hours after injection. (ii) Accumulation of radioiodine label in popliteal nodes reaches a plateau within 2 to 4 hours and then declines. The mechanisms by which radioisotope is eliminated from the node probably include dissociation of bound antibody, catabolism of antibody, and cell migration. (iii) For irrelevant control IgG, the percent of dose appearing in lymph nodes and in systemic organs is essentially independent of the dose administered (11,12). That is, the kinetics of distribution are linear. (iv) In contrast, the percentages of specific antibody found in various organs are strongly dependent on the absolute dose (11,12). In the case of antibody directed against H-$2K^k$, binding sites proximal to the point of injection appear to saturate first. Antibody then "overflows" to more distant lymph nodes along the chain, and finally into the bloodstream for systemic distribution. Doses on the order of 0.3 µg optimize the ratio of node image to body background. (v) At appropriately chosen doses, the selectivity ratio (specific/nonspecific) in the target lymph nodes exceeds 50:1 as early as 2 hours after injection, and more than 20% of the injected dose can be bound in the nodes. (vi) Antibody directed against a subset of T-cells in the lymph node (i.e., Lyt 2) appears to label all cells of that subset in the popliteal nodes. (vi) Delivery of antibody to regional nodes is faster and more efficient after injection in the footpads than in other subcutaneous sites tested.

These studies with anti-H-$2K^k$ and anti-Lyt 2 provided an experimental basis for imaging of radiolabeled monoclonals directed against normal cell types and lymphoma cells in human nodes. Such antibodies could provide an alternative to ^{99m}technetium-labeled colloids for use in "nonspecific" lymphoscintigraphy. More speculatively, they might provide a non-invasive way to assess or modulate the immune status of regional lymph nodes.

MONOCLONAL ANTITUMOR ANTIBODIES IN GUINEA PIGS

Studies with antibody to normal cell types also contributed to the design of protocols for lymphatic delivery of an antitumor monoclonal antibody to lymph node metastases of Line 10 hepatocarcinoma in guinea pigs (2). Line 10 cells were implanted intradermally in the flank and allowed to metastastasize to regional lymph nodes. ^{125}I-labeled D3 (a murine IgG_1 directed against a 290,000 dalton dimer on Line 10 cells) was then injected either into the forepaw or subcutaneously into the flank near the primary tumor. Gamma camera scintigraphy indicated accumulation of counts in the regional nodes, and double-label experiments with ^{131}I-labeled isotype-matched irrelevant IgG showed specificity of the uptake. Normal nodes and other organs showed no appreciable selectivity. As expected if antibody were reaching the nodes principally by local lymphatic flow, nodes outside of the regional drainage system contained very little radioactivity, even when they contained metastases.

It remained possible, however, that the accumulated antibody was not bound to metastatic cells, but rather to antigen shed from the primary tumor into lymphatic vessels and sequestered in the lymph nodes. This was thought to be the case in clinical studies (to be described later) with a polyclonal antibody directed against carcinoembryonic antigen (13,14). Such interference by shed antigen would tend to produce false positives whatever the route of administration. The possibility of binding to shed antigen was ruled out in the Line 10 system by autoradiographic studies and by experiments in which primary tumors were extirpated after metastasis (2).

The studies with Line 10 were done at a very early stage of tumor growth in the lymph nodes. At later stages, it appeared that antibody could reach only a portion of the mass. This finding indicated a major possible limitation of lymphatic delivery to metastases from solid tumors: an inability to "percolate" through, or even to reach, large metastases. Elsewhere (15,16), we have summarized our theoretical calculations on the "percolation" problem. The most interesting predictions from that analysis — which has application to delivery of antibody and other ligands whatever the route of administration — is that high affinity reagents may not be preferable if the aim is to achieve uniform distribution of immunoglobulin throughout the substance of a tumor. Figure 3 shows numerical simulations which illustrate that point.

BRONCHOSCOPIC ADMINISTRATION

A second problem of access requires attention. Many node groups of interest are not accessible from subcutaneous sites of injection. To reach the lymph nodes to which lung cancers metastasize, antibody could be injected through the bronchial wall in the area of a primary tumor using a cytology needle and fiberoptic bronchoscope (17). The antibody would then be expected to follow the same drainage pathway as do metastatic tumor cells, toward nodes of the carina and mediastinum. We conducted a feasibility study of this strategy using a murine monoclonal IgG reactive with mouse I-E^k. The anti-I-E^k was known to cross-react with determinants on antigen-presenting cells in a number of species, including the dog (18). Initial experiments with dogs showed uptake in hilar and carinal nodes that was antigen-specific and approximately 100-fold greater than in distant lymph nodes. Analogous endoscopic techniques could, in principle, be applied to colorectal carcinoma. In particular, the status of regional nodes can determine whether the anal sphincter is saved at surgery or whether a patient with low rectal carcinoma will require a permanent colostomy. However, the mixed results to be described in the next section render this a speculative possibility.

CLINICAL STUDIES

In 1975 Order, et al. (19,20) reported case studies using the polyclonal gamma globulin fraction of an antiserum raised against ferritin, a "tumor-associated" antigen. Patients with Hodgkin's disease were injected with antibody directly into cannulated lymphatic vessels. In 1980 DeLand, et al. (13,14) reported a series of patients injected subcutaneously with polyclonal IgG directed against carcinoembryonic antigen. Cancerous nodes were detected, but the investigators suggested that there was also interference by shed antigen.

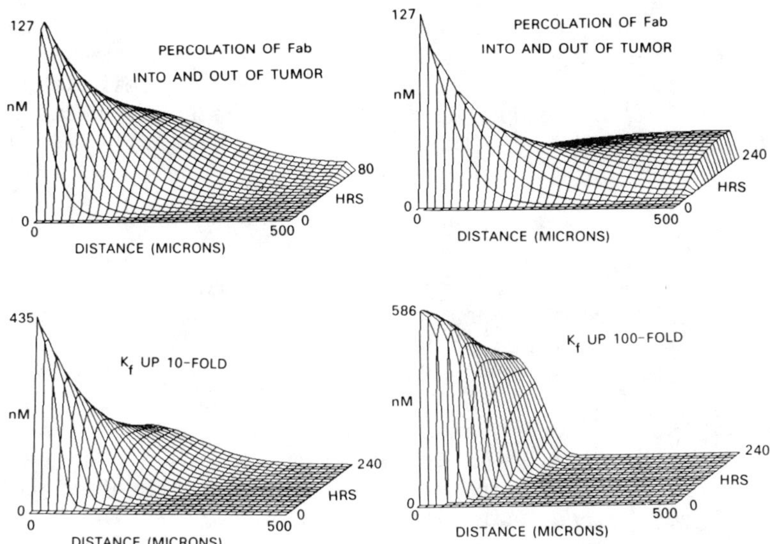

FIGURE 3. Sensitivity analysis of Fab diffusion through a tumor with different assumed forward rate constants (K_f) for binding to antigen. The lefthand boundary of each figure could correspond either to the surface of a nodule of tumor cells or to the outer surface of a capillary or lymph sinusoid. Fab concentration at that boundary rises from zero and then falls again as a function of time after injection. The top two panels are the same except for a change in the time scale. The bottom panels show the effect of increasing K_f. The distribution of Fab throughout the tumor becomes less uniform as K_f increases. Low K_f (and low affinity of binding) might be preferable for some therapeutic applications of monoclonal antibodies. On the other hand, high K_f might be desirable if the objective of therapy were to damage tumor vasculature. From ref. 15.

Thompson, et al. (21) injected a murine IgM monoclonal antibody (3E1.2) in the finger webs of patients with breast cancer in an attempt to image axillary nodes. A study of 9 patients showed generally positive correlation between antibody accumulation and the presence of metastases. The findings suggested specificity of localization, but surgical specimens were not available for verification.

Melanoma

At the National Institutes of Health, we have used several different preparations in an attempt to image metastases in patients with stage II melanoma: ^{131}I-labeled Fab fragments of antibody 96.5 (anti-p97) and antibody 48.7 (anti-gp250), as well as ^{111}In-labeled whole IgG 9.2.27 (anti-gp250). Some patients were injected subcutaneously near the site from which the primary tumor had been removed surgically. Others were injected in the web spaces between the fingers to reach axillary nodes or between the toes to reach node groups in the groin. A preliminary report (Lotze, et al., ref. 22) gives some suggestions of selective uptake but, overall, no reliable localization in cancerous nodes and no clinically useful imaging. Similar studies by Nelp, et al. (23) have, likewise, yielded no consistent, clinically useful imaging. It is not clear at this point whether optimal choice of reagent, label, and method of administration will change that picture. Lymph node metastases of melanoma may simply preclude effective access to (or "percolation" into) the affected nodes. Despite the reported success with breast cancer (21), these may be rather general problems with respect to metastases of solid tumors.

T-cell lymphoma.

The situation appears much more encouraging for T-cell lymphoma, as we have reported elsewhere (Keenan, et al., ref. 24). T101 is a murine IgG_{2a} directed against the T65 antigen on cells of chronic lymphocytic leukemia and cutaneous T-cell lymphoma (CTCL). The antigen is also expressed on normal T-cells. ^{111}In-labeled T101 was injected in the toe webs of patients with CTCL. Serial scintigraphic images showed large concentrations of radioactivity in the inguinal-femoral lymph nodes within 3 hours of injection. Activity in the nodes peaked within 24-48 hours, and then declined slowly. Figure 4 shows gamma camera images of the first two patients scanned. The first patient (left) had only minimal involvement (LN 2) of the nodes; the second (right) had bulky disease in the nodes. The 24 and 36 percent of dose imaged in nodes of those patients constituted the most efficient labeling achieved to date with monoclonal antibodies. Additional patients have shown similar uptake. The uptake appeared to be largely antigen-specific (24). However, it was not tumor-specific in that T101 reacts also with normal T-cells. Antibodies with greater selectivity for malignant cells would permit more sensitive and accurate assesssment of lymph node status.

DISCUSSION

Lymphatic delivery is clearly limited to regional lymph nodes. Metastases in the liver, lung, bone, brain, and other common sites cannot be detected (except in so far as antibody escapes the lymphatics and distributes via the blood). Since the lymphatic vessels must be patent if antibody is to reach its target, tumor may prevent access to the node or distort lymph flow. As suggested by the theoretical treatment of "percolation", access to antigenic sites buried within a tumor may not be easy. Lymphatic delivery is therefore best suited to small metastases. The lymphatic route would probably be used in combination with i.v. injection, the former to detect small masses, the latter to detect large, well-vascularized ones. As we have suggested elsewhere (10), it may also be helpful to use the antitumor antibody in combination with an antibody directed against normal lymph node cells, the latter to show the location of nodes and to signal disruption of the lymphatic chains.

Radiation toxicity at the injection site appears not to be a serious problem for diagnostic studies. However, local toxicity may be a major limitation for therapy with drug-, toxin-, or radionuclide-conjugates. Direct injection into a cannulated lymphatic vessel may be necessary for therapeutic applications.

The principal advantage of the lymphatic route is its efficiency. The mouse popliteal nodes weighed less than 1 mg, but they were imaged within a few minutes at doses of less than 1 μCi of ^{125}I (4). As little as a few mg of

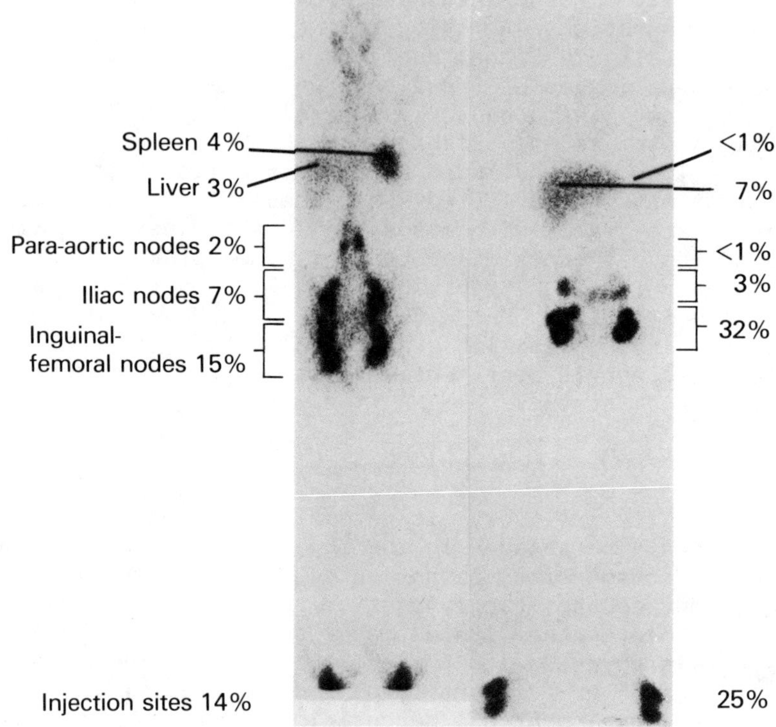

FIGURE 4. Anterior whole body scintigrams 2 to 3 days after administration of ^{111}In-labeled T101 in the toe webs of both feet in patients with cutaneous T-cell lymphoma. Numbers indicate percentages of injected activity in organs, nodes, and injection sites. Values for iliac and inguinal-femoral nodes represent the means of those obtained separately for left and right sides. The lefthand patient had minimal disease in the nodes. The patient on the right had bulky disease. Double foci in the feet of the righthand patient resulted from dorsiflexion and plantar flexion during imaging. The efficiencies of lymph node imaging were orders of magnitude higher than after i.v. injection (From ref. 24).

metastatic tumor was imaged in the guinea pig (2). For therapy with antibody conjugates, the lower dose and compartmentation in the lymphatics would presumably decrease systemic toxicity. Regional localization in the lymphatics should reduce cross-reactive binding of antibody to antigen on normal cells.

At this point it is not clear whether lymphatic delivery of antitumor antibodies will be effective for detection or treatment of solid metastases. The clinical data thus far have been mixed. However, both animal studies and clinical protocols suggest very efficient access and selective binding to lymphoma cells and to normal immune cell subsets in the lymph nodes.

ACKNOWLEDGEMENTS

We thank M.J. Talley for her technical assistance and K. Marconi for help in preparation of the manuscript. The opinions or assertions contained herein are the private views of the authors and are not to be construed as official or reflecting the views of the Department of Navy or the Department of Defense.

REFERENCES

1. Weinstein JN, Parker RJ, Keenan AM, Dower SK, Morse HC, III and Sieber SM (1982). Monoclonal antibodies in the lymphatics: Toward the diagnosis and therapy of tumor metastases. Science 218:1334-1337.
2. Weinstein JN, Steller MA, Keenan AM, Covell DG, Key ME, Sieber SM, Oldham RK, Hwang KM, and Parker RJ (1983). Monoclonal antibodies in the lymphatics: Selective delivery to lymph node metastases of a solid tumor. Science 222:423-426.
3. Parker RJ, Weinstein JN, Keenan AM, Dower SK, Steller MA, Holton OD, III, and Sieber SM. Targeting of radiolabelled monoclonal antibodies in the lymphatics. Cancer Research, in press.
4. Weinstein JN, Parker RJ, Holton OD, III, Keenan AM, Covell DG, Black CDV, and Sieber SM (1985). Lymphatic delivery of monoclonal antibodies: Potential for detection and treatment of lymph node metastases. Cancer Investigation 3:85-95.

5. Weinstein JN. Immunolymphoscintigraphy and other regional applications of monoclonal antibodies. In Zalutsky M (ed): "Antibodies in Radiodiagnosis and Therapy," Boca Raton, Florida: CRC Press, in press.
6. Ege GN (1976). Internal mammary lymphoscintigraphy. Radiology 118:101-107.
7. Weinstein JN, Keenan AM, Holton OD, III, Covell DG, Sieber SM, Black CDV, Barbet J, Talley MJ, and Parker RJ (1985). Use of monoclonal antibodies to detect metastases of solid tumors in lymph nodes. In Ceriani RL (ed): "Monoclonal Antibodies and Breast Cancer," Boston: Martinus Nijhof, in press.
8. Barsky SH, Baker A, Siegal GP, Togo S, and Liotta LA (1983). Use of anti-basement membrane antibodies to distinguish blood vessel capillaries from lymphatic capillaries. Am. J. Surg. Pathol. 7:667-677.
9. Leak LV (1971). Studies on the permeability of lymphatic capillaries. J. Cell Biol. 50:300-323.
10. Weinstein JN, Black CDV, Keenan AM, Holton OD, III, Larson SM, Sieber SM, Covell DG, Carrasquillo J, Barbet J, and Parker RJ (1986). In Reisfeld RA and Sell S (eds): "Monoclonal Antibodies in Cancer Therapy," New York: Alan R. Liss, pp. 473-488.
11. Steller MA, Parker RJ, Covell DG, Holton OD, III, Keenan AM, Sieber SM, and Weinstein JN (1986). Optimization of monoclonal antibody delivery via the lymphatics: The dose-dependence. Cancer Research 46:1830-1834.
12. Covell DG, Steller MA, Parker RJ, and Weinstein JN (1985). Delivery of monoclonal antibodies through the lymphatics: Characterization by compartmental modeling. In "Computer Applications in Medical Care," Vol. 10, pp. 884-888.
13. DeLand FH, Kim EE, Corgan RL, Casper S, Primus FJ, Spremulli E, Estes N, and Goldenberg DM (1979). Axillary lymphoscintigraphy by radioimmunodetection of carcinoembryonic antigen in breast cancer. J. Nucl. Med 20:1243-1250.
14. DeLand FH, Kim EE, and Goldenberg DM (1980). Lymphoscintigraphy with radionuclide-labeled antibodies to carcinoembryonic antigen. Cancer Res. 40:2997-3000.
15. Weinstein JN, Black CDV, Parker R, Eger RR, Sieber SM, Mozayeni B, and Covell DG (1986). Selected issues in the pharmacology of monoclonal antibodies. In

Tomlinson E and Davis SS (eds): "Site-specific Drug Delivery," New York: John Wiley, pp. 81-92.

16. Weinstein JN, Parker R, Holton OD, III, Black CDV, Sieber SM, and Covell DG. In Bonavid B and Collier RJ (eds): "Membrane-mediated Cytotoxicity," New York: Alan R. Liss, in press.
17. Mulshine JL, Keenan AM, Carrasquillo JA, Walsh T, Linnoila RI, Holton OD, III, Harwell J, Larson SM, Bunn PA, and Weinstein JN. Immunolymphoscintigraphy of pulmonary and mediastinal lymph nodes: A new approach to lung cancer imaging. Cancer Research, in press.
18. Watanabe M, Cezuki T, Taniguchi M, and Shinohara N (1983). Monoclonal anti-Ia murine allo antibodies cross reactive with the Ia-homologies of other mammalian species including humans. Transplantation 36:712-718.
19. Order SE, Bloomer WD, Jones AG, Kaplan WD, Davis MA, Adelstein SJ, and Hellman S (1975). Radionuclide immunoglobulin lymphangiography: A case report. Cancer 35:1487-1492.
20. Order SE (1977). Immunospecific radionuclide immunoglobulin lymphography. In Clouse ME (ed): "Clinical Lymphography" Vol. 7, Baltimore: Williams and Wilkins, pp. 316-322.
21. Thompson CH, Stacker SA, Salehi N, Lichtenstein M, Leyden MJ, Andrews JT, and McKenzie IFC (1984). Immunoscintigraphy for detection of lymph node metastases from breast cancer. Lancet, Dec. 1, pp. 1245-1247.
22. Lotze MT, Carrasquillo JA, Weinstein JN, Bryant GJ, Perentesis P, Reynolds JC, Matis LA, Eger RR, Keenan AM, Hellstrom I, Hellstrom K-E, and Larson SM (1986). Monoclonal antibody imaging of human melanoma: Radioimmunodetection by subcutaneous or systemic injection. Annals of Surgery 204:223-235.
23. Nelp WB, Eary JF, Jones RF, Kishore R, Krohn KA, Beaumier PL, Hellstrom K-E, and Hellstrom I (1986). Preliminary studies of radiolabeled monoclonal antibody lymphoscintigraphy in malignant melanoma. J. Nucl. Med. 26:66.
24. Keenan AM, Weinstein JN, Mulshine JL, Carrasquillo JA, Bunn PA, Jr., Reynolds JC, Foon KA, Perentesis P, Ghosh B, and Larson SM (1987). Evaluation of lymphoma by immunolymphoscintigraphy: Subcutaneous injection of indium-111-labeled T101 monoclonal antibody. J. Nucl. Med. 28:42-46.

The Pharmacology and Toxicology of Proteins, pages 91–94

DISCUSSION SUMMARY: PHARMACOKINETICS - GENERAL PRINCIPLES

Leslie Z. Benet and Carol A. Gloff

Department of Pharmacy
University of California, San Francisco
San Francisco, CA 94143-0446

Triton Biosciences Inc.
1501 Harbor Bay Parkway
Alameda, CA 94501

Following Dr. Benet's discussion, Dr. James Tam asked what particular metabolism studies would be appropriate for proteins, and is it possible to carry out metabolic studies that would be relevant? Dr. Benet responded that normally one would proceed using the same methodology as with classical drugs, in that a radiolabelled compound is administered and distribution of radioactivity in various sites of the body is measured in terms of administered compound and metabolic and catabolic products. He also indicated that it may be more relevant to carry out such metabolic degradation studies only with that portion of the molecule that contains the active site, as exemplified in Dr. Mark's talk, where restriction enzymes were used to isolate the active site(s). Also, for classical drugs, the FDA is usually only interested in metabolites which represent greater than 5% of the total dose. Since we are entering a new area with proteins as drugs, it will be necessary to convince the FDA that such an approach is also appropriate here. At this stage, when we have little information about metabolism of protein and peptide drugs, this may be a difficult question to address. Dr. Gabriela Nicolau commented that although metabolites of classical drugs are generally more polar, this may not in fact be the case with protein and peptide drugs. With these drugs, the amino acid residues giving the compound polarity may in fact be cleaved, resulting in a less polar molecule.

Dr. Lawrence Souza discussed the fact that many of the protein drugs may act in a manner similar to that of nitrogen mustard type compounds, with reversible binding leading to their action, rather than like classical drugs with well-defined receptor interactions. He suggested that it may be more appropriate to carry out studies related to the kinetics of effect rather than to the disposition kinetics of the molecule and that with such interactive processes, kinetics of the molecule may not lead to meaningful information. Dr. Benet responded that this is also a problem with classical drugs. An example of this is compounds used in anticlotting relationships, where a cascade of factors is involved between drug concentrations and effect. However, once these effect mechanisms are understood, it is possible to develop a pharmacokinetic/pharmacodynamic model which will allow one to predict effect measurements on the basis of concentrations of the drug. He emphasized again that there need not necessarily be a one-to-one relationship between a measured plasma concentration and an effect and that more likely integrated measures of concentration will be predictive of integrated measures of effect.

Following Dr. Young's talk, Dr. Benet, as moderator, made two comments. The first was to indicate that since the largest dose of IL-2 showed the highest clearance, i.e., 4.07 ml/min/kg for the 25 mg/kg dose versus a 2.97 ml/min/kg average clearance for all doses, the longer half-life observed by Dr. Young for IL-2 at higher doses is not a major determinant in predicting clearance for that compound. If it were, than the highest dose would have yielded the lowest clearance, rather the highest. Dr. Benet also commented that since the kinetics following the IP, IM and SC doses of IL-2 yielded the same plasma concentration time curve, it was apparent that the rate limiting step for this drug was the vehicle rather than the kinetics of the drug itself. If the rate limiting step were not the vehicle, then identical kinetics would not have been obtained with IP, IM and SC administration. Dr. John Holcenberg asked about the volume of the vehicle used in the IP injection, wishing to know whether belly bath-type increases in volume could have yielded greater absorption. Dr. Young responded that the volumes were small, no effect of volume was tested and that no attempt had been made to calculate the absorption rate constants from these various routes of administration.

Dr. Ruth Billings asked about the appropriateness of using rats as a model animal system for studying the pharmacokinetics of a human protein. Dr. Young responded that the pharmacokinetics of IL-2 in man and rats were very similar, and that receptor specificity may not be a major factor. Dr. John Farrar added that human IL-2 reacts with the rat receptor in a similar manner to the human receptor. Dr. Farrar also briefly described results of some experiments done by Dr. Steve Gillis at Immunex. If T-cells are put into the peritoneum of mice, and the mice are not treated with IL-2, the T-cells die relatively rapidly. However, if IL-2 is given up to 12 hours prior to T-cell injection, the T-cells survive 18-24 hours. This occurs even though IL-2 levels are no longer measurable for some time prior to T-cell administration.

Following Dr. Weinstein's talk, Dr. Alec Sehon asked if antibodies were formed following the injection of monoclonal antibodies. Dr. Weinstein responded that no antibodies were formed after a single injection but that after repeated injections, antibodies were seen. Dr. Sehon also asked about the intervals for the multiple injections. Dr. Weinstein indicated that the dosing intervals were not designed into the study and therefore were variable. Dr. William Joyner commented on Dr. Weinstein's conclusion that most of the antibody, after IV administration, was in the liver and spleen and not much was in the lymph nodes. Dr. Joyner thought that the slide showed significant quantities in the lymph nodes. Several other members of the audience echoed his comments. Dr. John Baynes asked what happened to antibody conjugates in the liver, e.g., were they denatured. He also asked about the patency of the Indium on the antibody, that is, is everything localized intact, or is only some of the Indium on the antibody and/or has some metabolism taken place? With the data currently available, this question is unresolved. Dr. Baynes also asked if free Indium could react with antibody in the blood or liver and then just accumulate at that site. Dr. Chris Black indicated that this was not a significant reaction with the free Indium.

Dr. Bennett Laguzza asked about the practical considerations for penetration of monoclonal antibodies to sites of action. Dr. Weinstein responded that this was related to the diffusion length for the monoclonal antibodies. Other factors involved include the number of binding sites, potential delivery into the perfusion space, (i.e., blood), the affinity for the binding site, the diffusion coefficient to the binding site, the concentration of the binding site and the size of the tumor. Dr. Robert Zimmerman asked if there was a difference in the microvasculature of tumors versus what is found in normal tissues. Dr. Weinstein responded that the vasculature of tissues varies as a function of the different tumors. It is a very complicated situation and in most cases there is a marked difference in deviation from that found in normal tissue.

Dr. Holcenberg asked whether it may be possible to change the delivery of the monoclonal antibodies to the tumor site by changing the distribution of flow, for example by using a vasodilator. Dr. Weinstein indicated that flow may be a fundamental barrier and that in some cases, changing flow may be useful. Dr. Sehon commented that Dr. Weinstein had discussed primarily the forward rate constants in his physiologic models. However, the reverse rate constant is the dominant feature in defining distribution to a peripheral site. Dr. Weinstein agreed and indicated that he had carried out such calculations. In response to another question, he also indicated that most of the studies to date are on F(ab) fragments, not on whole IgG. He pointed out that the amount injected is an important consideration in the development of these models and that when a greater amount of the compound is injected it can saturate the sites. However, a lot of compound is needed to do this and this may also increase the "noise" level, therefore the background separation would not be as good and one could not define the model as well. Dr. James Starling asked whether there had been a comparison of high and low affinity antibodies. Dr. Weinstein responded no, but that anecdotal reports indicate that high affinity antibodies do not necessarily penetrate the site and that in fact, very often compounds that are nonbinding, i.e., low affinity, may yield better results since these compounds distribute more evenly throughout the tumor. Obviously more information is needed. It was asked whether a dose comparison of penetration was made where a greater amount of antibody could be shown to further penetrate to the site. Dr. Weinstein responded that in the melanoma study, done by Dr. Paul Abrams and coworkers, there had been some escalation of dose and that they had seen a more even penetration at the higher dose. However, there were problems with that particular study so no definitive statement can be made.

The Pharmacology and Toxicology of Proteins, pages 95–111

CHEMICAL MODIFICATION OF α_2-MACROGLOBULIN AND ITS EFFECTS ON RECEPTOR BINDING

Salvatore V. Pizzo and Maureane Hoffman

Departments of Pathology and Biochemistry, Duke University Medical Center, Durham, North Carolina 27710

ABSTRACT Reaction of the plasma proteinase inhibitor α_2-macroglobulin (α_2M) with proteinases induces a conformational change which exposes a receptor recognition site on the inhibitor. High affinity receptors (K_d = 0.5 nM) for α_2M-proteinase complexes are present on many types of cells including macrophages, hepatocytes and fibroblasts. Subsequent to macrophage binding, many of the functions of these cells are significantly altered. In this report, chemical modification studies are described which probe the nature of the α_2M receptor recognition site. Data on macrophage regulation by α_2M-proteinase complexes are also presented.

INTRODUCTION

α_2-Macroglobulin (α_2M) is one of the major proteinase inhibitors of plasma (1). The inhibitor is produced by a variety of cells including hepatocytes and macrophages and it is present in most body fluids as well as plasma. α-Macroglobulins are found in all vertebrates as well as those invertebrates which have been surveyed (2-5). It is present in the hemolymph of the horseshoe crab, Limulus polyphenus. This animal is one of the most primitive creatures remaining on earth, its ancestors having evolved by the Middle Cambrian period some 550,000,000 years ago. It is likely, therefore, that α-macroglobulins evolved at least half a billion years ago.

This work was supported by NIH grants HL-24066, CA-29589 and Council for Tobacco Research Grant 1690. Dr. Hoffman is the recipient of a Career Development Award from The Veterans Administration.

The protein is composed of either 2 or 4 identical subunits, $M_r \sim 180,000$ (1-6). Invertebrate α-macroglobulins and the fish proteins appear to exist as a dimer (2-5). In reptiles, birds and mammals, the protein is tetrameric consisting of two pairs of disulfide bonded chains (1,3-4,6). In amphibians, two different α-macroglobulins exist, one a dimeric protein and one a tetramer (3,4). The complete amino acid sequence of the human subunit is known (7) and these studies demonstrate that the protein chains are homologous to the complement proteins C_3 and C_4 which consist of only a single chain. Presumably, the complement proteins and α-macroglobulins evolved from a common precursor of only one chain and $M_r \sim 180,000$.

α_2M also is unique in that it inhibits proteinases of any mechanistic class and any amino acid specificity (8). Each subunit contains a sequence of amino acids termed the "bait region" which is highly susceptible to proteolytic cleavage (8). When the "bait region" is cleaved, a conformational change occurs which irreversibly traps the proteinase (8-10). This conformational change results in a decrease in the Stokes radius (10). The more compact form of α_2M resulting from reaction with proteinase migrates faster when electrophoresed under nondenaturing conditions and is termed the "fast" form (8). There are considerable data from electron microscopy, small angle X-ray scatter and hydrodynamic studies which recently have been employed by our laboratory to produce a three dimensional model of α_2M (11,12). The molecule has D_2 symmetry with three C_2 rotational axes (Figure 1). The four identical subunits are organized into two domains or "half-molecules" which can be separated by limited reduction and carboxamidomethylation (13,14). These "half-molecules" retain proteinase binding activity. Recently, NMR studies have supported the proposed model, particularly its D_2 symmetry (15).

α_2M, C_3 and C_4 possess a unique internal thioester bond formed from Glx and Cys residues. This bond, located about 1/3 of the way from the carboxy terminus of the α_2M subunit (7), is broken when proteinases cleave the bait region. Proteinases then become covalently attached to the resultant Glu residue via a Glu-Lys crosslink. Both methylamine and ammonia react directly with the thioester to trigger the same conformational change (10). This conformational change is important because it exposes a receptor recognition site which is bound by a variety of cells including hepatocytes, fibroblasts and macrophages (16-18). The α_2M-proteinase complex or $\alpha_2M\text{-}CH_3NH_2$ adduct (α_2M "fast" forms) are rapidly removed from the blood by hepatocytes and Kupffer cells (macrophage-like cells) of the liver (17). Inhibitor-proteinase complexes produced in tissues bind to and are internalized by macrophages (16,18,19).

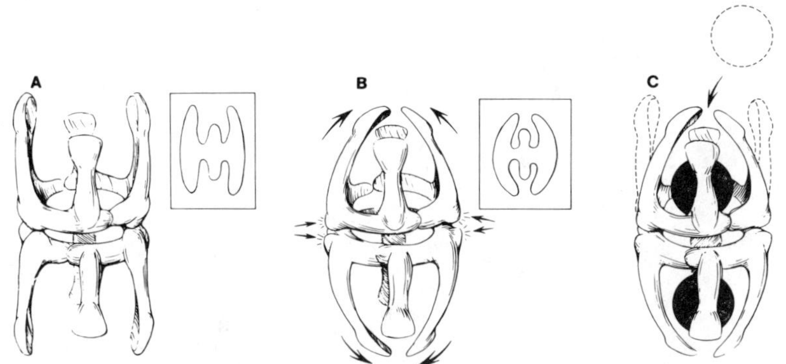

FIGURE 1. A model of α_2M. The native molecule (A) undergoes a conformational change when it reacts with methylamine (B) or proteinases (C). The small arrows denote the proposed placement of the receptor recognition sites.

RECEPTOR RECOGNITION OF α_2M.

The binding of α_2M-proteinase or α_2M-CH_3NH_2 to cells occurs with high affinity (Figure 2). By contrast, native α_2M has low affinity for the receptor. Human, murine, bovine, chicken and frog α-macroglobulins bind to murine peritoneal macrophages with a K_d of about 0.5 nM. The α_2M receptor on other types of cells seems to be identical to the macrophage receptor (Figure 2). It is evident from these studies that the α_2M receptor recognition site has been conserved for a very long period of time, since amphibians evolved some 200,000,000 years ago. This is particularly remarkable because there are about 1450 amino acids in the α_2M subunits (7) and the protein exhibits such a complex structure (11,12).

Because of a number of observations, we have suggested that the receptor recognition sites are located at the base of the long arms (11,12). The conformational change which occurs when α_2M reacts with proteinases or amine causes movement of the long arms about the base of the arm which functions as a hinge. Movement of the hinge exposes the receptor recognition sites.

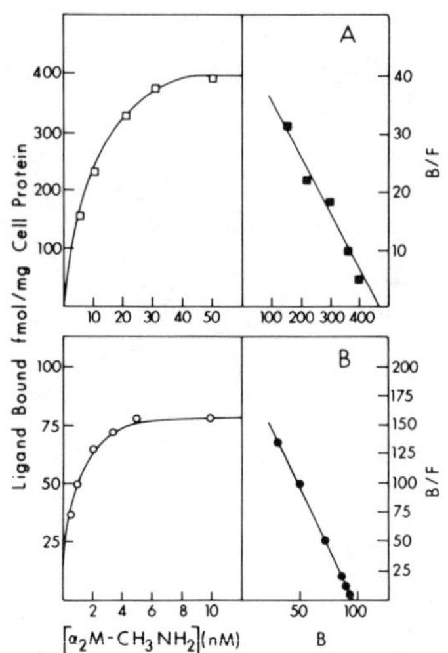

FIGURE 2. The binding of $\alpha_2M\text{-}CH_3NH_2$ to macrophages (A) and fibroblasts (B) at 4°C. The left hand panels show the binding isotherms while the right hand panels are the corresponding Scatchard plots.

The hinge region is also the location of a major portion of the carbohydrate chains of the α_2M molecule (4,6). These chains also become more exposed to solvent when the α_2M molecule undergoes conformational change (4,6). We have demonstrated that this carbohydrate is not involved in the receptor recognition of mammalian α-macroglobulins (19,21); however, when primitive α-macroglobulins are studied the data suggest that carbohydrate recognition could be a means by which these α_2M-proteinase complexes are removed from the circulation (4,6). Based on our data, it appears that the more specific receptor recognition systems for α-macroglobulins may have evolved only when amphibians appeared on the planet (4,6).

Chemical modification studies have shed considerable light on the nature of the specific α_2M receptor recognition site. Beginning in 1981, we demonstrated that α_2M reacts with the antitumor agent cis-dichlorodiammineplatinum (II) (cis-DDP) (21-25). The structure of cis-DDP is shown in Figure 3.

Amino acid analysis also indicated that oxidation modified two His residues per α_2M subunit; however, cis-DDP did not protect these residues from destruction. As a further test of the role of His in α_2M-CH_3NH_2 receptor recognition, diethyl pyrocarbonate (DEPC) (Figure 3) was employed to modify His residues in the inhibitor. This reagent modified a total of 53 of 156 His residues but did not alter recognition by the α_2M receptor (Table 1). Modification of these His residues, however, had an unexpected effect; namely, it was possible to demonstrate, by employing a direct binding assay, that DEPC-modified α_2M-CH_3NH_2 is recognized by the acyl-low density lipoprotein receptor (25). To our knowledge, this is the first example of a modification to a residue other than Lys which results in a preparation recognized by this scavenger receptor.

While these studies establish that a Met residue is at the α_2M receptor recognition site, they do not locate this residue among the 1450 amino acids in the primary sequence of the α_2M subunit (7). Other studies now in progress suggest that it will be possible to pinpoint the exact location of this Met residue. Recently, Sottrup-Jensen et al. obtained an α_2M fragment of about 20,000 molecular weight which demonstrated 0.1% receptor activity (26). We have also obtained a similar fragment by papain treatment of α_2M-CH_3NH_2 at low pH. However, our preparation demonstrates an $M_r \sim 30{,}000$. Moreover, in receptor binding studies it has much higher receptor binding activity. This fragment is being sequenced and it contains only three Met residues. Direct identification of the Met at the receptor recognition site should be possible by employing several approaches. These include use of cis-DDP treatment of the peptide and localization of the site of this modification. Initial studies demonstrate that reaction of the peptide with cis-DDP blocks receptor recognition just as observed with the parent α_2M-CH_3NH_2 molecule.

In addition to specific techniques to block the receptor recognition of α_2M-CH_3NH_2 or α_2M-proteinase, we have found that receptor binding is blocked by coupling polyethylene glycol (PEG) of various sizes to α_2M-proteinase complexes (27). Employing the chemistry shown in Figure 4, PEG-2, PEG-4 and PEG-20 were coupled to α_2M-trypsin. The effect of PEG coupling on receptor recognition was then evaluated in the murine clearance model. These studies were performed by injecting a small amount of ^{125}I-labeled protein into the lateral tail vein followed by repetitive blood sampling from the retroorbital venous plexus.

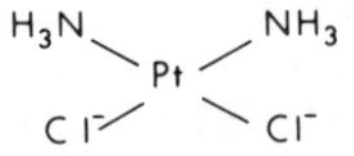

cis-Dichlorodiammineplatinum (II)

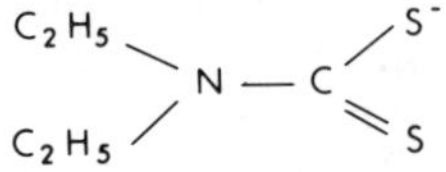

Diethyldithiocarbamate

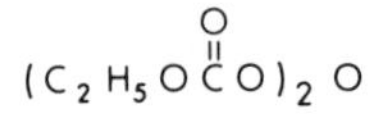

$(C_2H_5OCO)_2O$

Diethylpyrocarbonate

FIGURE 3. The structure of three compounds employed to chemically modify α_2M.

The reaction of α_2M-trypsin, α_2M-CH_3NH_2, or other α_2M "fast" forms with cis-DDP blocks receptor binding (21,23-25). This is best demonstrated by a displacement assay in which human ^{125}I-α_2M-CH_3NH_2 (0.3 nM) is allowed to bind to murine peritoneal macrophages in the presence of a competing concentration of a preparation to be tested for receptor recognition (Table 1). The concentration of unlabeled ligand required to decrease the binding of the ^{125}I-α_2M-CH_3NH_2 to cells by 50% is the I_{50}. The I_{50} is related to the apparent dissociation constant of the nonradiolabeled protein (K_I) by the equation: $I_{50} = K_I\,[1 + L/K_L]$ where L and K_L are the concentration and dissociation constant for ^{125}I-α_2M-CH_3NH_2, respectively (25).

cis-DDP treated unlabeled α_2M-CH_3NH_2 demonstrates a dose-dependent reduction in its ability to compete for the binding of ^{125}I-α_2M-CH_3NH_2. cis-DDP reacts preferentially with Cys and Met residues (24); however,the only Cys residues present in α_2M arise from cleavage of the four thioester bonds. When these residues are modified by carboxamidomethylation, there is no alteration in receptor binding (23) suggesting a role for Met in receptor recognition. If a Met residue were at the α_2M receptor recognition site, reaction with H_2O_2 might also alter receptor recognition of α_2M-CH_3NH_2. Table 1 demonstrates that this is the case. Amino acid analysis indicates that treatment of α_2M with H_2O_2 oxidizes 19 of 25 Met residues per subunit. Pretreatment of α_2M with cis-DDP blocks oxidation of two to four of these Met residues, consistent with the hypothesis that a Met residue is at the receptor recognition site of α_2M.

Reaction of cis-DDP with α_2M is reversible if the protein is treated with a potent nucleophile, diethyl dithiocarbamate (Figure 3), which displaces platinum from Met residues (25). When cis-DDP treated α_2M-CH_3NH_2 is reacted with DDC, receptor binding is restored (Table 1). This observation offered the possibility of a direct test of whether H_2O_2 and cis-DDP react with the same Met residues. The α_2M-CH_3NH_2 was first reacted with cis-DDP, then treated with very high concentrations of H_2O_2. The peroxide was destroyed and then DDC employed to remove the cis-DDP. The resultant preparation demonstrated a significant recovery of receptor binding activity (Table 1).

TABLE 1
RECEPTOR RECOGNITION OF MODIFIED FORMS OF α_2M

α_2M-CH_3NH_2	K_I (nM)	Receptor Recognized[a] (%)
Untreated control	0.3	100
Treated with cis-DDP[b] at:		
0.8 mM	5	6
1.2 mM	8	4
1.7 mM	11	3
Treated with H_2O_2	2	15
Treated with 1.7 mM cis-DDP, then 3.5 mM DDC	0.3	100
Treated with 1.7 mM cis-DDP, then H_2O_2, then 3.5 mM DDC	0.5	60
Treated with DEPC	0.3	100

[a]As determined with murine peritoneal macrophages *in vitro* where α_2M-CH_3NH_2 binding is taken at 100%.
[b]Abbreviations used are: cis-DDP, cis-dichlorodiammine-platinum (II); DDC, diethyl dithiocarbamate; DEPC, diethyl pyrocarbonate.

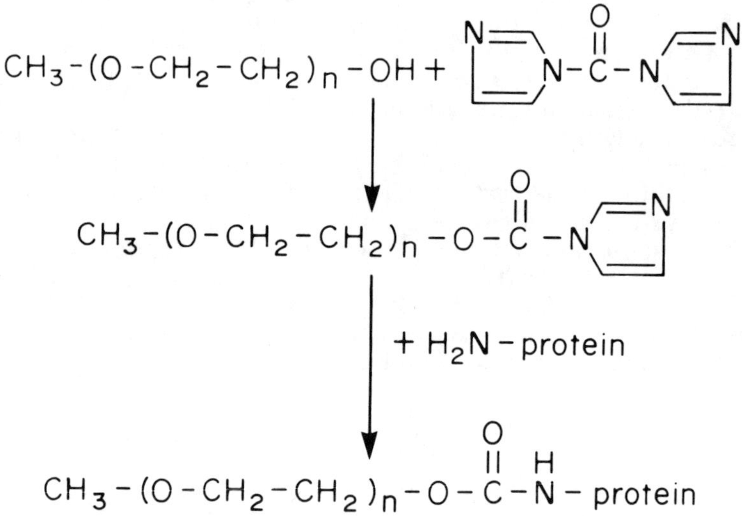

$$CH_3-(O-CH_2-CH_2)_n-OH + \text{N}{=}\!\!\diagdown\ \ \overset{O}{\overset{\|}{N-C-N}}\ \ \diagup\!\!{=}\text{N}$$

$$\downarrow$$

$$CH_3-(O-CH_2-CH_2)_n-O-\overset{O}{\overset{\|}{C}}-N\diagup\!\!{=}\text{N}$$

$$\downarrow + H_2N-\text{protein}$$

$$CH_3-(O-CH_2-CH_2)_n-O-\overset{O}{\overset{\|}{C}}-\overset{H}{N}-\text{protein}$$

FIGURE 4. Proposed mechanism for the formation of activated PEG employing 1,1′-carbonyldiimidazole as the activating reagent. The coupling of activated PEG to protein occurs via Lys residues (27) as also depicted in the figure.

α_2M-proteinase or α_2M-CH_3NH_2 clear from the murine circulation with a $t\frac{1}{2}$ of about 2 min and this process results from the specific receptor-mediated mechanism already discussed above (1,16-21,23,25,27). As is evident from Figure 5, coupling of PEG alters clearance. The change in clearance is dependent on the size of the PEG and the greatest effect is seen with the largest PEG, PEG-20. For this reason, we do not believe that the coupling of the PEG species to α_2M Lys residues (27) results in a specific block of the receptor binding process. Rather, this appears to be the result of steric hinderance with longer PEG chains producing greater obstruction to the interaction of the ligand with receptors on the cells. In our hands, similar results were obtained with other proteins whether these proteins clear via carbohydrate recognition or passively via kidney filtration (27). PEG coupling, therefore, is a nonspecific and general technique to block circulatory clearance mechanisms.

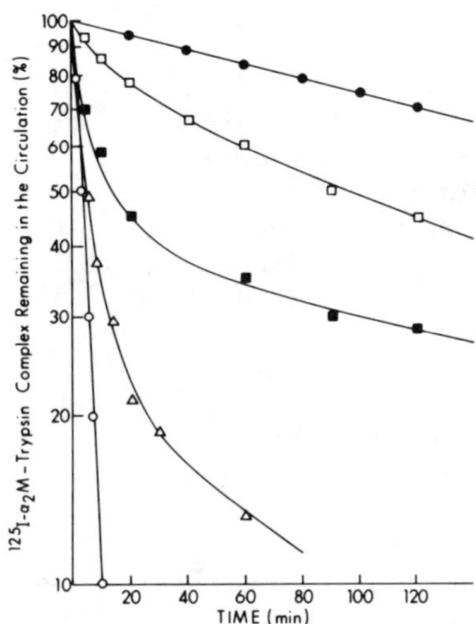

FIGURE 5. The clearance of ^{125}I-α_2M-trypsin before and after PEG coupling. The clearance of unmodified α_2M-trypsin (○) is compared to the clearance of α_2M-trypsin modified with PEG-2 (△), PEG-4 (■) or PEG-20 (□). The clearance of native α_2M (●) is shown for comparison.

The studies presented indicate that the α_2M receptor has changed little, if at all, over millions of years. This suggests that this receptor may have other functions than as a scavenger pathway to remove α_2M-proteinase complexes. The remainder of this report describes studies which indicate that α_2M plays a major role in regulation of immunity. It seems likely that the great importance of these functions accounts for preservation of the receptor recognition site over such a long evolutionary span.

IMMUNOREGULATORY ACTIVITY OF α_2M

As described above, α_2M probably evolved from the same progenitor as the immunologically important complement components C_3 and C_4 (7) and the structure of the α_2M receptor recognition site is highly conserved among a wide range of species. While a few cases of modest α_2M deficiency have been reported (28), no instances of severe deficiency of α_2M have been detected. In one study over 10,000 persons were surveyed (29). This suggests that a severe deficiency of α_2M is incompatible with life.

Several features make α_2M, particularly its complexes with proteinases, well suited for a role as a regulator of immune cell function. In addition to being present in large amounts in plasma, α_2M is present in significant concentrations in peritoneal, joint, and pulmonary lavage fluids. It is synthesized by macrophages and monocytes (30), and is available in the local tissue microenvironment to regulate cells of the immune system. As already noted, α_2M is unique among plasma proteinase inhibitors in its ability to inhibit proteinases from all major classes. This broad specificity makes α_2M-proteinase complexes suitable as feedback regulators of the wide variety of physiological processes in which proteinases are secreted. Once formed, the α_2M "fast" forms are rapidly cleared from the circulation, so that steady state plasma levels are very low. Native α_2M also has an unusual ability to bind other molecules, including lymphokines, mitogens, soluble Ia antigens, immune complexes and some metal ions. This property could allow α_2M to serve as a carrier or inactivator of immunoactive molecules. α_2M-proteinase complexes can inactivate other proteins in an additional way. Proteinases which are complexed to α_2M retain some activity and small proteins which can diffuse into contact with the active site of the trapped proteinase can be degraded. For example, a_2M-trypsin complexes degrade interleukin 2, an important modulator of lymphocyte function (31).

α_2M is structurally very similar to a pregnancy-associated protein termed pregnancy zone protein (PZP). Plasma concentrations of PZP increase dramatically during pregnancy, and it is synthesized at least partly by estrogen-exposed monocytes (32). PZP was initially studied as an immune regulator. It supresses rejection of mouse cardiac allografts *in vivo*, mitogenic responses of lymphocytes to antigens and lectins *in vitro*, mixed lymphocyte culture reactivity, and lymphocyte production of migration inhibition factor (32). It has been theorized that PZP is at least partially responsible for the non-specific immunosuppression present during pregnancy. PZP is a proteinase inhibitor and it undergoes a conformational change upon reaction with CH_3NH_2 similar to that of α_2M. It has recently been demonstrated that PZP-CH_3NH_2 and α_2M "fast" forms bind to the same high-affinity receptor on fibroblasts (33).

We are particularly interested in the effects of α_2M on the mononuclear phagocyte system (macrophages and monocytes). Macrophages play a central role in the activities of the immune system both as effector cells in microbial and tumor cell killing, and as regulators of the function of other immune cells. The two primary mechanisms by which macrophages kill invading microbes and tumor cells are by release of proteinases and production of superoxide anion. As described above, α_2M "fast" forms generated by reaction with any of a number of different proteinases or nucleophiles bind to a macrophage receptor (16). Preincubation of macrophages with α_2M "fast" forms results in a dose-dependent inhibition of the subsequent secretion of cytolytic factor (Figure 6), plasminogen activator (data not shown) and neutral proteinase (Figure 6) (34). Cytolytic factor is a serine proteinase which is able to selectively attack tumor cells and produce holes in the cell membrane. Preincubation with α_2M "fast" forms also decreases tumor cell killing in parallel with the decrease in proteinase secretion (34). Half maximal inhibition of proteinase release (ID_{50}) occurred at about 10-20nM α_2M "fast" form. α_2M "fast" forms generated by reaction with trypsin or with CH_3NH_2 were equally effective in inhibiting macrophage proteinase secretion and tumor cell killing. We found that α_2M "fast" forms also affect the second major mechanism used by macrophages to kill tumor cells. Preincubation of macrophages with α_2M "fast" forms inhibits their production of superoxide anion in response to subsequent stimulation with phorbol myristate acetate (Figure 6) (35).

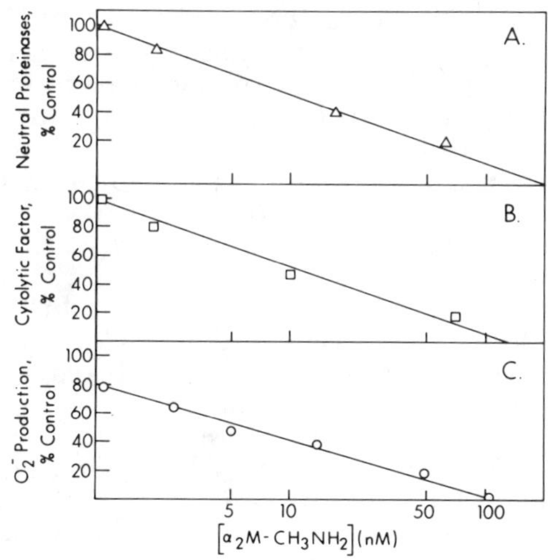

FIGURE 6. The effect of preincubation with α_2M-CH_3NH_2 on proteinase secretion and superoxide production by murine peritoneal macrophages. (A) secretion of neutral proteinases, (B) secretion of cytolytic factor, (C) production of superoxide anion following stimulation with phorbol myristate acetate.

The concentration of α_2M "fast" forms required for half maximal inhibition (ID_{50}) is about 14 nM, and native α_2M is about two orders of magnitude less effective an inhibitor of superoxide generation (data not shown). The ID_{50} determined in these studies is very similar to the K_d determined at 37°C for the binding of α_2M "fast" forms to the macrophage receptor (19,20,34). These studies strongly suggest that suppression of macrophage function is dependent on the binding of the ligand to the receptor.

We have also examined the effects of α_2M on a macrophage regulatory function, Ia antigen expression in mouse or HLA-DR expression in human. The ability of murine macrophages to regulate a wide range of immunologic events is a function of their expression of cell surface Ia antigens. The presence of membrane Ia is essential for macrophages to serve as accessory cells for mitogen induced lymphocyte proliferation (36), and to initiate specific T cell response to antigens (37). Macrophages can be immunologically activated by injecting mice with a number of different stimuli including Bacillus Calmette-Guerin (BCG) or peptone. The macrophages are then obtained by peritoneal lavage. Culture of these cells in the presence of recombinant murine γ-interferon (γIFN) (100U/ml) for three days results in a significant increase in the percent of Ia positive cells (Figure 7A). Culture in the presence of 100nM α_2M-trypsin alone has no effect on Ia expression, but α_2M-trypsin antagonized the Ia-inducing effect of γIFN (Figure 7A). A very similar pattern is seen for macrophages harvested from peptone-injected mice (Figure 7B). α_2M-trypsin also antagonized the γIFN-induced increase in HLA-DR expression on human peritoneal macrophages in the same manner. α_2M-CH_3NH_2 is as effective an antagonist of γIFN-induced Ia expression as is α_2M-trypsin. Thus, the effect could not be due to the proteolytic degradation of γIFN by α_2M-trypsin complexes.

The antagonism of γIFN-induced Ia expression is dose-dependent (Figure 8), with a maximal effect being achieved at about 100nM α_2M-trypsin. Both α_2M-trypsin and α_2M-CH_3NH_2 depressed the ability of murine macrophages to serve as accessory cells for lymphocyte proliferation in parallel with the depression in Ia expression (data not shown). The effects of α_2M "fast" forms on Ia expression are not due to endotoxin contamination (which is often difficult to eliminate from purified protein preparations), since macrophages from the endotoxin resistant C3H/HeJ mouse strain respond to α_2M "fast" forms in the same fashion as macrophages from the endotoxin sensitive C3H/HeN mice. In addition, boiling, which does not destroy endotoxin, abolishes the ability of α_2M-trypsin to antagonize γIFN induction of Ia expression.

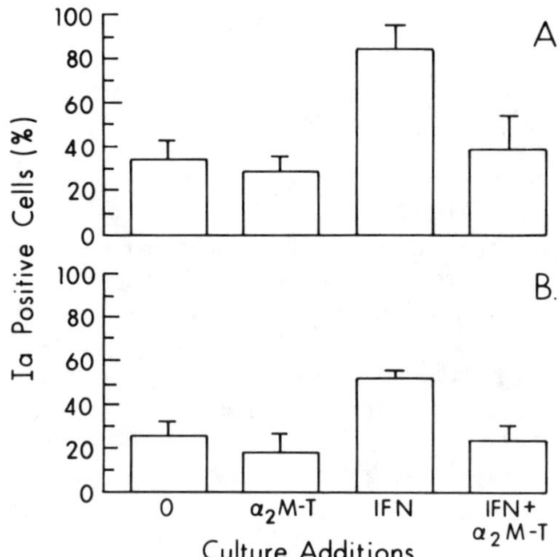

FIGURE 7. The effect of γIFN and α_2M-trypsin on Ia expression of BCG (A) or peptone (B) elicited peritoneal macrophages from CH3/HeN mice.

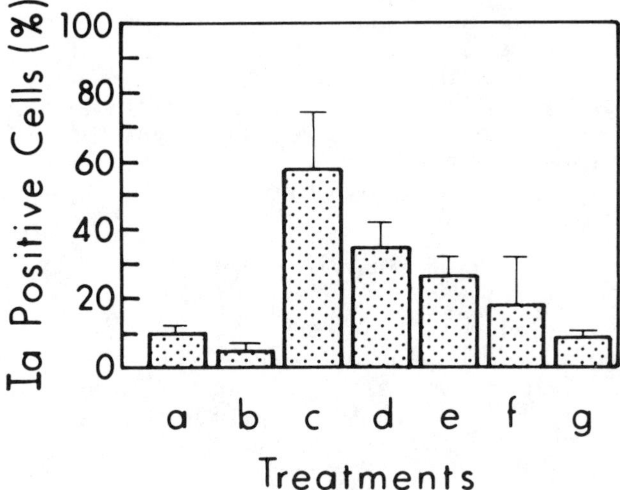

FIGURE 8. The dose-dependent effect of α_2M-trypsin on γIFN-induced Ia expression employing activated macrophages from C3H/HeJ mice. The symbols are: "a" control cells; "b" cells treated with 100nM α_2M-trypsin alone; "c" cells treated with 100 U/ml γIFN alone; "d"-"g" cells treated with 100 U/ml γIFN in the presence of 10, 25, 50 or 100 nM α_2M-trypsin, respectively.

In other experiments, murine macrophages were pretreated with α_2M "fast" forms for 36 hours, thoroughly washed, then cultured with γIFN. Pretreatment with α_2M "fast" forms still significantly reduced the increase in Ia expression in response to subsequent culture with γIFN. Therefore, α_2M does not directly inactivate the γIFN, but changes the responsiveness of macrophages to γIFN. It seems likely that the effects of α_2M "fast" forms on macrophage function are receptor-mediated, since the effectiveness of an α_2M preparation in inhibiting macrophage Ia expression, superoxide production and proteinase secretion is correlated with the known ability of the preparation to bind to the high affinity receptor. Proteinases can be liberated or generated as a consequence of a number of immune reactions, such as direct release by effector cells, release from lysed target cells, or production during complement activation. The released proteinases could then be bound by α_2M which would serve to protect host tissue from damage, and also to intiate local feedback inhibition of specific and nonspecific responses through its effects on macrophages. Since α_2M is present in concentrations as high as 5μM in the plasma and 1μM in peritoneal fluid, conversion of less than 10% of the α_2M in the local environment could produce concentrations of "fast" forms sufficient to cause maximal supression of macrophage functions.

In summary, modulation of macrophage functions by α_2M "fast" forms is specific, dose-dependent and receptor-mediated. The importance of these macrophage functions to regulation of the immune system, may in part explain why the α_2M receptor has been conserved over such a long period of time. Why this receptor is present on many other types of cells is at present unknown.

REFERENCES

1. Pizzo SV, Gonias SL (1984). Receptor-mediated Protease Regulation. In Conn PM (ed): "The Receptors", New York:Academic Press, Vol 1, Chapter 4, p 178.
2. Starkey PM, Barrett AJ, (1982). Evolution of α_2-macroglobulin. Biochem J 205:105.
3. Feldman SR, Pizzo SV (1984). Comparison of the binding of chicken α-macroglobulin and ovomacroglobulin to the mammalian α_2-macroglobulin receptor. Arch Biochem Biophys 235:267.
4. Feldman SR, Pizzo SV (1985) Purification and characterization of frog α-macroglobulin: Receptor recognition of an amphibian glycoprotein. Biochemistry 24:2569.

5. Quigley JP, Armstrong PB (1983). An endopeptidase inhibitor found in Limulus plasma: an ancient form of α_2-macroglobulin. Ann NY Acad Sci 421:119.
6. Feldman SR, Pizzo SV (1986). Purification and characterization of a "Half-molecule" α_2-macroglobulin from the Southern grass frog: Absence of binding to the mammalian α_2-macroglobulin receptor. Biochemistry 25:721.
7. Sottrup-Jensen L, Stepanick TM, Kristensen T, Wierzbicki DM, Jones CM, Lonblad PB, Magnusson S, Petersen TE (1984). Primary structure of human α_2-macroglobulin. J Biol Chem 259:8318.
8. Barrett AJ (1981). α_2-Macroglobulin. Methods Enzymol 80:737.
9. Barrett AJ, Starkey PM (1973). The interaction of α_2-macroglobulin with proteinases. Biochem J 133:709.
10. Gonias SL, Reynolds JA, Pizzo SV (1982). Physical properties of human α_2-macroglobulin following reaction with methylamine and trypsin. Biochim Biophys Acta 705:306.
11. Feldman SR, Gonias SL, Pizzo SV (1985) A model of α_2-macroglobulin structure and function. Proc Natl Acad Sci (USA) 82:5700.
12. Feldman SR , Pizzo SV (1986) A Three Dimensional Model of a Unique Proteinase Inhibitor: α_2-Macroglobulin, Seminars in Thromb and Hemostas, 12:223.
13. Gonias SL, Pizzo SV (1983). Characterization of functional human α_2-macroglobulin half molecules isolated by limited reduction with dithiothreitol. Biochemistry 22:536.
14. Gonias SL, Pizzo SV (1983). Reaction of human α_2-macroglobulin half molecules with plasmin as a probe of protease binding site structure. Biochemistry 22:4933.
15. Gettins P, Cunningham LW (1986). A unique pair of zinc binding sites in the human α_2-macroglobulin tetramer. Biochemistry 25:5004.
16. Feldman SR, Ney KA, Gonias SL, Pizzo SV (1983). *In vitro* binding and *in vivo* clearance of human α_2-macroglobulin after reaction with endoproteases from four different classes. Biochem Biophys Res Commun 114:757.
17. Ney KA, Gidwitz S, Pizzo SV (1984). Changes in the binding of "fast" form α_2-macroglobulin to 3T3-L1 cells after differentiation to adipocytes. Biochemistry 23:3395.
18. Feldman SR, Rosenberg MR, Ney KA, Michalopoulos G, Pizzo SV (1985). Binding of α_2-macroglobulin to hepatocytes: Mechanism of in vivo clearance. Biochem Biophys Res Commun 128:795.

19. Imber MJ, Pizzo SV (1981). Clearance and binding of two electrophoretic "fast" forms of human α_2-macroglobulin. J Biol Chem 256:8134.
20. Imber MJ, PIzzo SV, Johnson WJ, Adams DO (1982). Selective diminution of the binding of mannose by murine macrophages in the late stages of activation. J Biol Chem 257:5129.
21. Gonias SL, Pizzo SV (1983) Chemical and structural modification of α_2-macroglobulin: Effects on receptor binding and endocytosis studies in an *in vivo* model. Annals of the NY Acad of Science 421:457.
22. Gonias SL, Pizzo SV (1981). Inactivation of the plasma protease inhibitor α_2-macroglobulin by the antitumor drug cis-dichlorodiamineplatinum (II). J Biol Chem 256:12478.
23. Gonias SL, Pizzo SV (1981). Altered clearance of α_2-macroglobulin complexes following reaction with cis-dichlorodiamineplatinum (II). Biochim Biophys Acta 678:268.
24. Gonias SL, Oakley AC, Walther PJ, Pizzo SV (1984). Effect of diethyldithiocarbamate and nine other nucleophiles on the intersubunit protein crosslinking and inactivation of purified human α_2-macroglobulin by cis-dichlorodiammineplatinum (II). Cancer Res 44:5764.
25. Pizzo SV, Roche PA, Feldman SR, Gonias SL (1986). Further characterization of the platinum reactive component of the α_2-macroglobulin receptor recognition site. Biochem J 238:217.
26. Sottrup-Jensen L, Gliemann J, Van Leuven F (1986). Domain structure of human α_2-macroglobulin. FEBS Letters 205:20.
27. Beauchamp CO, Gonias SL, Menapace DP, Pizzo SV (1983). A new procedure for the synthesis of polyethylene glycol-protein adducts:Effects of function, receptor recognition and clearance of superoxide dismutase, lactoferrin and α_2-macroglobulin. Anal Biochem 131:25.
28. Stenbjerg S (1981). Inherited α_2-macroglobulin deficiency. Thromb Res 22:491.
29 Laurell CB, Jeppson JO. In "The Plasma Proteins" (F.W. Putnam, Ed.), pp 229-264. Academic Press, New York, 1975.
30. Hovi T, Mosher D, Vaheri A (1977) Cultured human monocytes synthesize and secrete α_2-macroglobulin. J Exp Med 145:1580.

31. Borth W, Teodorescu M (1986) Inactivation of human interleukin 2 (IL-2) by α_2-macroglobulin-trypsin complexes. Immunology 57:367.
32. Thomson AW, Horne CHW (1980) Biological and clinical significance of pregnancy-associated α_2-glycoprotein-a review. Invest Cell Pathol 3:295.
33. Van Leuven F, Cassiman J-J, Van den Berghe H (1986) Human pregnancy zone protein and α_2-macroglobulin. High-affinity binding of complexes to the same receptor on fibroblasts and characterization by monoclonal antibodies. J Biol Chem 35:15522.
34. Johnson WJ, Pizzo SV, Imber MJ, Adams DO (1982) Receptors or maleylated proteins regulate secretion of neutral proteases by murine macrophages. Science 218:574.
35. Hoffman M, Feldman S, Pizzo SV (1983) α_2-Macroglobulin "fast" forms inhibit superoxide production of activated macrophages. Biochim Biophys Acta 760:421.
36. Habu S, Raff MC (1977) Accessory cell dependence of lectin-induced proliferation of mouse T lymphocytes. Eur J Immunol 7:451.
37. Unanue ER (1981) The regulatory role of macrophages in antigenic stimulation. II. Symbiotic relationship between lymphocytes and macrophages. Adv Immunol 31:1.

The Pharmacology and Toxicology of Proteins, pages 113–129

LIKE CONGENERS AND CONJUGATES (INCLUDING PEPTIDYLATION) OF CATECHOLAMINES, BETA ANTAGONISTS AND HISTAMINE PRODUCE UNUSUAL PHARMACOLOGIC SELECTIVITY

Kenneth L. Melmon, Manzoor M. Khan and Murray Goodman*

Division of Clinical Pharmacology
Departments of Medicine and Pharmacology
Stanford University School of Medicine
Stanford, California 94305
and
*Department of Chemistry
University of California-San Diego
La Jolla, California 92093

ABSTRACT We hypothesized that the pharmacologic effects of catecholamines, their antagonists and histamine might be preserved by derivatizing them at a site distant from the pharmacophore. The purpose of the derivatization was to alter the pharmacokinetics of the resulting congener derivatives and conjugates based on the chemical qualities of the ligands attached to them. We did not expect that the derivatives would be effect or tissue specific; we expected only altered pharmacodynamics or tissue distribution based on altered *in vivo* pharmacokinetics and/or tissue distribution respectively.

A series of congener derivatives and conjugates of catecholamine has unusual pharmacologic effects. Several of the derivatives are more potent than their progenitor, and some have surprisingly unprecedented receptor, tissue and effect specificity. For instance, norepinephrine with a branched alkene chain and toluidide moiety attached to the amino end of the molecule has more inotropic and much less chronotropic effects than norepinephrine or isoproterenol. The derivative's effect on heart and fat cells is far more pronounced than the progenitor but few if any effects on peripheral sites of catecholamine action including the bronchus are produced by this congener. The analogous congener of propranolol becomes proportionally more able to bind to the beta 2 receptors of S-49 cells than to the same receptors of lung tissue. Finally an analogous compound of histamine loses its H_2 cardiac, and H_1 vascular activity and appears to become a specific H_1 agonist on murine natural suppressor lymphoid cells. The dipeptide conjugate of the norepinephrine congener derivative becomes orally active with prolonged *in vivo* cardiac effects when compared to norepineprhine or isoproterenol. Rather surprising is the finding that parallel series of unrelated pharmacophores (histamine and catecholamines) often have predictable parallels in potency, effect and tissue specificity that was seen with the catecholamine.

The changed tissue and effect specificity are so far not explained by altered *in vivo* pharmacokinetics. Nor can the difference of conjugates from progenitor be explained by anything other than the pharmacologically inert ligands (e.g., the peptidylation, etc.). The ligands do not interact with classical receptors, nor do they prevent the pharmacophore from recognizing the receptors. All the effects of the derivatives are competitively blocked by antagonists or agonists, but after washing an antagonist from the preparation that was stimulated by a catecholamine derivative, the response to the agonist returns. The unusual effects and their duration seem attributable to the effects of the congener portion and peptide moeity of the conjugate that may be attracted to some undefined receptor microenvironment. Perhaps such derivatization of other surface active pharmacophores could impart similar tissue and effect specificity.

INTRODUCTION

For some time, our laboratories have been involved in projects that attempt to modify endogenous catecholamines and histamine. Each amine has widespread (multiple tissue) and a spectrum of effects within tissues. We reasoned that if the drugs could be made tissue and/or effect specific, the derivatives could be used as probes to define the specific physiologic and pathologic roles of their progenitors. We also believed that effect and tissue specific catecholamines, histamine and/or their antagonists ultimately could have interesting therapeutic potential (1-5).

At first our approach was oriented towards modifying the amines and beta adrenergic antagonists by using the amino end of the molecule and adapting the molecule (congener) so that it could be linked to monodisperse oligopeptides, and even proteins such as albumin and monoclonal antibodies (conjugates) (Fig. 1). We assumed that the chemical characteristics of the pharmacologically inert ligands used in preparing the congeners and conjugates would produce major net chemical alterations of the new molecule as opposed to the chemical characteristics of the progenitor compound. We had planned to systematically synthesize and test the effects of the ligands on the alterations they would cause in the pharmacokinetics and tissue distribution of the congeners and conjugates compared to the unaltered progenitor. By this approach, we had hoped to develop drugs that would have their predominant effects in one tissue or another and for different durations than the native amines.

We found that we could chemically modify the amines by the strategy proposed. We obtained remarkable and unexpected tissue AND effect specificity when we tested the congener and conjugate derivatives. All of the effects we saw were those produced by the classical mechanisms of agonism or antagonism of the progenitor amines or antagonists. But the catecholamine agonists we studied almost exclusively affected the heart. Even in that tissue inotropic effects were greatly augmented while chronotropic effects were almost nill (6,7). The congeners that affected heart, had no systemic or hormonal effects, and were many magnitudes more potent than the progenitor drug. Surprisingly they also were orally active (unpublished data). Our original assumption, however, was wrong. The tissue and effect specificity could not be explained by alteration of the pharmacokinetics of at least the congeners and a peptide conjugate of catecholamine (8).

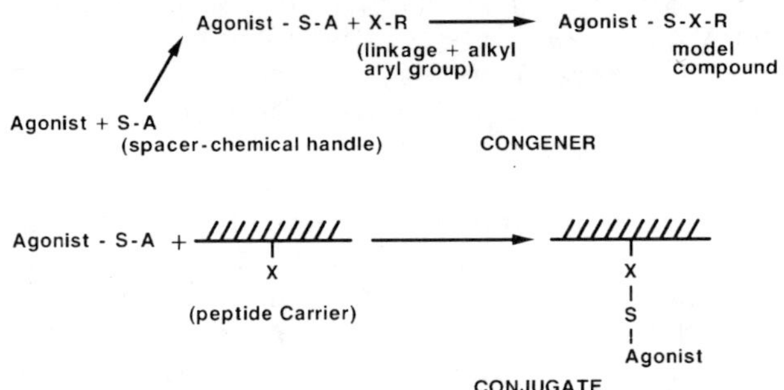

FIGURE 1: Congener Approach to Drug-Conjugate Design

Series of congeners and conjugates that paralleled the series made with agonists of catecholamines have been made with beta adrenergic antagonists (propranolol and practolol) and with a non-catecholamine agonist -- histamine. Surprisingly we often find that like derivatives of each antagonist and of histamine show remarkably predictable parallels in potency, effect and tissue specificity as was seen with catecholamine (9,10).

Thus, we now have congeners of propranolol that are several magnitudes more potent than propranolol, have selective anti-beta 1 effects, and have 400 fold more affinity to the beta 1 receptors of heart as opposed to adipose tissue. Likewise, histamine derivatives can be made to have their exclusive effects on lymphocytes. In lymphocytes their exclusive effects are H_1 agonism.

We believe that our work may ultimately prove that these autacoid agonists with diffuse and unselected effects when given systemically, can be modified to become very selective agonists. If we are right, we will have made pharmacologic tools to define hitherto unclear physiologic and pathologic roles of catecholamines and histamine. We also will be able to consider the more selective effects *in vivo* than we now think possible using the progenitor drug. Some of these selected activities could be clinically useful.

We believe that our work may be uncovering a fundamental principle of membrane pharmacology. We hypothesize that our ligands utilize the microenvironment of the receptor for their tissue and effect specificity. Proof of this theory and its extrapolation to larger drugs, including biologically active peptides, is now feasible to undertake. If this theory is true, it may be applicable to other membrane active drugs.

METHODS

The methods for synthesis of the catecholamine congeners and conjugates (up to octa peptides); propranolol and practolol congeners and conjugates, and histamine congeners and conjugates (up to dipeptides); their purification and characterization have been fully described (2,4,5,9,10-16) (Fig. 1).

The pharmacologic testing for each of the derivative of beta-adrenergic agonist and antagonist include binding and displacement assays on whole cell and membrane preparations of S-49 mouse lymphoma cells (beta 2 receptors), turkey erythrocytes (beta 1 receptors), rat adipocytes (beta 1 receptors), rat heart (beta 1 receptors), and rat lung (beta 2 receptors). Biologic activity of the progenitors and their derivatives were first measured by cAMP accumulation assays. Inhibition of isoproterenol induced cAMP accumulation was the assay for the antagonists. Later more complex effects were assessed. *In vitro* preparations of intact heart, papillary muscle from the heart, bronchial tissue, and tissues mentioned above were used to provide membranes. *In vivo* tests ranged from simple measurements of heart rate and blood pressure in anesthetized rats, to complex measurements of cardiac contractility, chronotropic actions, coronary vasodilation, and systemic metabolic and hormonal effects ordinarily expected of catecholamines on rats, rabbits and dogs.

The biologic functions of the derivatives of histamine were tested by accumulation of intracellular AMP in murine lymphocytes (H_2 receptor effect) (10) and in MLR preparations (H_1 receptor effects) (10). Other assays were the adenylate cyclase activation in the guinea pig myocardium (an H_2 receptor effect) and vasoconstriction of the rabbit aorta (H_1 receptor effect) (17).

RESULTS

Catecholamines

In all, about 150 derivatives of the catecholamine norepinephrine, the beta-adrenergic blockers propranolol and practolol, and histamine have been synthesized and characterized. Because we focused on the branched chain substitutions made on the amino end of norepinephrine, comparisons of the pharmacologic effects of the congeners and conjugates were made with isoproterenol, rather than norepinephrine. The lessons learned from the catecholamine congeners and conjugates were used in design of the derivatives of the other series. For instance, we have not yet synthesized derivatives of practolol, propranolol or histamine similar to those of the inactive derivatives of the catecholamines that were constructed by using sites other than the amino end of the molecule.

As can be seen in Table 1, the potency of the congeners tested on S-49 cells' intracellular accumulation of cAMP showed the most striking changes when the lengths of the methylene chain varied from 2-5 carbons, and when different substituents were added to the aromatic moiety. The most profound increase in potency was obtained from the C_4 methylene chain with a p-trifluoromethyl substitution on the toluidide group. This compound (Table 1, 3) and its p-methyl toluidide partner (Tables 1,2) became key derivatives for comparison with like derivatives using the beta blocking and histamine pharmacophores. The table also illustrates the important shifts in potency by simple position changes of the trifluoromethyl group on the toluidide moiety (Table 1-3, 1-4 and 1-5).

TABLE 1

IN VITRO BIOLOGICAL ACTIVITY OF CONGENERS AND MODEL DERIVATIVES OF ISOPROTERENOL

Biological activity was measured by cyclic AMP accumulation in S-49 cells. Relative activity is expressed as the ratio of the K_a for isoproterenol to K_a for the test compound.

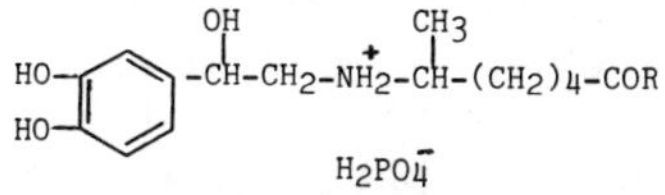

$$(HO)_2C_6H_3\text{-}CH(OH)\text{-}CH_2\text{-}\overset{+}{N}H_2\text{-}CH(CH_3)\text{-}(CH_2)_4\text{-}COR \quad H_2PO_4^-$$

Compound	R	n[a]	Activity relative to isoproterenol
1. Carboxylic acid	-OH	4	1.6×10^{-4}
2. p-Toluide	$-NH-C_6H_4-CH_3$	4	5.4×10^{1}
3. p-Trifluoromethylanilide	$-NH-C_6H_4-CF_3$	4	8.6×10^{2}
4. m-Trifluoromethylanilide	$-NH-C_6H_4-CF_3$ (meta)	4	9.8×10^{-1}
5. o-Trifluoromethylanilide	$-NH-C_6H_4-CF_3$ (ortho)	4	2.5×10^{-4}
6. p-Butylanilide	$-NH-C_6H_4-(CH_2)_3CH_3$	4	3.2×10^{3}
7. p-Methoxylanilide	$-NH-C_6H_4-OCH_3$	4	3.1
8. p-Toluide	$-NH-C_6H_4-CH_3$	2	7.1×10^{-3}
9. p-Toluide	$-NH-C_6H_4-CH_3$	3	3.6×10^{-1}
10. p-Toluide	$-NH-C_6H_4-CH_3$	5	1.9×10^{-1}

[a]n=Number of methylene groups in the spacer group.

Table from reference (7).

Since there are four enantiomers for each of these preparations, stereospecific synthesis was done to test the pharmacologic effects of each enantiomer. Using the S-49 cell assay, the RR enantiomer was found to be the most potent enantiomer followed by RS and SR enantiomers.

The SS compound was generally inactive, i.e., it was 85 times less potent than the RR.

Peptide conjugates ranging from a dipeptide with phenylalanine-glycine that was four times more potent than isoproterenol to an octapeptide that still maintained classical beta adrenergic actions were also studied (18).

In general, the compounds were selective in tissues they would effect and even within a single tissue effects were skewed when compared to those of the progenitor pharmacophore. For instance, despite the fact that chronotropic and inotropic effects of beta mimetic amines seem to go hand in hand, compound 3 (Table 1) was close to 5 times more potent as an inotropic agent, but only 2.4 times more potent as a chronotropic agent when tested on the guinea pig atrial muscle (7,14). The dichotomous effects on inotropism vs chronotropism was also seen in rabbits and dogs _in vivo_ (data unpublished).

The _in vitro_ data indicated that compound 3 could not be easily washed from the heart tissue, an action that corresponded to, but may not have fully explained, the very long duration of the drug effect when given _in vivo_ as compared to isoproterenol (7). Furthermore the compound was a much more potent and efficacious coronary vasodilator than isoproterenol, but had no effect in reversing histamine induced broncho contriction in guinea pig bronchi (7) or provoking systemic metabolic changes as isoproterenol did. Finally, a dipeptide conjugate of the norepinephrine that had most of the selective effects shown by compound 3, as well as compound 3 were pharmacologically active when given orally (unpublished data) -- an action not seen with isoproterenol.

Perhaps not surprisingly after obtaining these data, we should have anticipated that the _in vivo_ pharmacokinetics of compound 2, another derivative with very long term _in vivo_ effects, would have a pattern that was generally indistinguishable from isoproterenol (8). Furthermore, while early studies do show much higher relative affinity of compound 3 for beta 1 receptors of turkey erythrocytes than isoproterenol, the tissue selectivity of the compound may not be explainable on this single feature of altered affinity for the receptor itself

(19). In fact when compound 3 was incubated *in vitro* with guinea pig atrial muscle, its effects were competitively blocked by propranolol. But when the bath was changed and the propranolol was washed free of the preparation, the compound 3 induced changes in contractility returned without adding additional drug (unpublished data). Thus, simple binding to receptors *per se* could not have accounted for the complex response described.

In summary, we had synthesized congeners and conjugates of "isoproterenol" all of whose properties acted via classical beta adrenergic receptors. But the potency of the derivatives were different from that of the progenitor; the tissues they affected were not the same group as those affected by the progenitors; and effects within a given tissue were more specific than was seen when the progenitor was used.

We next asked whether the potential value of making tissue and effect specific beta-agonists could also apply to beta-antagonists and whether the ligands that had been used to create the most interesting skews in potency and effect with the agonist would do the same with antagonists.

Beta Adrenergic Antagonists

Table 2 shows a sampling of propranolol derivatives analogous to those made with isoproterenol. While a number of comparisons between the agonist and antagonist series show analogy as well as differences in the changes of potency and tissue effect specification in the two series, note compounds 2-4 and 2-6 correspond with compound 1-2 and 1-3 of the agonists (9). The table reveals the remarkable tissue and effect specificity of the two antagonist derivatives. A full description of these effects can be found in a recent report (9). In that report we also identified a peptide conjugate of propranolol that became one of the most potent selective beta 1 antagonists of the series and was far more potent in this selective effect than practolol or any of its congeners and conjugates (9).

TABLE 2

RELATIVE POTENCIES OF BETA ADRENERGIC ANTAGONIST DERIVATIVES*

	S-49/Lung	Fat cells/Heart	S-49/Fat cells	Lung/Heart**
Propranol	.54	1.0	0.1	.77
1 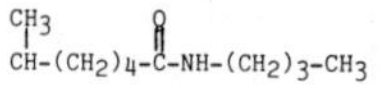 $CH(CH_3)-(CH_2)_4-C(=O)-NH-(CH_2)_3-CH_3$	0.1	25	.003	10
4 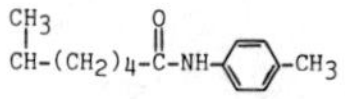 $CH(CH_3)-(CH_2)_4-C(=O)-NH-C_6H_4-CH_3$	.0054	15	.0005	1.3
6 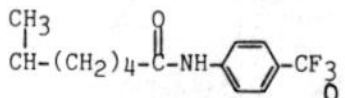 $CH(CH_3)-(CH_2)_4-C(=O)-NH-C_6H_4-CF_3$	.05	414	.0003	2.5
9 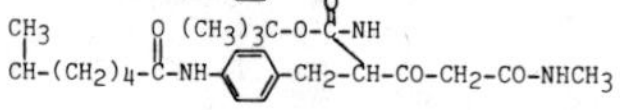 $CH(CH_3)-(CH_2)_4-C(=O)-NH-C_6H_4-CH_2-CH(NH-C(=O)-O-C(CH_3)_3)-CO-CH_2-CO-NHCH_3$	.26	3.0	.13	1.5
Practolol	.25	.81	.91	14.5
14 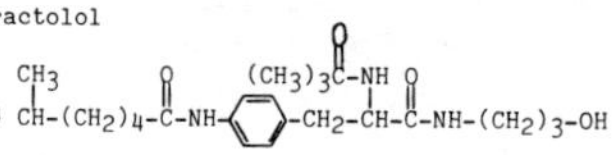 $CH(CH_3)-(CH_2)_4-C(=O)-NH-C_6H_4-CH_2-CH(NH-C(=O)-C(CH_3)_3)-C(=O)-NH-(CH_2)_3-OH$	.048	1.45	1.39	42.0

*Taken from inhibition constants (Ki).

**Rat lung and heart contain 15-20% beta 1 and beta 2 receptors respectively.

Taken from reference (9).

The data confirmed the fact that the same pharmacologically inert ligands when linked to beta antagonists had imparted altered potency, and allowed only selected effects of the progenitor to be expressed by the new congener or conjugate. Yet as with the catecholamine, all the effects of these congeners or conjugates were mediated by classical drug receptor action.

Although the progenitor agents in the two series were related pharmacologically, they were not chemically similar. The next question to ask was whether pharmacologically and chemically unrelated progenitors would develop unusual changes in potency and selection of effects. We chose histamine as the progenitor in making this third series of congeners and conjugates in parallel with the catecholamine agonists and antagonists. Now, the only remaining obvious property shared by the four progenitors used in this work was that each of their actions is mediated by well defined interactions with receptors on the surface membrane of cells.

Histamine

Once again the alkyl chain length of the congeners and conjugates was a key determinant of their altered potency, as well as their dominant H_1 and H_2 stimulatory properties. The H_2 effects of selected derivatives of histamine on lymphocytes can be seen in Table 3.

TABLE 3

COMPARATIVE POTENCIES OF SIMILAR ISOPROTERENOL AND HISTAMINE CONGENER DERIVATIVES ON S-49 LYMPHOMA AND NATURAL SUPPRESSOR CELLS

	Activity Relative to Isoproterenol	Activity Relative To Histamine
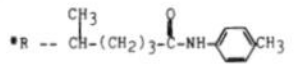 *R -- $CH(CH_3)-(CH_2)_3-C(=O)-NH-C_6H_4-CH_3$	3.11×10^{-1}	3.0×10^{3}
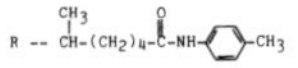 R -- $CH(CH_3)-(CH_2)_4-C(=O)-NH-C_6H_4-CH_3$	4.40×10^{1}	$.30 \times 10^{1}$
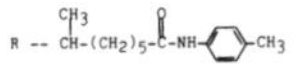 R -- $CH(CH_3)-(CH_2)_5-C(=O)-NH-C_6H_4-CH_3$	1.11×10^{-1}	$.20 \times 10^{1}$
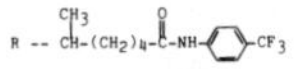 R -- $CH(CH_3)-(CH_2)_4-C(=O)-NH-C_6H_4-CF_3$	8.60×10^{2}	4.2×10^{4}
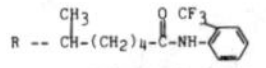 R -- $CH(CH_3)-(CH_2)_4-C(=O)-NH-C_6H_4(CF_3)$	2.6×10^{-4}	Inactive
R -- $CH(CH_3)-(CH_2)_4-C(=O)-NH$-BOC-Phe-Gly-$NHCH_3$	0.4×10^{1}	Inactive
R -- $CH(CH_3)-(CH_2)_4-C(=O)-NH-(CH_2)_3-CH_3$	2.5×10^{-1}	Inactive

Biological activity was measured by cAMP accumulation in a mouse natural suppressor cell and in S-49 cells. Relative activity is expressed as the ratio of Ka for histamine or isoproterenol to Ka for the compound.

*R= (imidazolyl)-CH_2-CH_2NH or (3,4-dihydroxyphenyl, HO, HO)-$CH(OH)$-CH_2-NH

When chain length of the histamine congener ranged from C_1-C_3, the effects on lymphoid cells were predominantly H_1 with the C_3 trifluoromethyl derivative being most potent and lacking any ability to affect the rabbit aorta or the guinea pig myocardium (17,10). The paratrifluoromethyl derivative of histamine with C_4 methylene chain possessed both H_1 and H_2 effects. The deletion of a single methyl group resulted in a total loss of H_2 activity and the resulting derivative became an exclusive H_1 agonist of lymphocytes. The H_1 receptors on aorta did not respond to this agonist. When the trifluoromethyl group was moved from the para position, the potency of the histamine congener fell dramatically as it had in the "isoproterenol" congener.

Only when we made a dipeptide conjugate of histamine (analogous to the dipeptide conjugate of "isoproterenol" that had selective cardiac effects) did we see some retention of the cardiac stimulatory actions of histamine. This cardiac effect was the only agonist effect that we were able to detect using the dipeptide.

Once again we had found that although there were dissimilarities in changes in potency, tissue and effect specificity when similar congeners of the four series of agonists and antagonists were compared, there were enough similarities in a narrow band of derivatives. Therefore, it appeared that employing ligands to create the small number of derivatives might successfully restrict the action of surface membrane active pharmacophores.

DISCUSSION

Extrapolation from our data must be supported by the observation that similar derivatives of all pharmacophores had similar changes in their potency or tissue selectivity profiles. In fact, there were a number of pharmacologic changes that were consistent within a small group of parallel derivatives in each of the four series (see Table 4). Futhermore, it was interesting to see the work of Jacobson et al that used the congener-conjugate approach described in this paper. They created different functional classes of adenosine antagonists (20) by using the same approach. These authors suggested, and we agree, that the ligands we have described may be used to alter potency and receptor selectivity of these antagonists and conceivably other drugs (20).

We now have experience with the derivatives from five different membrane active agents. Perhaps a similar strategy could be adopted to derivatize pharmacophores with higher molecular weights. Based on our previous observations, we could use similar ligands and could predict the tissue and effect selectivity of the derivatives of the higher molecular weight pharmacophores.

TABLE 4

STRUCTURE PREDICTS PHARMACOLOGIC NOVELTY

STRUCTURE	ISOPROTERENOL (Type I)	PROPRANOLOL (β_2) (Type II)	PRACTOLOL (β_1)	HISTAMINE (Type III)
METHYLENE CHAIN				
C_2	⬇	⬇	—	↓
C_3	↓	⬇	↓	⬆ H_1-Lymph
C_4	⬆	⬆	⬆	⬆ H_1,H_2-Lymph
C_5	↓	↓	↓	↓
BRANCHED METHYLENE CHAIN	↑	↑	↑	↑
TOLUIDIDE MOIETIES				
P-methyl	↑ β_1 (Heart vs Lung-TS)	↑ β_1 (Fat vs Heart) RS & TS	↑ β_1 RS & TS	⬆ RS-H_1 & TS-lymph
P-trifluoromethyl	⬆ OA; TS [Peripheral vs Coronary Metabolic, Skeletal Muscle (Glycogen)]	↓ β_1 (Fat vs Heart)	↑ β_2,RS & TS	⬆ RS-H_1 & TS-lymph
Ortho	↓	—	—	↓
Meta	↔	—	—	—
AMINO ACID	↑; β_1 (TS, RS)	↑; β_1 (TS, RS)	↑; β_1 TS, RS	—
DIPEPTIDE	↑; OA; (TS, RS)	↓ Fat vs Heart	—	TS-Heart vs lymph RS-H_2
LONGER PEPTIDE	↔ or ↓	—	—	—

* Histamine derivatives not active on heart (H_2) can be vasoconstrictive (H_1), III-1 or vasorelaxing with H_2III-6 & 9

OA = Oral activity, TS = Tissue selectivity, NT = No Tachyphylaxis in heart

RS = Receptor selectivity, —— = Not tested

The mechanisms of altered potency, tissue, and effect specificity of the derivatives are not clear, but must relate to the ligands that are pharmacologically inert. Although our leads towards explaining the altered pharmacologic effects of the derivatives are quite limited, early data point more towards some altered interaction with the microenvironment of the receptors than an interaction with the receptor alone or simply in altered pharmacokinetics. Defining the exact mechanisms of alteration of the pharmacologic effects of our derivatives is necessary because such information might help us design tissue and effect specific derivatives of other pharmacophores with ubiquitous effects.

In the meanwhile, our approach might apply to the concerns of molecular biologists who are grappling with means to focus other pharmacologic effects of the new larger molecular weight substances. Among the strategies to limit the immunogenicity of the new large molecules, regulation of the molecules is being tried and small peptides are being sought that possess most of the pharmacologic actions of the progenitor protein. But both approaches are likely to lead to drugs that still possess most of the wide spectrum of biological effects that the unmodified proteins had. As long as the drugs are widely distributed and have an inherent spectrum of pharmacologic potential, then the modification will not modify their diffuse effects. Application of our approach may focus the pharmacologic profile of these drugs.

In addition to the data we have emphasized on modification of tissue and effect specificity of low molecular weight amines, we have shown that drugs can be attached to macromolecules without compromising the *in vitro* functions of the pharmacophore or of the macromolecule. Molecular biologists in this meeting ask for vasodilators to be given at the site they wish to concentrate a macromolecule. We ask why not test the effect of macromolecules conjugated to a vasodilator such as histamine via the ligands we have discussed?

In addition to seeking combined effects of two molecules by linking them, perhaps it would be fruitful to use our approach to determine whether some peptides could be made more potent, tissue and effect specific or even orally active by our approach. It might even be possible to affect intracellular transport of macromolecules by conjugating them to impotent high affinity agonists for which there are receptors on the targeted cells. Such an approach could theoretically lead to internalization of the whole molecule as the receptor-agonist complex is internalized.

ACKNOWLEDGEMENTS

The authors are grateful to our collaborators, particularly Drs. Brian Hoffman, Roberto Rosenkranz, Quan Ming Zhu, Ms. Leslie Krasney, Moon Choo, Lynne Anderson, Julie Bauer, Andrea Wilson, and Karen Keaney at Stanford and Drs. M. Verlander, K. Jacobson, and D. Leisy at the University of California, San Diego. We also thank Ms. Sara Fisher for her excellent editorial assistance.

This work was supported in part by NIH HL 26340 and AI 23463.

REFERENCES

1. Melmon KL, Weinstein Y, Bourne HR, Poon T, Shearer G, Castagnoli N Jr (1976). The pharmacological effects of conjugates of pharmacologically active amines to complex or simple carriers: a new class of drug. Mol Pharmacol 12:701.
2. Melmon KL, Verlander MS, Krasny L, Goodman M, Kaplan N, Castagnoli N Jr, Insel P (1979). The chemistry and pharmacology of catecholamines and other low molecular weight ligands conjugated to carriers. In IV International Catecholamine Symposium: Catecholamines: Basic and Clinical Frontiers, Pergamon Press, Vol 1, p 474.
3. Rosenkranz RP, Jacobson KA, Goodman M, Verlander M, Melmon KL (1982). *In vitro* and *in vivo* betamimetic activity of congeners of isoproterenol. Proc West Pharmacol Soc 25:19.
4. Jacobson KA, Rosenkranz RP, Verlander MS, Melmon KL, Goodman M (1982). Monodisperse conjugates of catecholamines. Proceedings of the 17th European Peptide Symposium. In Blaha K, Malon P (eds) Peptides, Prague, p 337.
5. Verlander MS, Jacobson KA, Rosenkranz RP, Melmon KL, Goodman M (1983). Some novel approaches to the design and synthesis of peptide-catecholamine conjugates. Biopolymers 22:531.

6. Rosenkranz RP, Jacobson KA, Verlander MS, Goodman M, Melmon KL (1983). Betamimetic activity of peptide catecholamine conjugates. Proc West Pharmacol Soc 26:381.
7. Rosenkranz RP, Hoffman BB, Jacobson KA, Verlander MS, Klevans L, O'Donnell M, Goodman M, Melmon KL (1983). Conjugates of catecholamines II. Pharmacologic activity of N-alkyl functionalized carboxylic acid congeners and amides related to isoproterenol. Mol Pharmacol 24:429.
8. Ferraiolo BL, Halldin MM, Asscher Y, Akita Y, Melmon KL, Benet L, Castagnoli N. Pharmacokinetics and excretion of unique beta adrenergic agonists. Pharmacology (in press).
9. Khan MM, Leisy D, Verlander M, Hoffman B, Goodman M, Melmon KL (1987). In vitro pharmacologic activity of congener derivatives and model conjugates of propranolol and practolol. Biochem Pharmacol (in press).
10. Khan MM, Leisy D, Verlander M, Bristow MR, Strober S, Goodman M, Melmon KL (1986). The effects of derivatives of histamine on natural suppressor cells J of Immun 137:308.
11. Jacobson KA, Verlander MS, Rosenkranz RP, Melmon KL, Goodman M (1983). Conjugates of catecholamines III. Synthesis and characterization of monodisperse oligopeptide conjugates related to isoproterenol. Int J Pept Protein Res 22:284.
12. Goodman M, Verlander MS, Melmon KL, Jacobson KA, Reitz AB, Taulane JP, Avery MA, Kaplan NO (1983). Characterization of catecholamine-polypeptide conjugates Eur Polym J 19:997.
13. Rosenkranz RP, Jacobson KA, Verlander MS, Klevans L, O'Donnell M, Goodman M, Melmon KL. Conjugates of catecholamines IV. In vitro and in vivo pharmacological activity of monodisperse oligopeptide conjugates. J Pharmacol Exper Ther 227:267.
14. Verlander MS, Jacobson KA, Reitz AB, Rosenkranz RP, Melmon KL, Goodman M (1984). Application of the congener approach to the design and synthesis of peptide-catecholamine conjugates. Proceedings of the International Symposium on Polymers in Medicine, Porto Cervo, Sardinia, Italy, 1982. Polym Med p 57.
15. Reitz AB, Sonveaux E, Rosenkranz RP, Verlander MS, Melmon KL, Hoffman BB, Akita Y, Castagnoli N, Goodman

M (1985). Conjugates of catecholamines. 5. Synthesis and beta-adrenergic activity of N-(aminoalkyl) norepinephrine derivatives. J Med Chem 28:634.
16. Reitz AB, Avery MA, Rosenkranz RP, Verlander MS, Melmon KL, Hoffman BB, Akita Y, Castagnoli N, Goodman M (1985). Conjugates of catecholamines. 6. Synthesis of beta-adrenergic activity of N-(hydroxyalkyl) catecholamine derivatives. J Med Chem 28:642.
17. Khan MM, Leisy D, Verlander M, Egli M, Lok S, Goodman M, Melmon KL (1987). Congener derivatives and conjugates of histamine: Synthesis and tissue and receptor selectivity of the derivatives. (manuscript in preparation).
18. Fakuda M, Beechan CM, Verlander M, Goodman M, Khan MM, Melmon KL (1985). Conjugates of catecholamines VII. Synthesis of beta adrenergic activity of peptide catecholamine conjugates. Int J Peptide and Protein Res (in press).
19. Schramm M, Eimerl S, Goodman M, Verlander MS, Khan MM, Melmon KL (1986). High potency congeners of isoproterenol: binding to beta-adrenergic receptors, activation of adenylate cyclase and stimulation of intracellular cyclic amp synthesis. Biochemical Pharmacology 35:2805.
20. Jacobson KA, Kirk KL, Padgett WL, Daly JW (1986). A functionalized congener approach to adenosine receptor antagonists: amino acid conjugates of 1,3-dipropylxanthine. Molecular Pharmacology 29:126.

The Pharmacology and Toxicology of Proteins, pages 131–148

DELIVERY OF THERAPEUTIC AGENTS BY OSMOTIC PUMPS

Alfred A. Amkraut, John W. Fara,
Kirstin C. Nichols, and Nigel P. Ray

ALZA Corporation, 950 Page Mill Road,
Palo Alto, CA 94304

ABSTRACT The availability of osmotic devices that allow for long-duration control over the rate of peptide administration in laboratory animals has opened up numerous methods, protocols, and models for delivering these substances to animals and potentially also to man. We present studies that have been done utilizing two of these devices: the implantable osmotic pump (ALZET®) for studies in animals, and the OSMET™ module for oral, vaginal, or rectal administration of drugs in clinical research.

In animals, osmotic pumps have been used extensively to deliver proteins, peptides and other short half-life agents which are difficult to deliver by other methods. These pumps allow for prolonged, constant delivery for up to 4 weeks following a single implantation procedure. With a short half-life agent, they assure drug presence without the need for repeated injections which cause stress-related physiological changes. Complete implantation of the infusion device permits free animal movement throughout the period of infusion. Osmotic pumps have been utilized as artificial organs, replacing or supplementing endocrine gland secretions, such as insulin. Additionally, they have been modified via catheter attachment for micorperfusion into soft tissue such as brain, into hollow organs, as the intestinal tract and the uterus, and into the systemic circulation as well as CSF.

Another important capability is the adaption of these osmotic systems to deliver at preprogrammed rates that vary over time. It is not yet clear which proteins or peptides are best given by a constant rate regimen and which might require a pattern of input. Researchers have devised various approaches to mimic circadian rhythms, to infuse agents to alter existing patterns, or to study what existing patterns may do to the test agent.

The same osmotic technology has been adapted to clinical studies with the OSMET™ modules. Designed for delivery over 8, 12, or 24 hours, these modules have been used in clinical investigative studies to deliver anti-inflammatories, antihypertensives, various receptor-blocking agents, and are now finding usefulness in evaluating peptide absorption across various mucosal surfaces.

INTRODUCTION

An implantable osmotic pump for animal studies, the ALZET® (1,2), allows investigators to control both temporal and spatial aspects of drug delivery by determining both the rate of delivery and the target organ or tissue, respectively. The ALZET® is being used routinely in the early stages of drug research for drug screening, toxicology, and pharmacology. Applications in these areas are creating new therapeutic opportunities to select appropriate drug delivery patterns and the effect of drug targeting in initial animal studies during drug development.

ALZET OSMOTIC PUMPS

In preclinical studies, the ALZET® provides rate-controlled, unattended administration of a wide variety of bioactive agents for 1 to 4 weeks. Pumps currently available have delivery rates and durations of 1 or 10 µl/hr or 7 days, 0.5 or 5 µl/hr for 14 days, or 2.5 µl/hr for 28 days (Fig. 1). Smaller pumps and lower release rate pumps are expected to be available in the near future. These osmotic pumps have the following components (Fig. 2): an impermeable, inert, flexible drug reservoir open to the exterior through a single portal; a thin sleeve containing osmotic agent surrounding the reservoir; and a semipermeable membrane surrounding the sleeve of the osmotic agent. A researcher fills the reservoir through the portal with a solution or suspension of the drug he is studying and then inserts a flow moderator so osmosis will control the delivery rate and the solution will not diffuse from the portal. After the filled pump is implanted, water from the surrounding tissue moves osmotically through the membrane at a rate controlled by the membrane's permeability. Because the membrane is rigid, the osmotic sleeve swells to displace the liquid drug formulation within the reservoir out through the portal in a continuous, controlled manner. The precise delivery profile and the excellent in vitro /in vivo correlation of pumping rate is illustrated in Figure 3.

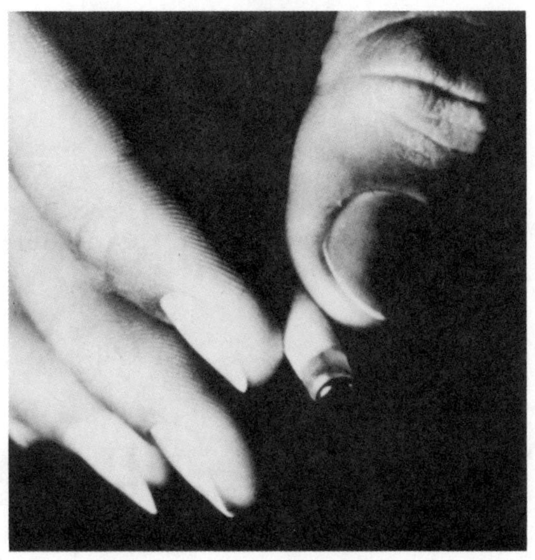

FIGURE 1. ALZET® osmotic pump

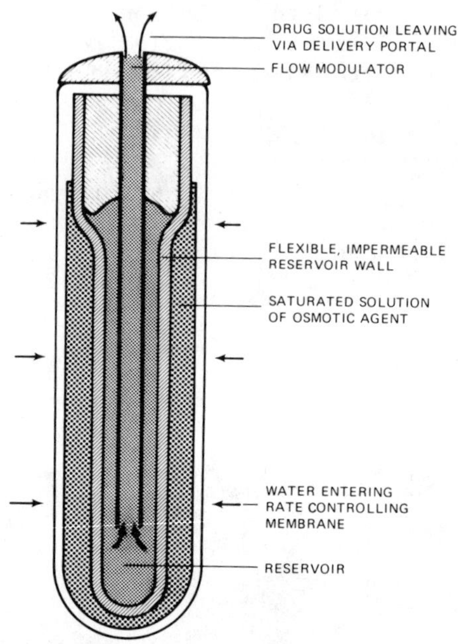

FIGURE 2. Cross-section of the osmotic pump

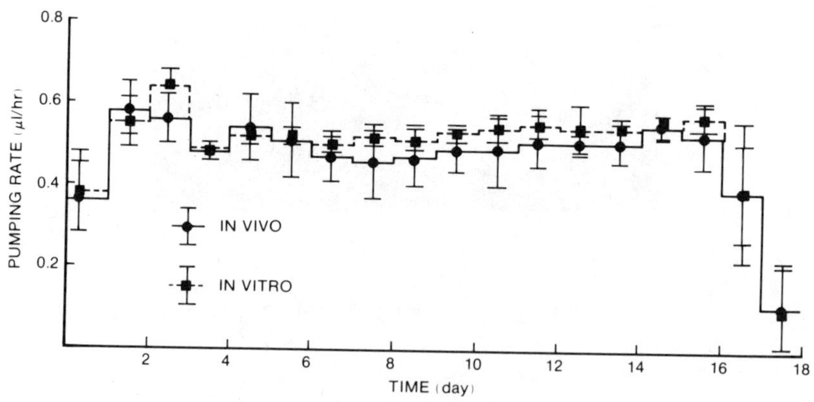

FIGURE 3. *In vitro* and *in vivo* pumping rates of an osmotic pump designed to deliver 0.5 μl/hr of an agent for 14 days.

ALZET® pumps have the added advantage of providing data from unstressed animals for the duration of the experiment after the first day (3,4). In addition, the animals remain untethered and unrestrained and do not require night or weekend dosing. Many investigators have demonstrated absence of toxic or stress effects on experimental animals following s.c. or intraperitoneal implantation.

ALZET® has been implanted in all common laboratory animals as well as in sheep, baboons, cows, cheetahs, horses, and even cold blooded animals (reptiles, fish, and crustaceans).

Of all the agents (over 400) that have been delivered with ALZET®, those with a short half-life (minutes) are particularly suited to delivery with ALZET®. In the body, peptides are often released continuously or in short pulses that are not feasible to mimic with bolus delivery. Bolus delivery of peptides produces only a short period of systemic exposure which makes it necessary to make observations at precisely the right moment to recognize many of the effects. The peptides that have been delivered with the ALZET® are listed in Table 1. This list continues to grow as more researchers discover the advantages of ALZET® delivery.

TABLE 1

PEPTIDES DELIVERED BY OSMOTIC PUMPS IN RECENTLY REPORTED EXPERIMENTS

ACTH	LH
Angiotensin	LHRH
Atrial Natriuretic Factor	LHRH analog, agon/antag
Bombesin	α-MSH
Bungarotoxin snake venom	Muramyl dipeptide
Calcitonin	Nerve Growth Factor
Cholecystokinin	Neuropeptide Y
Colony - Stimulating Factor	Neurotensin
Dermorphin	Oxytocin
Endorphins	Parathyroid hormone
Enkephalins	Pentagastrin
Erythropoietin	Pituitary extract
FSH	Prolactin
Gastrin	Somatomedins
Glucagon	Substance P
Growth hormones	Teprotide
IGF	Tetragastrin
Insulin	Vasoactive Intestinal Polypeptide
Interferons	Vasopressin
Interleukins	

CONTINUOUS DRUG DELIVERY

The ALZET® osmotic pumps have been used extensively to compare the effects of continuous drug delivery with those produced by conventional therapy. Such comparisons illustrate regimen dependent drug actions--information that is very important to have prior to initial human trials with a new agent to optimize dosing regimens. The following examples, selected from an extensive literature, are examples of the effects of dosage regimens of drugs and peptides on physiologic responses.

Nau et al. (5) administered the same dose of valproic acid (VPA) by two different regimens to pregnant mice from gestation days 7 to 15: by injection once daily and by continuous, constant-rate infusion from implantable pumps. Injections caused drug plasma concentrations to peak and decline quickly with long periods between injections in which the drug was not detectable.

The continuous infusion maintained a narrow range of drug and metabolite concentrations. In addition, Nau discovered that the infusion regimen shifted the embryotoxicity curve to the right--a ten-fold higher dose was necessary with the infusion regimen to yield the same resorption rate observed with the injection regimen (Fig. 4).

Sikic et al. (6) found that efficacy, in addition to toxicity, was regimen dependent. These investigators delivered the anticancer drug, bleomycin, in three different five-day regimens: injections twice daily, alternate day injections; and continuous infusions by implanted osmotic pumps (Figs. 5 and 6). Peng et al. (7) confirmed enhanced bleomycin efficacy by continuous infusion. Clearly, the therapeutic index for bleomycin is widened by using the infusion regimen and narrowed by the injection regimen.

The effects of hormones have long been recognized as rate dependent rather than dose dependent. The ALZET® osmotic pump has been used to explore the action of parathormone, human growth hormone, and triiodothyronine.

Tam et al.(8) demonstrated that the net effect of parathryroid hormone on bone formation and resorption depended on the regimen used. They administered parathyroid hormone by either daily s.c. injections or

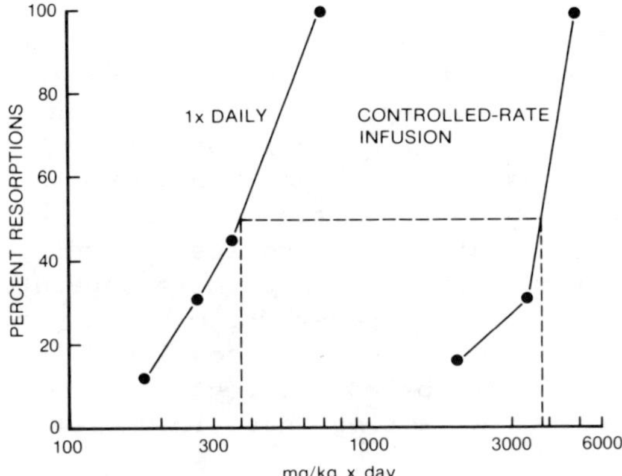

FIGURE 4. Fetal resorption rate observed when valproic acid is administered as a single daily injection or by continuous infusion. (Adapted from Nau, H. et al., Metabolism of Antiepileptic Drugs, Levy, R.H. et al., Eds, Raven Press, 1984, 85, with permission.)

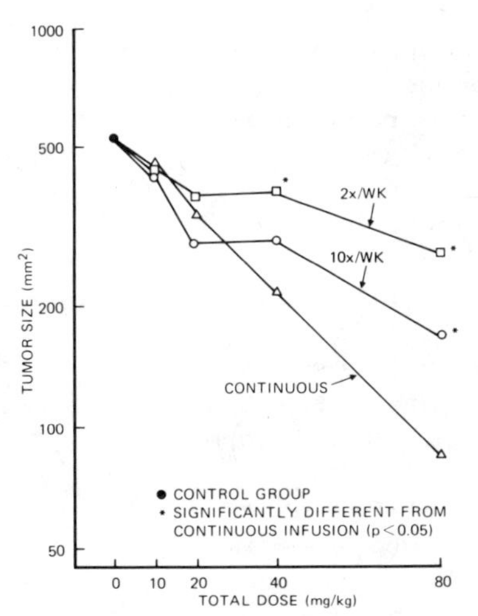

FIGURE 5. Dose-response curve of bleomycin given in 3 regimens against Lewis lungs carcinoma. Measurements shown are on the 15th treatment day but are representative of all other treatment days.[6]

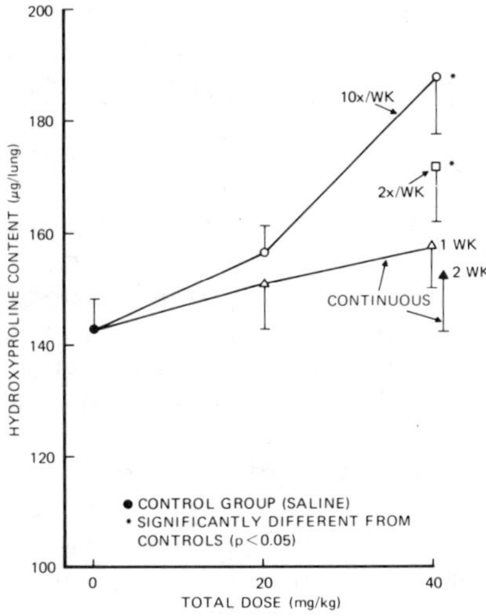

FIGURE 6. Effect of various regimens and doses of bleomycin on pulmonary toxicity in normal mice, as measeured by the lung hydroxyproline content after 10 weeks of treatment.

by continuous infusion in thyroparathyroidectomized rats. The infusion resulted in a net decrease in trabecular bone by increasing bone apposition and increasing both bone formation and resorption surfaces. Equal doses by s.c. injection increased bone volume by increasing bone apposition rate and bone formation surface without increasing resorption surfaces. Because the s.c regimen permits separation of the resorptive effects of parathyroid hormone from its effects on apposition rate, they concluded that intermittent dosing would be more effective than continuous infusion in promoting anabolic skeletal effects.

Connors and Hedge (9) used ALZET® to deliver physiologic amounts of T3 to unanesthetized, thryoidectomized rats to explore the quantitative relationships in the control of thryoid hormone secretion. Continuous infusion maintained near normal T_3 plasma levels while TSH levels rose steadily over 144 hours; at higher T_3 levels, T_3 was elevated and TSH was in the normal range. When equivalent amounts of T_3 were given by s.c. injection, animals had nonphysiologic peak-and-trough T_3 levels and TSH supression to pre-thyroidectomy levels. Injection also reduced reduced TRH responsiveness to a greater degree than the infusion.

Cotes et al. (10) administered growth hormone (hGH) in various dose regimens and vehicles to hypophysectomized rats to determine the optimal way to administer the limited amounts available (Fig. 7). hGH given continously induced a greater growth response than did higher doses of hGH given by intermittent daily injection, despite the fact that rats and normal children have episodic secretion of GH.

Chang et al. (11), in their study of the most effective routes of administration and dosage regimens of interleukin-2 (RIL-2), determined that osmotic pumps placed subcutaneously or intraperitoneally in mice --providing continuous administration of RIL-2 over a 4-day period--resulted in optimal _in vivo_ generation of lymphokine-activated killer cells in the spleen and

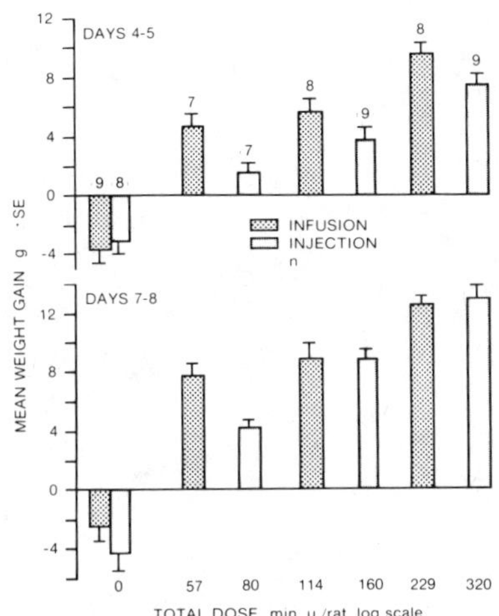

FIGURE 7. Mean+SEM weight gain at days (a) 4-5 and (b) 7-8 after treatment started on day 1 in hypophysectomized rats treated with human growth hormone by continuous infusion from implanted ALZET® pumps (shaded bars), and by intermittent injection with control infusion from implanted ALZET® pumps (solid) (with permission).[10]

peritoneal exudate (Table 2). Nishimura et al. (12) also demonstrated that continuous infusion of interleukin-2 by ALZET® pumps in mice significantly prolonged serum IL-2 levels. A single s.c. injection was not able to achieve this effect.

Atrial natriuretic factor has been successfully delivered with osmotic pumps. In one three day study (13), ANF infusion with osmotic pumps reduced blood pressure in angiotensin II-induced hypertension in rats.

Continuous administration of antigen by ALZET® enhanced both immunogenic and tolerogenic effects as measured by antibody production (14). Groups of 10 Swiss-Webster Mice were used in these experiments. Three injections of 10 µg of lysozyme spaced at weekly intervals did not give rise to measurable antibody levels 28 days after initiation

TABLE 2

Comparison of different routes of administering recombinant interleukin-2 (RIL-2) and generation of natural killer (NK) activity

Expt No.	Days of Treatment	NK Activity (LU/Spleen)[a]											
		Intraperitoneal		Intraperitoneal with gelatin		Intraperitoneal pump		Subcutaneous		Subcutaneous with gelatin		Subcutaneous pump	
		A[b]	B[c]	A	B	A	B	A	B	A	B	A	B
1[d]	2	<100	<120	475	<140	496	<110	255	<45	168	<45	550	<35
	4	126	<95	580	<130	2,090	<90	420	<90	392	<65	906	<60
	6	570	<145	<190	<125	731	215	504	<85	264	<60	1,704	<55
2[e]	2	162	--	<80	--	923	--	532	--	540	--	1,460	--
	4	<75	--	377	--	4,264	--	214	--	1,170	--	5,838	--
	6	136	--	<190	--	900	--	337	--	1,298	--	1,150	--

a One lytic unit (LU) = 33% lysis of labeled 10^4 YAC cells in a 4-h ^{51}Cr release assay

b With RIL-2

c Without RIL-2

d 80,000 units RIL-2 was delivered daily by the pump or given by injection

e 120,000 units RIL-2 was delivered daily by the pump or given by injection

(with permission from A. Chang et al. J. Biol. Resp. Modif. 3(5): 561, 1984)

of the procedure. However, continuous administration of only 10 µg for 28 days (15 ng/hr)--with or without E. Coli lipopolysaccharide--achieved levels comparable to those reached following injection of the same amount of antigen in complete or incomplete Freund's adjuvant (Table 3). Antibody was determined by capacitation of ^{125}I-labeled lysozyme with gamma globulin.

TABLE 3

ANTIBODY RESPONSE TO LYSOZYME
LYS Bound by Serum (ug/ml)

Mode Immzn	14d	28d	60d	*	90d	150d
10 µg LYS 4W AZ	0.87	5.50	1.13	*	39.30	15.6
10 µg LYS IFA	0.73	4.10	4.45	*	27.8	3.6
10 µg LYS 100 µg LPS 4W AZ	1.65	16.2	3.48	*	53.8	12.8
10 µg LYS CFA	1.87	7.82	5.13	*	42.0	15.5

* Challenge on day 83 (s.c.)
LYS=lysozyme; 4W AZ=4 weeks ALZET®; LPS=lipopolysaccharide; IFA=incomplete Freund's adjuvant; CFA=complete Freund's adjuvant

ALZET® osmotic pumps can also be used as implantable delivery systems to replace substances secreted by organs removed by surgery or other means. As an artificial pancreas, ALZET® was implanted to infuse insulin for 6 days to rats with experimentally (streptozotocin) induced diabetes. Large pulses of insulin release, typical of meal-eaters, do not normally occur in rats because they nibble instead of eating meals. Therefore, continuous infusion permits assessment of the total daily need for insulin. Patel et al. (15) filled ALZET® pumps with various concentrations of unmodified crystalline bovine insulin in isotonic saline and implanted them subcutaneously to infuse 2, 4, and 10 U per 200 gram rat. Pumps filled with isotonic saline served as controls in diabetic rats. Experiments spanned 60 to 80 days, with replacement of osmotic pumps every 2 weeks. A dose of 2 U/day was sufficient to return the diabetic rats to normoglycemic levels of 90 to 120 mg%.

DISCONTINUOUS DRUG DELIVERY

ALZET® osmotic pumps can be adapted to deliver substances at a pre-programmed rates that vary over time. This potential to alter the delivery pattern may be particularly significant in determinining the true physiological effect of some naturally occurring hormones and peptides that can only be observed when the agent is given in an on/off or phasic/tonic pattern. An excellent example of this principle is Knobil's work demonstrating the significance of delivery pattern on the physiological effects of LHRH (16).

An adaptation of the osmotic pump has permitted researchers to mimic the circadian rhythm of melatonin (Fig. 8). Before implanting ALZET® pumps, Lynch et al. (17,18) attached a coiled polyethylene catheter to the pumps. The coil was filled with an alternating sequence of vehicle/drug solution/vehicle/drug solution, etc. allowing investigators to slowly and intermittently infuse melatonin according to a predetermined pattern of 6 hr on and 18 hours off over 6 days. To document the 24-hr excretion of melatonin, they measured urinary excretion of melatonin or of a mixed solution containing a dye.

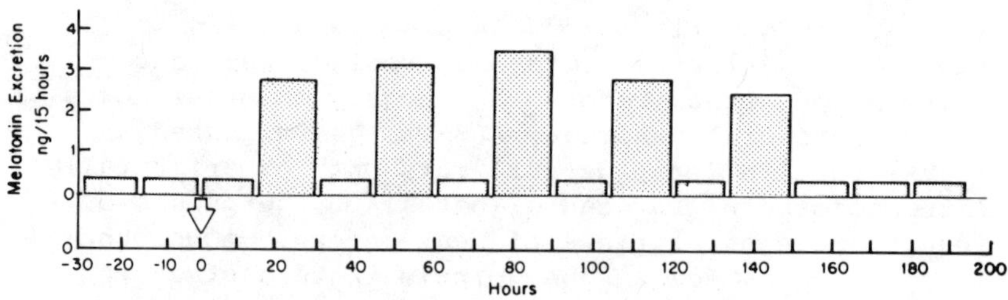

FIGURE 8. Melatonin intermittently infused with ALZET®[18] (with permission).

TARGETED ORGAN- AND TISSUE-SPECIFIC DELIVERY

The osmotic pump has permitted chronic delivery of drugs into the cerebral ventricles and brain tissue and into the kidney via the renal artery. Selected areas of the visual cortex have also been microperfused with ALZET®.

Targeted infusions can produce relatively high drug concentrations in one organ or in a specific site within an organ--with low systemic levels. For example, Smits et al. (19) infused the right kidney by implanting a cathether attached to an osmotic pump into the right suprarenal artery of unrestrained, conscious, uninephrectomized rats (Fig. 9). By investigating kidney function by several methods, he determined that the catheter did not induce changes in kidney function. This technique was later used for the intrarenal administration of vasoactive drugs. Ruers et al. (20) continuously infused prednisolone, an immunosuppressive drug, directly into rat renal allografts for 13 days. They discovered that the immunosuppressive effect of continuous intrarenal infusion is superior to continuous systemic infusion (Fig. 10).

Kasamatsu et al. (21) continuously infused norepinephrine to study the effects of this neocortical catecholamine on cortical plasticity. To measure effect, they used well known visual cortical changes in ocular dominance that follow monocular deprivation in kittens. Two separate osmotic pump/cannula systems delivered norepinephrine and control solutions, respectively, to appropriate sites in the left and right visual cortex of kittens deprived of susceptibility to the effects of monocular lid suture because of earlier treatment with 6-hydroxy-dopamine. As shown by a shift in ocular dominance, norepinephrine restored cortical plasticity. Norepinephrine treatment also decreased binocularity in older animals who were no longer susceptible to ocular deprivation. The authors stated that using the ALZET® was advantageous to the intraventricular infusion used previously because it (1) allowed precise localization of drug effect, (2) made it possible to use an aminal for its' own control (in a distant part of the same hemisphere or in the opposite hemisphere), and (3) eliminated side effects that often occur with repeated intraventricular injections.

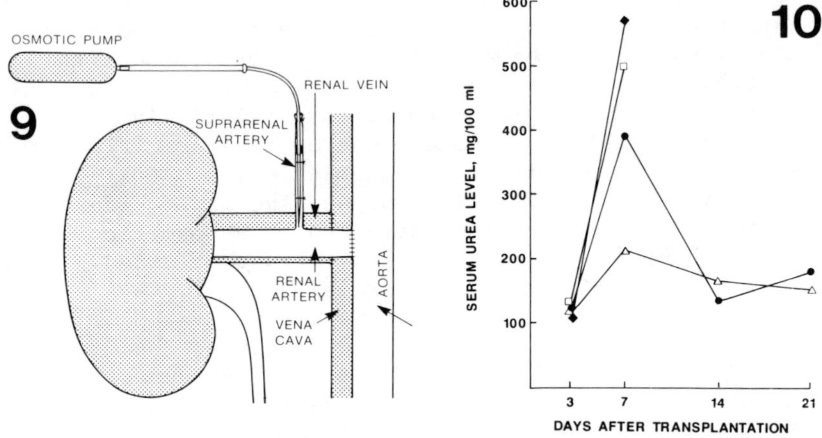

FIGURE 9. Local perfusion of the kidney using an ALZET® pump.[18]

FIGURE 10. Mean serum urea levels in Lewis rats receiving BN renal allographs. Lewis recipients were untreated or received various prednisolone treatments ◆ (untreated), □ 12 mg/k/day i.p. bolus, ● 8 mg/kg/day i.p. continuous, △ 4 mg/kg/day intrarenal continuous[20] (with permission).

Recently Parkinsonism has been treated by transplanting dopamine-producing tissue (chromaffin tissue) into the brains of affected individuals. Stromberg et al. (22) used ALZET® pumps to infuse nerve growth factor (NGF) into brains of mice who had received autotransplants of chromaffin tissue from the adrenal medulla. They modified the pumps by gluing stainless steel and dialysis tubing with epoxy resin to form a cannula. The cannula was attached to an ALZET® and implanted in the backs of rats. To maintain the infusion over 28 days, the investigators removed the spent pump at the end of two weeks and attached a fresh pump to the cannula-catheter assembly. NGF increased not only increased graft survival; it induced a transformation in graft cell morphology and enhanced adrenergic fiber outgrowth into host brain tissue.

OSMET DRUG DELIVERY MODULES

The osmotic technology described for use in preclinical studies with ALZET® has been adapted to clinical studies. OSMET™ can administer a suspension or solution of an agent of interest either orally, rectally, or vaginally. Despite being slightly larger than conventional tablets, the capsule-shaped modules go through the human gastrointestinal tract in the same manner as conventional tablets. The smaller 0.2 ml capacity modules deliver agents continuously at a near-constant rate of 8, 15, or 25 µl/hr. For vaginal or rectal administration, the 2.0 ml capacity modules deliver 60 µl/hr for 30 hr or 120 µl/hr for 15 hr.

De Leede and colleagues (23) used the modules in clinical trials to study rectal administration of the choline salt of theophylline (Fig. 11). Use by each subject of two modules sequentially resulted in extended periods of virtually constant plasma levels of theophylline and good agreement of *in vitro* and *in vivo* functionality.

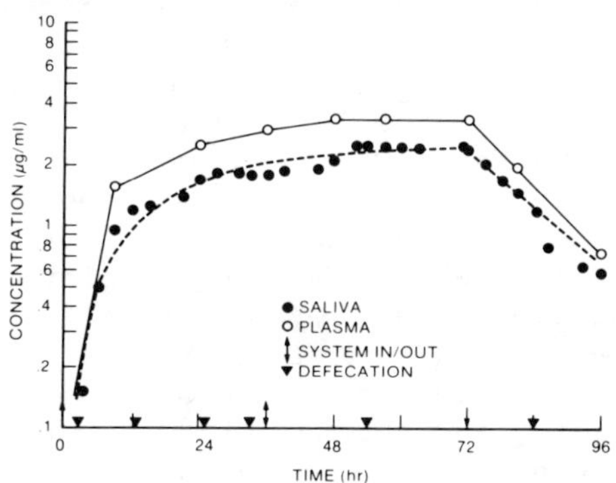

FIGURE 11. Theophylline concentrations in plasma and saliva after administration of an osmotic rectal delivery system (delivering 11 mg/hr theophylline) in one subject.[23]

OSMET™ been used in clinical studies (24) to deliver antiinflammatories, antihypertensives, vitamins, and receptor-blocking agents.

CONCLUSIONS

We have presented various examples of the use of osmotic pumps for continuous, discontinuous, and local delivery of drugs, peptides, and proteins in animals and in man. The studies discussed demonstrate the profound influence of the mode of delivery on the physiological effect of these agents. These studies emphasize the need to explore various delivery patterns to determine the optimal regimen for administration of novel therapeutic agents. These osmotic pumps also offer a powerful tool for the clinical application of these agents.

References

(1) Theeuwes F, Yum SI (1976). Principles of the design and operation of generic osmotic pumps for the delivery of semisolid or liquid drug formulations. Ann Biomed Eng 4:343.

(2) Eckenhoff B, Theeuwes F, Urquhart J (1981). Osmotically actuated dosage forms for rate-controlled drug delivery. Pharm Technol 5:35.

(3) Jarnagin K, Brommage R, DeLuca HF, Yamada S, Takayama H (1983). 1-but not 24-hydroxylation of vitamin D is required for growth and reproduction in rats. Am J Physiol 244:E290.

(4) Akhtar FB, Marshall GR, Wickings EJ, Nieschlag E (1983). Reversible induction of azoospermia in rhesus monkeys by constant infusion of a gonadotropin-releasing hormone agonist using osmotic minipumps. J Clin Endocrinol Metab 56:534.

(5) Nau H, Zierer R, Spielmann H, Neubert D, Gansau C (1981). A new model for embryotoxicity testing: teratogenicity and pharmacokinetics of valproic acid following constant-rate administration in the method using human therapeutic drug and metabolite concentrations. Life Sci 29:2803.

(6) Sikic BI, Collins JM, Mimnaugh EG, Gram TE (1978). Improved therapeutic index of bleomycin when administered by continuous infusion in mice. Cancer Treat Rep 62:2011.

(7) Peng YM, Alberts DS, Chen HS, Mason N, Moon TE (1980). Antitumour activity and plasma kinetics of bleomycin by continuous and intermittent administrations. Br J Canc 41:644.

(8) Tam CS, Heersche JNM, Murray TM, Parsons JA (1982). Parathyroid hormone stimulates the bone apposition rate independently of its absorptive action: differential effects of intermittent and continuous administration. Endocrinology 110:506.
(9) Connors JM, Hedge GA (1980). Feedback effectiveness of periodic versus constant triiodothyronine replacement. Endocrinology 106:911.
(10) Cotes PM, Barlett WA, Das REG, Flecknell P, Termeer R (1980). Dose regimens of human growth hormone: effects of continuous infusion and of a gelatin vehicle on growth in rats and rate of absorption in rabbits. J Endocrinol 87:303.
(11) Chang AE, Hyatt CL, Rosenberg SA (1984). Systemic administration of recombinant human interleukin-2 in mice. J Biol Response Mod 3:561.
(12) Nishimura T, Uchiyama Y, Yagi H, Hashimoto Y (1986). Administration of slowly released recombinant interleukin 2. Augmentation of the efficacy of adoptive immunotherapy with lymphokine-activated killer (LAK) cells. J Immunol Methods 91:21.
(13) Yasujima M, Abe K, Kohzuki M, Tanno M, Kasai Y, Sato M, Omata K, Kudo K, Takeuchi K, Hiwatari M, Kimura T, Yoshinaga K, Inagami T (1986). Effect of atrial natriuretic factor on angiotensin II-induced hypertension in rats. Hypertension 8:748.
(14) Martins A, Amkraut A (unpublished).
(15) Patel DG (1983). Rate of insulin infusion with a minipump required to maintain a normoglycermia in diabetic rats (41529). Proc Soc Exp Biol Med 172:74.
(16) Knobil E (1980). The neuroendocrine control of the menstrual cycle. Recent Prog Horm Res 36:53.
(17) Lynch HJ, Rivest RW, Wurtman RJ (1980). Artificial induction of melatonin rhythms by programmed micro-infusion. Neuroendocrinology 31:106.
(18) Lynch HJ, Wurtman RJ (1979). Control of rhythms in the secretion of pineal hormones in humans and experimental animals. In Suda M, Hayaishi O, Nakagawa H (eds): Biological Rhythms and their Central Mechanism," Amsterdam:Elsevier/North Holland, 117.
(19) Smits JFM, Kasbergen CM, van Essen H, Kleinjans JC, Struyker-Boudier HAJ (1983). Chronic local infusion into the renal artery of unrestrained rats. Am J Physiol 244:H304.

(20) Ruers TJM, Buurman WA, Smits JFM, van der Linden CJ, van Dongen JJ, Struyker-Boudier HAJ, Kootstra G. (1986). Local treatment of renal allografts, a promising way to reduce the dosage of immunosuppressive drugs. Transplantation 41:156.

(21) Kasamatsu T, Pettigrew JD, Ary M (1979). Restoration of visual cortical plasticity by local microperfusion of norepinephrine. J Comp Neuro 185: 163.

(22) Stromberg I, Herrera-Marschitz M, Ungerstedt U, Ebendal T, Olson L (1985). Chronic implants of chromaffin tissue into dopamine-denervated striatum: Effects of NGF on graft survival, fiber growth and rotational behavior. Exp Brain Res 60: 335.

(23) DeLeede LGJ, DeBoer AG, Breimer DD (1981). Rectal infusion of the model drug antipyrine with an osmotic delivery system. Biopharm Drug Dispos 2:131.

(24) Wilson CG, Hardy JG, Davis SS (1983). Controlled release dosage forms. Pharm J 231:334.

The Pharmacology and Toxicology of Proteins, pages 149–158

COMMERCIALIZATION OF LIPOSOMES

Annie Yau-Young and Frank Martin

Liposome Technology Inc.
1050 Hamilton Court
Menlo Park, California 94025

ABSTRACT The development of liposome formulations which satisfy the regulatory requirements of pharmaceutical products has been an area of intense effort in the last few years. The attributes of liposome formulations aside from their biocompatibility are that they can be designed for sustained release of drugs, targeted delivery to organ of action or site avoidance of organs sensitive to toxic effects of the drug. The commercial issues confronting the industry include raw materials, dosage form development, production scale-up and regulatory issues. In the selection of raw materials, the source, qualifications of suppliers, purity, cost, stability and safety have to be considered. With regard to the development of liposome dosage forms, lipid composition, particle size, drug loading, dose, in vivo release rates and disposition, stability of the formulation during preparation, storage and use, and final package form have to be optimized. Pharmacokinetics and efficacy must also be evaluated and optimized. The process used for the production of liposome formulations in large scale for pre-clinical and clinical tests has to yield reproducible preparations according to specifications and has to be performed in accordance with FDA's Good Manufacturing Guidelines. The FDA regulations require quality control of the potential product, standardization and validation of analytical assays and production process, stability data and extensive documentation systems. A number of parenteral applications of liposomes are currently in clinical trials.

INTRODUCTION

Liposomes are thermodynamically stable microspheres bounded by thin membranes formed by amphiphiles, such as lecithin, organized in lamellar bilayers. The membrane separates the internal aqueous compartment from the external aqueous medium. Liposomes were originally developed and used in academic laboratories as model biological membranes[1,2]. They have proven to be very useful systems to study transport of ions and other molecules across the phospholipid bilayers and the functions of membrane proteins[3,4]. They have also been used to introduce drugs and macromolecules into cells[5]. More recently, considerable effort has been made to develop and commercialize liposome-based products for diagnostic and therapeutic applications.

One of the features that makes liposomes particularly useful as drug carriers is their biocompatibility - the basic components are the same as those found in biological membranes. The applications of liposomes as drug carriers have been reviewed recently[6,7,8,9]. They can be used to deliver both hydrophilic as well as hydrophobic drugs. Water soluble drugs are encapsulated in the internal aqueous compartment of the liposome while hydrophobic drugs are usually intercalated in the fatty acyl region of the bilayer. By encapsulating a drug in liposomes, the release kinetics can be altered in a manner which enhances its therapeutic index. This is typically done by lowering the peak plasma level while sustaining plasma concentration[10]. Other approaches include reduction of the hepatic first pass and lowering undesirable side effects by directly delivering drugs to their site of action or by avoiding sites of toxicity.

Much has been published on the mechanism underlying many of the therapeutic applications of liposomes. In this discussion, we will focus on the commercial issues confronting the development of liposome formulations. These major issues include:

- Raw Materials (sources, purity requirements, cost)
- Dosage Form Development (pharmaceutical attributes)
- Production Scale-up (Compliance with FDA manufacturing guidelines)
- Regulatory Issues

RAW MATERIALS

The raw materials used in any product are of utmost importance to its commercial success. The key raw material issues include sources, qualifications of suppliers, purity, cost, stability and safety. It is critical that more than one supplier be identified for each key material to minimize the probability of encountering shortage. Equally important, the competition generated among several suppliers tends to reduce the cost of raw materials. The lipid raw materials typically used to form liposomes are extracted from natural sources, such as, egg yolks and soy beans. Currently, these lipids of pharmaceutical quality are available in bulk quantities at reasonable prices. These natural phospholipids have traditionally been used as emulsifiers in products used for total parenteral nutrition. These nutritional products call for intravenous infusion of up to 12 g of egg phosphatides daily. Patients can be maintained by such fat emulsions for many months without noticeable undesirable effects.[11]

Synthetic lipids also can be considered as raw materials for liposomes. They have the advantage of being well-defined chemically and are generally more stable to peroxidation. However, relatively few companies manufacture these specialty lipids in quantities sufficient for commercial products. Moreover, these lipids are quite expensive relative to the natural ones. If synthetic lipids are to become a reasonable choice as liposome raw materials, significant capital will be necessary to develop new and scale-up processes to produce them in bulk quantities.

Reproducible quality of the lipids supplied is very important. Current purification processes appear to remove the protein and endotoxin in the natural lipids. Similarly, it is necessary to monitor the purity of synthetic lipids to assure the absence of any potentially undesirable by-products in the synthetic process. Regardless of which lipid source is used, appropriate panels of assays have to be established and validated to assure quality and batch to batch reproducibility.

Phospholipids are susceptible to two major degradative processes: hydrolysis of the ester bonds linking the fatty acids to the glycerol moiety and peroxidation of double bonds in the fatty acid chains. Natural enzymes hydrolyse the ester bonds in vivo and this represents an important safety advantage in using liposomes as drug carriers. However, the ester bond also undergoes non-enzymatic hydrolysis during storage in aqueous suspensions. The rate at which the hydrolysis occurs is pH and temperature dependent. The pH at which the lowest rate of hydrolysis occurs is about 6.5[12] and may vary slightly depending on the specific drug formulation. The temperature dependence obeys Arrhenius' Law with the exception of lipid mixtures which exhibit thermal transitions. There is evidence that when hydrolysis occurs, the products, namely, lyso-compound and fatty acids, remain embedded in the bilayer and do not adversely affect stability of the bilayer structure below levels of 56%[13]. Whether the presence of these hydrolytic products affects the leakage of encapsulated drug or leads to aggregation or fusion of the liposomes will depend on the specific drug and its interaction with the lipid components. The leakage of water soluble macromolecules such as peptides and proteins may be less sensitive to the pressure of lysocompounds in the bilayer. However, in the case of small molecules, especially those which partition into both aqueous and lipid phases, accumulation of relatively low amounts of lyso compounds in the lipid bilayer may cause changes in the performance of the liposome formulation.

In the situation where the formulation cannot tolerate the projected extent of hydrolytic products, dried liposome formulations can be considered. This can be accomplished by lyophilization or spray drying techniques which remove most of the water from the preparation. These drying processes frequently require the addition of cryoprotectants and bulking agents to the formulation. The development of optimal freeze-drying cycles for liposome formulations is a new challenge in the technology. The integrity of the lipid bilayer has to be maintained throughout the drying and reconstitution steps. The reconstituted formulation ideally should have the identical physical and biological properties of the original formulation.

Most natural phospholipids contain polyunsaturated fatty acids. These fatty acids are susceptible to peroxidation propagated by free radicals. Although there is no evidence that the oxidative state of the phospholipids will affect the liposome formulation's general properties, the presence of free radicals may affect the stability of the drug or may cause variability in interactions in biological fluid. This type of oxidation is frequently initiated by trace amount of transition elements such as iron in glassware or water[14]. To minimize this problem, one can use synthetic lipids which have saturated fatty acyl chains, natural lipids which have been catalytically hydrogenated, or antioxidants can be included in the formulation. The disadvantage of the first approach is the increase in cost as discussed above and the requirement for elevated temperatures in some cases during the formation of the liposomes (in order to avoid crystallization of the phospholipids in the encapsulation solution, the liposome formation process must be conducted above the liquid crystalline phase transition temperature). High temperatures may accelerate the degradation of drugs and thus pose an additional problem. Another added risk is the lack of a safety data base to support the biocompatibility of synthetic lipids, particularly with respect to their metabolism. The preferred solution to reduce peroxidation of natural lipids, is hydrogenation or the incorporation of antioxidants in the formulations. The antioxidants may be incorporated in the lipid bilayer, in the aqueous solution or both. Two commonly used antioxidants in liposome formulations are vitamin E (a-tocopherol) and butylated hydroxy-toluene (BHT). These antioxidants retard but do not inhibit completely the peroxidation process[15]. It is also advisable to include chelating agents for transition metals to reduce the probability of initiating the formation of free radicals.

DOSAGE FORM DEVELOPMENT

With regard to liposome dosage form development, it is essential to engineer into the formulation the pharmaceutical attributes desirable for a particular route of administration and clinical indication. Important considerations include the lipid composition, size of the particles, drug loading (drug to lipid ratio), dose, in vivo release rates and disposition, stability of the formulation during preparation, storage and use, compatibility with preservatives (if required), and cryoprotectants included for freeze dried formulations. In the course of developing the formulations, pharmacokinetics and efficacy must be evaluated and optimized. Other questions such as loss of potency of the drug during encapsulation, possible immunogenicity of the formulation and localized tissue response to the formulation at injection sites also need to be addressed.

Using a peptide hormone such as salmon calcitonin as an example in our dosage form development strategy, the following approach is taken. The clinical objective is to develop a sustained release dosage form able to deliver the peptide over a seven day period following subcutaneous or intramuscular administration. Since calcitonin is extremely potent, drug loading is not a problem in this case. A bioassay of calcitonin activity in rats was used to evaluate possible loss of activity due to encapsulation in liposomes. The results were quite encouraging, the hormone was found to retain its full activity after lysis of the liposomes with a mild detergent. A series of in vivo experiments were carried out to optimize the rate of release from the site of injection[16]. This information was used to optimize the liposome formulation with respect to the lipid composition and size requirement. The selection of the lipids took into consideration the added cost to the dosage form and stability issues. Furthermore, the histopathological response of the tissue following multiple injections of the calcitonin containing liposomes was evaluated. The level of tissue response to the hormone alone, liposomes alone and drug containing liposomes were no more irritating than that observed in tissues receiving multiple doses of saline injections[17].

Since salmon calcitonin is a macromolecule and may be antigenic, a pilot study was conducted to monitor the production of humoral antibodies to calcitonin-liposomes, free calcitonin and calcitonin dispersed in complete Freund's adjuvant. Following the first boost after priming, an antibody titer of 10-4 was measured for the Freund's group while the other groups had no detectable antibody. After five boosts an antibody titer of 1:8 was detectable in the case of both free and calcitonin liposomes immunized animals while those immunized with calcitonin in complete Freund's adjuvant developed antibody titer at 10-6. These data suggest that calcitonin presented in liposome form is no more antigenic than free calcitonin. It should be noted that all samples injected were sterile and the level of endotoxin was below the limit of detection by the Limulus Amebocyte Lysate test. These are essential precautions when evaluating both the tissue and humoral antibody responses. The dosage form development stage of this potential product is now at the point of studying the long-term stability profile and production scale-up.

PRODUCTION SCALE-UP

Production scale-up is a new and challenging area confronting the biotechnology industry. Most of the new technological advances in the last decade occur first in the research laboratory where only small quantities are necessary to conduct the research work. In the product development phase, liter or kilogram quantities of the drug are required and the task of making this material reproducibly and economically is not a trivial matter. In the case of liposomes, the traditional methods of preparing milliliter quantities generally do not lend themselves to proportionate scale-up. In fact some of the methods require the use of solvents, such as ether, which are not acceptable in an industrial environment. Therefore, novel approaches to prepare multi-liter quantities have to be explored. This type of work requires extensive co-operation and collaboration between liposome formulators and process engineers. Our general approach has been to use commercially available equipment rather than custom equipment when possible. After the technical feasibility of the process is established, multiple runs are required to validate the reproducibility of the process. In-process control and monitoring systems are important to help identify and modulate the critical parameters. Manpower, capital equipment and chemicals costs are just as important as the choice of raw materials in developing the production process. In the case of proteins and peptide drugs which are themselves extremely expensive, not only does the process have to be optimized to maximize the encapsulation efficiency, but additional steps in the production process are needed to recover the unencapsulated drug. The recovered drug will need to be further processed to conform to its original specifications. For injectable formulations, the production process must be designed to produce sterile, pyrogen-free products. Depending on the particle size requirement of the formulation, terminal filtration may not be feasible to sterilize the product. In such cases, aseptic processing and containment are required.

REGULATORY ISSUES

Regulatory issues are extremely important and must be addressed throughout the product development process. The United States Food and Drug Administration must approve the safety and efficacy of a pharmaceutical product in human clinical trials prior to marketing. In addition, pre-clinical animal studies which typically include pharmacokinetic studies and acute and chronic safety studies to evaluate potential side effects, must be submitted to the FDA for approval prior to beginning clinical studies. Pharmaceutical products such as new drugs or formulations must be produced in compliance with FDA's Good Manufacturing Practices regulations. These require quality control data on a potential product, standardization and validation of analytical assays for raw materials, in-process sampling and finished product; validation of production processes, facilities and equipment; and extensive documentation systems. Stability data are also required for preclinical and clinical studies and the product shelf life must be determined. The ability to comply with these regulatory requirements can provide opportunities for innovation and significant technical achievements.

At present, clinical trials are in progress to test intravenous liposome formulations of antineoplastic (doxorubicin) and antifungal (amphotericin B) drugs. Other I.V. trials are evaluating the ability of liposomes to deliver an immunomodulator (muramyl dipeptide derivative) to the reticulo-endothelial systems and imaging agents to tumors. The ability to deliver liposome encapsulated water soluble bronchodilators to their site of action in the lung has been demonstrated in animals and is now the subject of a clinical trial. These trials provide evidence that the pharmaceutical engineering issues regarding stability, sterilization and scale-up have been successfully addressed.

ACKNOWLEDGEMENTS

We would like to thank Drs. D. Mufson, T. Parkinson, A. Huang and T. Smith for helpful criticism of this manuscript. The excellent assistance of Mrs. J. Neuner in typing this manuscript is gratefully acknowledged.

REFERENCES

1. Bangham AD (1968). Membrane models with phospholipids. Prog Biophys & Mol Biol 18:29.
2. de Gier J, Block MC, van Dijck PWM, Mombers C, Verkley AJ, van der Neut-Kok ECM, van Deenen LLM (1978). Relations between liposomes and biomembranes. Ann NY Acad Sc 308:85.
3. Montal M, Lindstrom J (1982). Reconstitution of the acetyl-choline receptor in lipid vesicles and in planar lipid bilayer. In Martonosi AN (ed.): "Membranes and Transport," New York: Plenum Press, p 331.
4. Klaerke DA, Karlish SJD and Jorgensen PL (1987). Reconstitution in phospholipid vesicles of calcium-activated potassium channel from outer renal medulla. J Membr Biol 95:105.
5. Fraley R, Papahadjopoulos D (1982). Liposomes: the development of a new carrier system for introducing nucleic acids into plant and animal cells. Curr Topics in Microbiol Immunol 96:171.
6. Patel HM, Ryman BE (1981). Systemic and oral administration of liposomes. In Knight CG (ed): "Liposomes: From physical structure to therapeutic applications," New York: Elsevier/North Holland Biomedical Press, p 409.
7. Mayhew E, Papahadjopoulos D (1983). Therapeutic Applications of Liposomes. In Ostro MJ (ed): "Liposomes," New York: Marcel Dekker, Inc., p 289.
8. Mufson D (1985). The application of liposome technology to targeted delivery systems. Pharm Tech Suppl, 16.
9. Ostro MJ (1987). Liposomes. Sci Amer 256:102.
10. Arakawa E, Imai Y, Kobayashi H, Okumura K, Sezake H (1975). Application of drug-containing liposomes to the duration of the intramuscular absorption of water-soluble drugs in rats. Chem Pharm Bull 23:2218.
11. Jeejeebhoy KN, Zohrab WJ, Langer B, Phillips MJ, Kuksis A, Anderson GH, (1973). Total parenteral nutrition at home for 23 months without complication and with good rehabilitation. Gastroenterology 65:811.
12. Frojkaar S, Hjorth EL, Warts O (1982). Stability and storage of liposomes. In Bundgaard H, Hansen AB,

Kofod H (eds): "Optimization of Drug Delivery," New York: Raven Press p 394.

13. Sundler R, Alberts AW, Vagelos PR (1978). Phospholipases as probes for membrane sidedness. J Biol Chem 253:5299.
14. O'Connell J, Ward RJ, Baum H, Peters TJ (1985). The role of iron in ferritin- and haemosiderin-mediated lipid peroxidation in liposomes. Biochem J. 229: 135.
15. Konings AWT (1984). Lipid peroxidation in liposomes. In Gregordiadis G (ed): "Liposome Technology," vol 1, Boca Raton, Florida: CRC Press Inc., p 139.
16. Yau-Young A, Chow J, Law M (1986). Liposome delivery of a biologically active peptide. Papers presented before the Am Pharm Assoc 16:105.
17. Yau-Young A, Law M, Chow J, Lin JP (1987). Sustained release of a peptide from liposome formulations. J Cellular Bioch S11B:188.

SECTION IV: PROTEIN TOXICOLOGY

J. Kopplin, Cetus Corporation, introduced this session and summarized the major issues. D. Johnson, International Research and Development Corporation, showed that nephrotoxicity of alpha-2u globulin in male rats may be a model for protein-induced renal damage. H. Lewis, Purdue University presented strategies for toxicologic testing of gene products.

Finally, E. Esber, Biologics Division, FDA discussed the unique regulatory problems of biologic drugs. She stressed the needs for strong in-house science and ongoing relationships with pharmaceutical companies during development of these agents. These products may have increased heterogenicity due to instability of coding sequences, variations in N-termini, folding, disulfide bond formation and glycosylation, and incomplete chemical and physical modifications.

The Pharmacology and Toxicology of Proteins, pages 161–164

INTRODUCTION: TOXICOLOGY SESSIONS

Joanne R. Kopplin, DVM, PhD
Cetus Corporation
Emeryville, CA 94608

Good Evening. I am Dr. Joanne Kopplin, Director of Toxicology for Cetus and Convener of this evenings' session on toxicology. As you now know, from Dr. Winkelhake's opening remarks, the original organizers of this symposium were unaware of the early start date for the Society of Toxicology Meeting this year. That meeting is occuring at this moment and has had a significant impact on the participation of toxicologists in this symposium. A workshop discussion session was originally scheduled for this afternoon to serve those interested in the untoward effects of recombinant protein therapeutics. Rescheduling the workshop to this evenings' session has created a lengthy program on top of a very long day. Both of the workshop leaders, Dr. H. Lewis and Dr. Dale Johnson, have agreed to redesign their discussions into a presentation format and to keep the presentations short so that we may proceed rapidly to the main sessions.

The agenda for the toxicology portion of the program will consist of the following:

1. Overview: J. Kopplin
2. Safety Assessment Strategy; A Position Paper: H. Lewis
3. A Possible Generic Renal Toxicity of Protein Therapeutics: D. Johnson
   "Nephropathy in Male Rats Induced by Modulation of an Endogenous, Low Molecular Weight Protein"

The issues of concern in the field of recombinant protein toxicology center in three areas.

1. Regulatory Issues
2. Toxicity Profile Determination "Testing Strategy"
3. *In Vivo* Testing Results

The regulatory issues will be amply covered later by Dr. Elaine Esber of the FDA. They are refreshingly flexible and provide wide latitude for creative toxicity testing. Today it is possible for a specific recombinant cytokine, manufactured by three different pharmaceutical companies, to reach Phase I clinical trials by three quite different paths. This flexibility has lead to increased secretiveness among toxicologists since efficient, pertinent and sound testing strategy can provide a major competitive edge to an innovative company. Very few people are revealing specific testing strategy and very few people are revealing specific animal test results for the same reasons. Our speakers this evening will address strategy and testing results from a general or generic viewpoint and I will briefly address the issues of concern expressed by the toxicologists themselves at the moment.

WHAT ARE THE CURRENT ISSUES IN PROTEIN TOXICOLOGY?

During the 1970's, the major issues in toxicology were regulatory. The Good Laboratory Practice Regulations were codified and product development took a back seat to science administration. Toxicologists worried about staying out of jail, not for bad scientific judgement but for inadequate documentation and the flagrant use of "White-Out." In the 1980's, the major issues are centered on the environment and the emerging discipline of risk assessment. As the 80's are ending, concerns about toxicity evaluation of recombinant proteins are growing with each new player in this field. Since the players are all so different, they each have different concerns.

Key Players in the Recombinant Protein Field

1. FDA Office of Biologics
2. FDA Office of Drugs
3. Very Small Biotechnology Companies (1+ Products)
4. Larger Biotechnology Companies (5 Products)
5. Old Line Pharmaceutical Companies
6. Universities/Government
7. Individual Academic Scientists
8. Industry Toxicologists

The FDA Office of Biologics may have the most difficult job of all and the most valid concerns. They must learn to make wise decisions in the midst of a developing data base on a totally new class of compounds and they must do so in a manner that will shepherd small companies without establishing guidelines while still being evenhanded and fair to all. Their answer has been to use a case by case approach based upon the Points To Consider. The opposite case exists for their counterpart, the Bureau of Drugs. There seem to be no unusual issues for them. They are evaluating drugs of every class, every day and each of these new compounds is just another drug. They are conducting business as usual for products that fall in their jurisdiction.

Small companies and individual inventors share many of the same problems relating to preclinical toxicity testing. Drug discovery has been completed but drug development is blocking their path to Phase I. They generally have no toxicologists and often no *in vivo* biologists. The issues of concern are centered on how they can turn their recombinant molecules into parenteral products and how to find the money to pay for the preclinical and clinical testing. Their present strategy seems to be to encourage friendly takeovers or joint ventures and to hire consultants to guide them.

The larger biotechnology companies have made the transition into drug development and the preclinical toxicologists are most concerned about competition. They are focused on developing a strategy for the best and most pertinent science conducted in the shortest time for the least amount of money to reach an IND. They are trying to do this by recruiting each others' experienced toxicologists and/or using the major contract laboratories.

The toxicologists at many of the older established pharmaceutical companies are concerned about testing strategy. They are literally suffering from a lack of guidelines. Symposia are arranged around a general theme asking how to get together as a group and decide how to test these compounds. Those who fought guidelines in the 70's seem to be begging for them now. And finally, an interesting concern that seems to be resident primarily with companies which have already experienced such toxicities with xenobiotics; will recombinant protein therapeutics cause an epidemic of anaphylaxis or serum sickness? While they are searching for answers, many of the established firms are using standard FDA Bureau of Drugs strategy; do one of everything in each of the usual animal species, do it well and spend whatever it takes.

I am sure that in the coming months, more data will appear on the specific preclinical toxicities of these new products. It will be slow because the competitive stakes are very high for companies operating primarily on venture capital. However, I suspect that very little will be revealed regarding precise preclinical strategy and the guessing-game will continue for many toxicologists. Thank you.

And now, may I present Dr. Dale Johnson, Director of Experimental Toxicology and Associate Director of the Toxicology Division of International Research and Development Corporation. He will present toxicity data, generated by the administration of a xenobiotic, which may possibly have considerable relevance to the new recombinant protein therapeutics.

The Pharmacology and Toxicology of Proteins, pages 165–171

PROTEIN INDUCED NEPHROPATHY IN MALE RATS

Dale E. Johnson

Division of Experimental Toxicology,
International Research and Development Corporation,
Mattawan, Michigan 49071

ABSTRACT

A variety of xenobiotics (xb) can alter the renal deposition of alpha-2u globulin (AG), an endogenous, low molecular weight protein ocurring predominately in male rats. Excessive AG accumulation in proximal tubule cells (PTC) leads to hyaline droplet formation, PTC degeneration, granular casts, tubular dilation, PTC regeneration and chronic inflammation. Morphological, biochemical and xb-AG binding studies suggest that AG nephropathy may be an interesting model for studying potential protein-induced renal toxicity.

INTRODUCTION

The kidney is highly susceptible to toxic chemicals or altered physiological events which can lead to decreased function or total organ failure (1). It has been estimated that three fourths of all cases of acute renal failure in humans are associated with acute tubular necrosis of which approximately ten percent may be nephrotoxin-induced. While tubular alterations comprise the highest incidence of morphological manifestations of nephrotoxicity by chemicals (2), they can also be induced by physiologic responses to an altered protein load in proximal tubular cells (PTC). In humans, Bence-Jones protein in multiple myeloma patients and lysozyme in mononuclear cell leukemia patients are examples where PTC reabsorption of protein which appears in the glomerular filtrate at abnormal levels lead to nephropathies characterized by granular cast formation (3). In male rats, a similar type of nephropathy has been induced by a variety of xenobiotics (xb), primarily hydrocarbons, which alter PTC deposition of alpha-2u globulin (AG), an endogenous, low molecular weight protein (4,5). AG (18-20,000 daltons) is secreted in the liver under androgenic control and appears in urine as a pheromone (4). The protein is rapidly filtered at the glomerulus and reabsorbed primarily in the mid portion of the proximal tubule (6). Chronic AG accumulation in PTC leads to a nephropathy characterized by PTC degeneration and regeneration in the mid portion of the tubule, hyaline droplets, granular casts, tubular dilatation and chronic inflammation (4,7). Evidence that xb *per se* do not induce the nephropathy comes from the lack of any similar response in female rats (no AG) or in any other species where AG does not occur (4,5,7).

Since the kidney is a primary site for low molecular weight protein catabolism, it was surmised that an alteration of lysosomal function, as seen with several nephrotoxic chemicals, may contribute to AG-induced nephropathy. Recent work in our laboratory was conducted to define the initial event in AG nephropathy induced by a model xb, 1,3,6-tricyanohexane, and to study effects on PTC lysosomal function at varying AG loads. Other investigators have studied the potential binding of xb to AG which may affect recognition for catabolism leading to accumulation.

METHODS

All methods and materials used in these studies including chemicals and xb-dosing regimens have been previously described (7-9). Male CD® rats (Charles River Breeding Laboratories, Inc., Portage, MI) were used exclusively. Morphological changes were studied using standard light and electron microscopy techniques (7,8). In studying the catabolism of low molecular weight proteins, we used an *in vivo*/*in vitro* system similar to the one described by Cojocel, *et al*. (10). ^{125}I-human growth hormone (HGH) was used as a probe protein by giving intravenous injections in animals with altered AG loads. Ten minutes later, the peak time of HGH accumulation in renal tissue (3), kidneys were removed and cortical slices prepared and incubated for various periods of time. The percent TCA-soluble radioactivity increasing per incubation time was considered to be an estimate of the degradation of HGH in renal cortical slices (9,10). Lysosomal enzyme preparations and Cathepsin D activity, a marker enzyme inducible by high protein load, were prepared and assayed according to Fowler, et al. (11). Protein concentration was determined by the Biorad method (12). Statistical comparisons were made between various treatment groups using one way analysis of variance and Student-Newman-Keul's test at the 0.05 level of significance.

RESULTS

Morphological Changes

The initial lesion detected morphologically was the accumulation of protein droplets or crystalloid bodies in PTC with little, if any, other discernable cellular change. Ultrastructurally, these crystalloid inclusions were viewed as large, angular bodies with some evidence of membrane enclosure, presumably lysosomal. This protein deposition was confirmed with histochemical staining and occurred after five days of xb treatment (8). Immunochemical analysis of these crystalloid bodies in similar studies have shown them to be AG (13). These findings and the time course involved have been consistently seen regardless of the xb used to induce AG accumulation (4,14-17).

Biochemical Changes

Initial xb treatment for five days produced an overall decrease in catabolism of HGH. However, when xb treatment was increased to ten days, the catabolic curve returned to control levels. There was also a trend, but not statistically significant, that suggested Cathepsin D activity was higher in lysosomes of animals treated for ten days (9). This is the opposite effect that would be expected if the xb itself was nephrotoxic since altered protein degradation in general is an early indicator of chemically-induced nephrotoxicity (10). When the biotransformational state of the animal was altered by modulation of cytochrome P-450 mediated xb-metabolizing enzymes in the liver, the HGH degradation curve was also altered. Pretreatment of xb-animals with phenobarbital (induced P-450) resulted in an elevation of HGH catabolism above control levels. In addition, Cathepsin D activity and lysosomal enzyme protein were both induced ($p<0.05$) (9). With the lack of any other correlating morphological changes, these results suggest that protein load, presumably AG, increases in PTC under different experimental conditions, and physiological mechanisms adapt to accomodate for the higher levels of protein.

DISCUSSION

The results of the research together with findings by other investigators using different model xbs (13,14) suggest a probable sequence of events in the AG nephropathy syndrome. The initial event occurs in the liver where AG is produced. A recent study using trimethylpentane suggests binding, but not covalently, of the xb and AG (18). If true, conformational changes could be induced in the protein which could alter recognition for catabolism and lead to excess accumulation. In our research, pretreatment with phenobarbital then treatment with the xb led to induced Cathepsin D and increased HGH catabolism. Phenobarbital itself does not induce Cathepsin D, but presumably acts to increase the metabolism of the xb in the liver but not kidney (19,20). This gives evidence that hepatic metabolism of the xb either increases the amount of AG eventually transported to the kidney or binding occurs which alters the structure of the protein. Due to the great variety of structures of xb known to induce AG nephropathy in male rats (4) a similar binding mechanism for all compounds to AG does not seem feasible. The second important event is the excess accumulation of AG in PTC. The catabolism of AG by lysosomes is of high capacity, but saturable (6). As AG accumulates, the level of hydrolytic activity increases as shown by the induction of Cathepsin D (9). It is easy to speculate that hydrolytic enzymes "leak" out and eventually lead to cellular death. The next event seen morphologically is cellular degeneration and sloughing into the tubular lumen. Granular casts are seen near the corticomedullary junction at the thin portion of the loop of Henle (7). Blockage of the tubule can result which would alter intratubular pressure, glomerular filtration rate and eventually lead to ischemic effects. As the disease progresses, regeneration of tubular cells occurs (7,14). It is the cell turnover that apparently results in other chronic effects (21).

This sequence of events is consistent with a proposed mechanism of protein toxicity by induction of unwanted physiological effects (22). It is not known if exogenously administered proteins could induce similar effects, but the kidney must be viewed as a potential target organ in these instances for the following reasons. Kidneys comprise less than one percent of total body mass, yet receive nearly twenty-five percent of the cardiac output. Consequently, low molecular weight proteins are rapidly cleared at the glomerulus and the kidney, in general, plays a major role in plasma turnover and catabolism of these proteins (3). In addition, the kidney can be the site of injuries resulting from the formation or deposition of antigen-antibody complexes, e.g. glomerulo-nephropathies, which could lead to an alteration in quantity and substance of proteins appearing in the glomerular filtrate (3,23).

AG nephropathy in male rats may be an interesting model to study conformational changes in endogenous proteins that lead to altered physiological events, or effects of over-saturating physiological processes with normal proteins.

REFERENCES

1. Hook JB (1981). "Toxicology of the Kidney" New York: Raven Press, p vii.
2. Racusen LC, Solez K (1986). Nephrotoxic tubular and interstitial lesions: morphology and classification. Tox Path 14:45.
3. Maack T, Johnson V, Kau ST, Figueiredo J, Sigulem D (1979). Renal filtration, transport, and metabolism of low-molecular-weight proteins: a review. Kidney Int 16:251.
4. Alden CL (1986). A review of unique male rat hydrocarbon nephropathy. Tox Path 14:109.
5. Halder CA, Warne TM, Hatoum NS (1984). Renal toxicity of gasoline and related petroleum naphthas in male rats. In Mehlman MA, Hemstreet GP, Thorpe JJ, Weaver NK (eds): "Renal Effects of Petroleum Hydrocarbons," Princeton: Princeton Scientific, p 73.
6. Roy AK, Raber DL (1972). Immunofluorescent localization of alpha-2u globulin in the hepatic and renal tissues of rat. J Histochem Cytochem 20:89.
7. Barnett JW, Johannsen FR, Levinskas GJ, Boothe AD, Johnson DE (1986). Hydrocarbon nephropathy induction in male rats by crude tricyanohexane. Toxicologist 6:173.
8. Barnett JW, Johnson DE, Boothe AD, Johannsen FR (1987). Protein accumulation as the initial event in tricyanohexane induced nephropathy in male rats. Toxicologist 7:27.

9. Johnson DE, Barnett JW, Farnum LC, Johannsen FR (1987). Effect of tricyanohexane on renal lysosomal function. Toxicologist 7:27.
10. Cojocel C, Smith JH, Maito K, Sleight SD, Hook JB (1983). Renal protein degradation: a biochemical target of specific nephrotoxicants. Fund Appl Tox 3:278.
11. Fowler BA, Lucier GW, Hayes AW (1982). Organelles as tools in toxicology. In Hayes AW (ed): "Principles and Methods in Toxicology," New York: Raven Press, p 635.
12. BioRad Laboratories Richmond CA.
13. Alden CL, Kanerva RL, Ridder G, Stone LC (1984). The pathogenesis of the nephrotoxicity of volatile hydrocarbons in the male rat. In Mehlman MA, Hemstreet GP, Thorpe JJ, Weaver NK (eds): "Renal Effects of Petroleum Hydrocarbons," Princeton: Princeton Scientific, p 107.
14. Short BG, Burnett VL, Swenberg JA (1986). Histopathology and cell proliferation induced by 2,2,4-trimethylpentane in the male rat kidney. Tox Path 14:194.
15. MacNaughton MG, Uddin DE (1984). Toxicology of mixed distillate and high-energy synthetic fuels. In Mehlman MA, Hemstreet GP, Thorpe JJ, Weaver NK (eds): "Renal Effects of Petroleum Hydrocarbons," Princeton: Princeton Scientific, p 121.
16. Busey WM, Cockrell BY (1984). Non-neoplastic exposure-related renal lesions in rats following inhalation of unleaded gasoline vapors. In Mehlman MA, Hemstreet GP, Thorpe JJ, Weaver NK (eds): "Renal Effects of Petroleum Hydrocarbons,: Princeton: Princeton Scientific, p 57.
17. Phillips RD, Cockrell BY (1984). Effect of certain light hydrocarbons on kidney function and structure in male rats. In Mehlman MA, Hemstreet GP, Thorpe JJ, Weaver NK (eds): "Renal Effects of Petroleum Hydrocarbons," Princeton: Princeton Scientific, p 89.
18. Lock EA, Charbonneau M, Strasser J, Bus JS (1987). The reversible binding of 2,2,4-trimethylpentane (TMP) to renal alpha2u-globulin (alpha2u) in male Fischer 344 rats. Toxicologist 7:27.
19. Orrenius S, Das M, Gnosspelius Y (1969). Overall biochemical effects of drug induction on liver microsomes. In Gillette JR, Conney AH, Cosmides GJ, Estabrook RW, Fouts JR, Mannering GJ (eds): "Microsomes and Drug Oxidations," New York: Academic Press, p 251.
20. Anders MW (1980). Metabolism of drugs by the kidney. Kidney Int 18:636.

21. Loury DJ, Smith-Oliver T, Butterworth BE (1987). Assessment of unscheduled and replicative DNA synthesis in rat kidney cells exposed in vitro or in vivo to unleaded gasoline. Tox Appl Pharm 87:127.
22. Galbraith WM (1986). New challenges in the safety evaluation of drugs and biologics. Presented at Preclinical safety of biotechnology products intended for human use: Satellite Symposium to IUTOX, Tokyo.
23. Lebish IJ, Hurvitz A, Lewis RM, Cramer DV, Krakowka S (1986). Immunopathology of laboratory animals. Tox Path 14:129.

The Pharmacology and Toxicology of Proteins, pages 173–184

STRATEGIES FOR THE TOXICOLOGIC TESTING OF GENE PRODUCTS

Hugh B. Lewis, DVM, PhD

School of Veterinary Medicine, Purdue University
Lafayette, Indiana 47907

There is a lot of confusion regarding what constitutes an appropriate toxicity evaluation of gene products. I think this either results from lack of understanding of the purpose of toxicity testing or differences in the philosophy behind it. Many people, including some toxicologists, are unsure of the purpose behind toxicity testing, so it is worth reviewing.

In general, our society subscribes to the notion that new drugs must not be given to humans without some prior knowledge of their toxic potential. The purpose then of the toxological evaluation of a drug is to define its toxic potential in animals, thereby permitting the formation of an informed judgement regarding the safety of administering compounds to humans. "Safe" is a relative term; it does not imply absolute safety. Clearly, the toxic potential of the drug must first be defined before a judgement regarding safety can be made. The initial toxicity studies in animals should attempt to define the toxic potential of the drug, as far as that is possible.

For toxicity studies to be justified, there must be some expectation that animal studies can provide useful toxicity information. The important word here is "can". Prospectively, there are no guarantees that animal studies will generate relevant information; however, that does not invalidate them.

It is worth reviewing a few definitions because a lot of the confusion that exists probably reflects misunderstandings of terminology. If we define a few things, now we'll all know what we're talking about for the discussion to follow:

(a) Toxicity study:

This is a study usually done in animals, rarely in humans. Toxic doses are used and toxicity is detected. In fact, toxicity is the desired endpoint of the study; doses may have to be escalated until toxicity is produced. The establishment of a highest non-toxic dose is also important in these studies. Fundamentally, one is trying to define and understand the toxicity of a compound in a toxicity study. This type of study is not usually performed with human subjects, except when anti-cancer drugs are being tested. With the latter, doses are usually titrated until limiting toxicity is achieved, the assumption being that the maximum anti-cancer effect in the individual coincides with the maximum tolerated toxic dose. However, most drugs are not developed as aggresively as this; toxicity is to be avoided in people, even in the early clinical trials.

(b) Safety study:

In a safety study, doses are pre-selected, often being multiples of the anticipated therapeutic dose in humans. If toxicity is observed, the study may be equivalent to a toxicity study. If no toxicity is found, then it is a safety study. The only conclusion possible from the latter is that the dose used was safe in that species of animal. Nothing useful can be learned about the toxic potential of a drug in this sort of study. However, this approach is appropriate for studies in humans.

Many toxicologists and others in industry use the terms "toxicity study" and "safety study" interchangeably, and this leads to much confusion. Clearly, there is little justification for performing animal safety studies, except, of course, the drug is intended for therapeutic use in animals. Even then it's not a very informative approach. Little or nothing is learned, and animal life is therefore wasted. It is important to understand that ignorance is not bliss when it comes to developing new therapeutic agents for use in humans. Although there clearly must be a philosophy behind the toxicity testing, oneu does not often hear discussions on the subject.

However, the philosophy behind testing is reflected in the several existing approaches and attitudes towards toxicity testing.

There are three major approaches and attitudes.

1. Toxicity testing is viewed as a regulatory hurdle. Adherents to this philosophy naturally feel a need to have the hurdle clearly defined; this is manifested as pleas or demands to the FDA for guidelines detailing every facet of testing required. They want to know exactly what that hurdle is, so that they can run the compound through the process as quickly as possible, keeping costs to a minimum. The results of such a testing approach are not viewed as particularly interesting or important, so long as they don't impede drug development. These sort of studies are non-challenging, and demand only technical skills. There is little need of scientific expertise. Companies with this philosophy frequently contract the studies out to the cheapest bidder. These are generally minimal studies, done with a cookbook type of approach. Under these circumstances, toxic findings tend to produce bewilderment and alarm and often viewed as major stumbling block to drug development. In fact, toxicity is very unwelcome views to people who view toxicity studies as just a regulatory hurdles.

2. Toxicity testing viewed as a necessary hurdle. Groups with this philosophy agree that new drugs for human use should not be introduced without a full evaluation of their safety. All drugs are evaluated in a comprehensive series of tests and safety studies, a lockstep approach, in which routine or standard study protocols are used. Drugs found to be toxic are usually not developed any further. Since toxicity is such bad news, groups that have this approach tend to emphasize safety studies. In "successful" safety studies, no toxicity is found. Little scientific expertise is required for this type of "routine" approach. Toxicity is passively accepted as bad news and usually results in withdrawl of the drug from development. Some of

the bigger companies fit this mold and they tend to defend the status quo regarding toxicity testing. They're very comfortable with this kind of approach and apparently have the muscle to cope with product wastage that's involved.

3. Toxicity is viewed as important and worthy of investigation and understanding. This approach recognizes an important responsibility of drug developers, i.e. to minimize the possibility of toxic effects occurring in humans given new therapeutic agents. Toxicology groups employing this approach recognize that the purpose of toxicity studies in animals is indeed to detect and define potential toxicities for humans. They view this as a worthwhile scientific challenge requiring a flexible and rational approach. Drugs are tested using a case-by-case approach, each study being designed specifically for the particular drug in hand.

   Toxicity studies are viewed as clinical investigations of iatrogenic disease and are designed to obtain the maximum relevant information for the animal life expended. Emphasis is placed on understanding the pathophysiology of the lesion inflicted, the mechanism of toxicity, its reversability, and so on. The information obtained from such studies contributes to the design and monitoring regime for subsequent studies in humans. Toxic effects are aggressively sought and an attempt is made to define them and understand the basis for them. Toxic effects are not considered to preclude human studies. This approach is one of inducing iatrogenic disease, studying it and understanding it, so that steps can be taken to determine whether or not humans may be at risk too.

This is a very demanding task and requires scientists who are capable of mounting sophisticated clinical investigations. It is understood that toxicity is most destructive when it occurs in human patients, and when it is unexpected and/or unexplained. Toxicity is least destructive when it occurs in animals and can be explained; this is therefore the time to find and understand it. The following is an example illustrating the third approach, clearly the one I subscribe to:

> Several high dose dogs in a toxicity study exhibited a mild anemia. Evidence of a compensated anemia was found in dogs from the other dose groups, including the low dose, which was similar to the intended human therapuetic doses. There are basically two ways of reporting this sort of information: (a) simply as a statistically significant decrease in hemoglobin levels in high-dose dogs, or, (b) as a well-compensated red-cell loss or destruction, affecting all dose groups and needing to be understood. In this particular study, all the dogs in all the drug groups showed evidence of splenic reactivity, indicating that the anemia probably was hemolytic in origin. In all the groups there was evidence of marrow compensation for red cell destruction, even in the low dose. A review of the red cell morphology revealed there were a number of red cells that looked as though something had been torn from them; a few of them even had Heinz bodies (fragments of precipitated hemoglobin), usually only seen in dogs after oxidant damage to hemoglobin. It thus looked as though this was a mild case of Heinz body hemolytic anemia. THis was confirmed by an in vitro study in which the parent drug and its known canine metabolites were incubated with whole dog blood. The parent compound did not induce Heinz bodies in the test tube, but its n-oxide metabolite did. The latter metabolite also induced Heinz bodies in human blood in vitro. Thus, the advantage and importance of defining the cause of mild compensated "anemia" in dogs induced by this drug is that it allowed one to deliberately design the Phase I clinical study in humans. Firstly, G6PD-deficient people were screened out, because presumably, they'd be at a great risk of red-cell destruction should the n-oxide metabolite be formed in humans. Volunteers in the Phase I study were monitored very carefully, and no anemia was found. Indeed, it was later discovered that the n-oxide metabolite was not produced in people.

On reflection, one could either say that this animal study was a wasted effort, or, it was absolutely what a toxicity study in animals is all about, i.e., to forewarn of potential toxicity. The fact that the individuals in the Phase I did not respond with the same sort of toxicity as dogs does not mean to say that all people will so respond; if in the future, some individuals developed such a hemolysis it would not be totally unexpected, and it would not be unexplained. Thus, the value of animal toxicity studies is not necessarily in their predictive value, but that they provide the basis for continuing the responsible development of the drug.

Predictibility is a major point of confusion. As noted above, not all human toxicities will be predicted by findings in animal studies, but that does not invalidate them. Currently they are without doubt the best alternative available. In Phase I human studies, normal healthy volunteers are used, and these studies are basically safety studies because they rarely result in any toxic effect. They are therefore not very useful for predicting toxicity in humans

Phases II and III clinical studies in people usually involve patients of a variety of ages. By definition, these people are sick. As a group, they are likely to be heterogenous, with variable sensitivies, metabolic capabilities, and susceptibilities. They're also more likely to be exposed to more other drugs and environmental stresses than the general population, simply because they are sick. In contrast, animal toxicity studies involve healthy normal young animals kept in an ideal environment and provided an ideal diet. This is why toxic doses are used in animal studies; one needs to provoke a toxic response, which may mimic the response seen in human subjects that are particularly susceptable, for one reason ot another, to the drug.

In my experience, this approach, when viewed in retrospect, is reasonably predictive of human toxicity. The problem is, one can only judge predictiveness retrospectively, following extensive clinical trials using humans. Until that time, the relevance of animal toxicity is uncertain, but forewarned is forearmed! For this approach to be successful, capable scientists who are skilled in animal clinical investigative work and able to design studies and stratagies on a case-by-case basis, must be employed.

What are some of the points that must be considered during the preclinical development of gene products?

Many toxicologists are of the opinion that the traditional appraoch to toxicity testing of xenobiotics (as embodied in the various regulatory guidelines) is inappropriate for products of human and animal genes and their close analogues. They feel that the risk associated with exposing humans to substances entirely foreign to them should be viewed perhaps as different from exposure to naturally occuring endogenous substances. Exactly where the line is drawn between naturally occuring and foreign is highly debatable. In some cases, the route of administration of an endogenous substance may constitute foreign-ness. For example, the natural exposure to a substance released continuously in minute amounts (well within the body's capacity to transport, utilize, degrade, and excreted it) is quite different from exposure to a large parental bolus that overwhelms the normal containment and receptor target systems, resulting in abnormal concentrations of secondary or even tertiary target sites.

A second concern relates to the interpretation of findings of toxicity studies on endogenous substances; in some instances at least, this should be different from findings on xenobiotics. For example, both glucocorticoids and thyroid hormones are highly toxic when administered to animals in large doses. Although this is clearly important and highly relevant information, it sheds light on the biology of these compounds rather than indicating any atypical effect which should preclude their development and therapeutic use in humans. Such information is pertinent to decisions regarding dose selection , patient selection, continuation of treatment, etc., but it is not an appropriate basis for any go/no go development decisions.

Thirdly some concerns appropriate for xenobiotics may not be appropriate for human gene products, e.g., carcinogenic or mutagenic potential. Given the current dogma that there is no official room for uncertainty regarding a compound's carcinogenic or mutagenicity, and that there is no "clean" dose for a carginogen or mutagen, then there is no way to evaluate the carcinogenicity or mutagenicity of a natural product. If the test is positive, what does one conclude, and what can be done about it?

DISCUSSION

HBL: Heman gene products, proteins, peptides are capable of producing toxic effects as diverse and severe as any class of xenobiotic. Do people in the audience generally agree or diagree with that statement?

HBL: Most people seem to agree that potentially these products are just as toxic as xenobiotics. Certainly, when one considers the hormones, immunomodulators, growth factors, and thrombolytics, I would have to agree. to me, this possibility is sufficient justification to pursue a policy of careful evaluation of their toxic potential in animals prior to their therapeutic use in man.

HBL: Statement: Target organs of human gene product toxicity may sometimes be predicted on the basis of toxicity studies in animals. Does the audience generally agree or disagree with the statement, in light of their own experience?

Audience response: Most poeple agreed with the statement.
HBL: Just as is seen with xenobiotics, minor modifications in the primary structure of proteins and peptides may produce alterations in the inherent toxic profiles, e.g., a single amino acid difference in the antitrypsin molecule has been shown to account for a clinically expressed fibrinolytic condition in a child. Thus, even a minor modification of polypeptide or protein would auger for a full toxicologic evaluation. There is no doubt that in the future, many new products will be analogues of endogenous substances; strictly speaking, they should not even be considered endogenous. The argument that endogenous products must be treated very differently to xenobiotics from the standpoint of toxicologic evaluation will be increasingly difficult to maintain.

HBL: What is the general experience to date with the

predictiveness of animal studies to man for gene product toxicity. My own experience is very limited, but so far, the animal studies I'm familiar with have been highly predictive.

There was agreement from members of the audience that animal toxicity studies with proteins were generally (but not always) predictive for findings in man. No details or examples were provided.

HBL: That's very useful to know, but I think that there will be many exceptions. It is remarkable how different the opinions and experiences of toxicologists are on this matter. I saw a report from a European toxicology working group, claiming that animal studies were totally non-predictive; indeed they proposed that instead of doing animal toxicity studies, limited animal safety studies should be done because in their collective experience, toxicity studies lacked predictive value. Their experience appears to be quite different from the experience of many groups in this country, including this group. Perhaps it reflects different philosophies behind toxicity testing, different types of study, different products, and/or different definitions of toxicity.

Question: ??

HBL: I think that's true, but one of the dogmas of toxicology holds animal studies can't predict idiosyncratic effects in humans. That would appear to be axiomatic, but in my experience, if toxicity is vigorously sought in animal studies, it is surprising how often so-called "idiosyncratic drug effects" (e.g., immune mediated hemolytic anemia, thrombocytopenia, leukemia, thyroiditis, marrow aplasia, etc.) are seen.

Question: Immunogenicity can be a source of major toxicity, at least in our experience, how would you address that problem?

HBL: Yes, I think immunogenicity is one of the major problems we face in testing gene products. I believe it is a very significant and, perhaps in some cases, insurmountable obstacle to conducting meaningful multiple-dose toxicity studies. Clearly, we have enough information indicating the importance of looking for neutralizing antibodies in repeat-dose studies. It certainly makes sense to evaluate this potential before embarking on subchronic studies. It is noteworthy that heterogenic antibody responses have been observed in humans given human gene products. I presume this reflects genetic variability within the human species. It follows that there may be heterogenic responses in pharmacologically relevant species used for toxicity testing. This has been reported, again in studies employing human interferon, where neutralizing antibodies were found in the monkeys. Maybe this is a problem that we just have to accept as a major limitation to toxicity testing with these products.

Concern has been expressed regarding species selection. Some toxicologists have proposed that all studies be done in monkeys; others vehemently object to this. I think that one should, obviously, do one's best to select a species that is pharmacologically relevant for the particular product under evaluation, but this might have major limitations. It is sometimes very difficult to identify an appropriate species and to find enough of them to do a toxicity study. if the "ideal species" is very large, then, from a practical standpoint, drug supply might become a major problem. I think the old dogma that the use of more than one species increases the probability of detecting a toxic effect relevant to humans, probably also holds true for this sort of product. The genetic makeup of the human population is very diverse. No one species is likely to be representative or predictive for all humans. There are certainly more similarities than there are differences between mammalian species. Animal studies have to be done, but sometimes we may have to accept the limitations of species selection.

Joy Cavagnero, Hazelton Laboratories: I agree with your point about traditional toxicology testing not being appropriate for the compounds. More to the point, what we end up with in terms of a toxicology evaluation is often determined by the fact that we're asked to develop a toxicology protocol based on very limited pharmacology and/or pharmacokinetics. And, I think that's why traditional toxicology testing doesn't really fit these products. Regarding toxicity testing and your comment about selecting a toxic dose, I think your third approach is really the way to go, but how would you go about selecting the doses that distinguish between a safety study and a true toxicicty study if, for example, you limited by the formulation. With proteins, we cannot get them as concentrated as we want to, so that we cannot administer enough to produce toxicicty. What are your recommendations under those circumstances?

HBL: There are times when practical reasons, e.g. inability to administer the large volumes necessary to produce toxicity or simply having insufficient material, limits what can be done. It is indeed possible that under these circumstances, one may not be able to perform a toxicity study. However, in general, this is only likely to be a problem in the single dose acute studies. For the most part, we are dealing with potent, biologically active substances. If insufficient drug is available to perform a toxicity study, but such a study must be done for development to progress, the answer is simple and applies equally to gene products an xenobiotics. More compound must be made available! Toxicity studies cannot be done by fiat. I suspect that this is the real reason behind the push to perform safety studies -- much less compound is needed. This is merely getting around the problem, not addressing it.

I noticed in data yesterday that in some cases, up to ten thousand times the human dose did not induce acute toxicity in animals. Rather than blasting away blindly, it might be better to re-evaluate the end-point of toxicity studies. For instance, in acute toxicity studies the traditional end-point is death. Little information of use is gained by this crude type of study. But maybe we are looking at the wrong end-point. With tissue plasminogen activator, very high doses can be administered without producing death in rats. However, if one looks closely at the hemostatic mechanism of those animals, even at doses only ten or fifteen times the human dose, there is total ablation of the fibrinolytic and coagulation systems. This could/should be viewed as a severe, albeit "latent" toxicity. These rats are walking hemophiliacs! Thus, I think that if one selects a better or more relevant end-point for evaluating toxicity, we will see plenty of it. I think we should re-examine the kind of toxicity studies that are being done on these new products rather than doing the knee-jerk type of study.

Experience to date.

Audience?

-- Animal models have failed to predict the adverse effects subsequently observed in man.

-- Animal models often yield information that retrospectively was predictive for man.

European toxicology panel report:

-- Animal toxicology studies nto viewed as predictive of toxicology of DNA products in humans.

  -- but most of the "toxicity" studies that I have seen from Europe are in fact safety studies and by definition are not designed to produce toxicity.

Therefore, how can they be predictive?

SECTION V: IMMUNITY, IMMUNE INTERFERENCE, AND ADVANCEMENT

J. Thomas, Baylor College of Medicine, illustrated the complexity of the immune response to insulin in man. For example, more than one Class II, HLA-DR antigen is involved in the restriction of the same epitope and there is evidence for immune regulatory networks. The majority of Type I diabetics express polyclonal antibodies even to human insulin; 40% percent have intradermal skin sensitivity but few have reactions to subcutaneous insulin injections. Continuous administration of other peptides and proteins may be expected to lead to cellular and humoral immunity but like insulin the clinical consequences are hard to predict due to idiotype anti-idiotype interactions, T suppressor cells, tolerance or other regulation.

The complexity of immune response was also shown by P. Ehrlich, Sandoz Corporation. Rhesus monkeys generally are resistant to human monoclonal antibodies but a monkey resistant to one monoclonal antibody rapidly cleared a monoclonal antibody with non-cross-reacting idiotypes.

A. Sehon, University of Manitoba, showed that monomethoxypolyethylene glycol conjugates of proteins (like PEG conjugates discussed by A. Abuchowski) can suppress antibody responses in mice. In addition, proper timing can lead to a tolerant state to the unmodified proteins. W. Hubbard, Biotherapeutics Inc., discussed the immunosuppressive effects of proteinase-complexed alpha-2 macroglobulin on antigen presentation in monocytes.

The Pharmacology and Toxicology of Proteins, pages 187–197

REGULATION OF THE IMMUNE RESPONSE TO THERAPEUTIC INSULINS IN MAN[1]

J.W. Thomas*, L.J. Nell*, and G.P.G. Miller+

Department of Medicine, Baylor College of Medicine* and Program in Infectious Diseases, The University of Texas Health Science Center+, Houston, Texas 77030

ABSTRACT The cellular and humoral mechanisms that regulate the immune response to insulin in man are examined. T cells are shown to recognize distinct epitopes on the insulin molecule in association with class II, HLA-DR antigens. More than one DR antigen may restrict the response to the same epitope. In addition to genetic restriction of T cell activation, studies of idiotypic determinants on anti-insulin antibodies are consistent with a regulatory network modulating anti-insulin antibodies. Together, these studies demonstrate the complexity of the regulatory events that modulate immunity to an exogenously administered protein.

INTRODUCTION

The natural course of "juvenile" or type I diabetes was dramatically altered by the introduction of insulin therapy. Not suprisingly, injection of crude extracts from animal pancrea resulted in allergic reactions. As the purity of pancreatic extracts improved, the incidence of hypersensitivity reactions declined. However, the highly purified and recombinant insulins used today may produce immunological reactions in some subjects (1). The majority of insulin treated diabetics develop antibodies to the molecule and in most cases these antibodies are of the IgG class. Competitive binding studies and isoelectric

[1]This work supported by NIH grants AM32329, AI20911 and P01 AI21289.

focusing demonstrate that human anti-insulin antibodies crossreact extensively with insulins from several species, including man (2). In addition, recent data confirm that anti-insulin antibodies may be present in the prodrome or autoimmune phase of type I diabetes (3,4). Understanding the mechanisms that regulate these immune responses will aid our management of allergic subjects and may provide insight into the immunopathology of type I diabetes.

In inbred animals the immune response to insulin has been shown to be linked to genes in the major histocompatibility complex (MHC) (5,6). The immune response genes or class II MHC antigens (Ia) determine the epitopes on the insulin molecule that induce insulin immunity in different strains of inbred mice and guinea pigs. For example, the immune response to the A loop of beef insulin can be detected in Strain 2 guinea pigs and $H-2^b$ mice (sequence differences in insulins are shown in Table 1). $H-2^d$ mice and strain 13 guinea pigs in contrast may recognize determinants associated with the B chain of beef and pork insulin. *In vitro* studies demonstrate that these immune response genes operate at the level of macrophage/T lymphocyte interaction and require co-recognition of insulin and Ia antigens by the T cell receptor (7).

TABLE 1
DIFFERENCES IN AMINO ACID SEQUENCES OF INSULINS

	A Chain			B Chain
Species	A8	A9	A10	B30
Human	Thr	Ser	Ile	-Thr-
Pork	-*	-	-	-Ala-
Beef	Ala	-	Val	-Ala-

*Identical residues

In man, immune response gene control of the response to insulin has been limited by the polyclonal nature of human T cell responses to insulin. As a result, most human studies have relied on statistical associations insulin immunity with the presence of serologically determined HLA antigens. Some studies have found associations with different DR antigens (8), but genetic restriction at the cellular level has not been demonstrated.

To identify genetic control at the cellular level in man, we have developed T cell lines that recognize distinct epitopes on the insulin molecule. We show that of T cell recognition of epitopes on the insulin molecule is genetically restricted in man. In addition the recognition of these epitopes is complex and different Ia antigens may present the same epitope. Further complexity in the human immmune response to insulin was identified when idiotopes on anti-insulin antibodies were examined. These studies show that the human immune response to insulin is complex both in its regulation by MHC antigens and by the dynamic expression of the anti-insulin antibody repertoire.

METHODS

Peripheral blood lymphocytes and sera were obtained from a single diabetic donor who was previously treated with beef/pork insulin mixtures. The procedures and protocols were approved by the Institutional Review Board of Baylor College of Medicine in accordance with guidelines set by the National Institutes of Health.

Highly purified beef, pork, and human insulins were the generous gift of Dr. Ron Chance (Lilly Research Laboratory, Indianapolis, IN). The amino acid difference in these insulins are shown in Table I. Lymphocyte proliferative studies were carried out in RPMI 1640 media (Hazelton, Denver, PA) containing 5% human AB serum (KC Biological Corporation, Kansas City, MO). Peripheral blood lymphocytes were prepared by density gradient centrifugation and responses to different insulin were measured in 96 well U bottom microtiter plates containing $5x10^5$ cells/ml in the presence or absence of insulin in a final volume of 200ml (9). Responses of the cells were determined by the addition of luCi of ^{3}H thymidine.

To produce T cell clones, blast cells were isolated over a discontinuous Percol gradient (Pharmacia, Piscataway, NJ) as previously described (10). The blast cell enriched population was then plated at limiting dilution in 96 well microtiter plates according to established procedures (11). Growth positive wells were expanded by intermitent stimulation with insulin in IL2 every 7 days. T cell lines produced in this fashion were tested for response to insulin by ^3H thymidine incorporation. Results are reported as mean of triplicate determinations. Antigen presenting cells (APC) used in this study were peripheral blood mononuclear cells obtained from the same donor. The HLA type of individual donors was determined by the 2 stage microcytotoxicity on B lymphocyte fractions as previously described (12).

Anti-insulin antibodies were determined in a solid phase enzyme immunoassay (ELISA). The binding of anti-insulin antibodies in the sera is detected by the activity of an enzyme conjugated second antibody. The results are reported as optical density (OD_{405}) as described for the alkaline phosphatase substrate (13). The presence of idiotypic determinants on anti-insulin antibodies was determined by competitive inhibition assay. In this assay idiotype (anti-insulin antibody) binding to plates coated with monoclonal anti-idiotypic is determined. Antibody fractions containing idiotype positive molecules are able to inhibit this interaction and the results are expressed as per cent inhibition (14).

RESULTS

The results of peripheral blood lymphocytes of several insulin treated diabetics is summarized in Table II. The results show the ^3H thymidine incorporation of T cells in the presence of beef, pork, and human insulin. As can be seen, the proliferative responses are detected to both animal and human insulin, and in some individuals the responses to human insulin are as strong as that to animal insulin. We have previously observed that these extensively crossreacting T lymphocyte responses are most common in type I diabetics (9). These results suggest that multiple determinants on the insulin molecule may be recognized by human lymphocytes, and in some cases the determinants are shared between human and animal insulins. In contrast to inbred animal studies where recognition of a single epitope dominates, these results indicate that the human peripheral blood response is polyclonal.

TABLE 2
PERIPHERAL BLOOD T CELL RESPONSE TO INSULIN

Subject	Insulin[a]	Response (SI)[b]
M.L.	Beef	54,300 (15)
	Pork	17,800 (5)
	Human	10,700 (14)
T.S.	Beef	45,400 (4)
	Pork	30,700 (11)
	Human	13,200 (5)
J.H.	Beef	8,700 (10)
	Pork	10,100 (13)
	Human	2,400 (2)

[a] Peripheral blood T cell response determined by ^{3}H-thymidine incorporation, (SI,stimulation index) as described (9).
[b] Insulins were at 100ug/ml.

To further examine human T cells reactive to insulin, we developed T cell clones that recognize the distinct epitopes on beef insulin. The procedures for developing these T cell clones are well described in the literature (10, 11). Table III shows the response of T cells highly specific for the 2 amino acid differences between beef and pork insulin. Another T cell clone is able to crossreact with pork and human insulin. These findings further support the polyclonal nature of human T cells reactive with insulin.

The availability of T cells that recognize discrete amino acid residues on a protein antigen provide a unique opportunity to examine genetic restriction of human T cell responses to insulin. To determine the role of class II or

TABLE 3
SPECIFICITY OF HUMAN T CELL CLONES[a]

Clone	Beef	Pork	Human
B1	11,060±601	1020±95	300±179
P1	20,500±550	8260±1300	10,040±1600
B2	12,400±1200	1170±90	940±40

a ^{3}H-thymidine incorporation of T cells to insulins at 100ug/ml. Background was 590±90.

HLA/DR antigens in the response of these clones, a panel of irradiated mononuclear cells was employed. Table IV summarizes the response of two beef insulin specific clones in the presence of several different accessory cell populations. As shown the response of clone B1 is detected only in the presence of accessory cells that carry the HLA/DR1 serological determinants. In contrast the response of the clone B2 is restricted to the HLA/DRw6 related determinants. Studies are in progress to determine if other class II antigens such as DP or DQ are important for these responses. Nonetheless, these data show that different genetic elements may restrict the T cell responses to the same epitope on the insulin molecule. Notably, the DR antigens that restrict the response of these T cells are not reported in previous population studies of T cell or antibody responses to insulin.

Since multiple genetic elements appear capable of restricting human immune response to insulin, we have initiated experiments to identify additional immunoregulatory events in man. The repetitive administration of "antigen" required for daily hormonal therapy might be expected to induce regulatory anti-idotypic responses (15). To test this possibility, we produced monoclonal antibodies that recognize determinants present on anti-insulin antibodies in approximately 40% of insulin treated diabetics (14). Because the donor of our T cell clones had multiple

TABLE 4
GENETIC RESTRICTION OF T CELL CLONES[a]

DR	Clone B1	Clone B2
1,w6 (donor)	9,000	11,000
1,w8	8,900	0
1,1	14,000	590
w6/7	290	17,000
3,4	0	0
5,5	0	0

a ^{3}H-thymidine incorporation of two T cell clones in the presence of irradiated MNC of the indicated DR type. Beef insulin is used at 100ug/ml.

phlebotomies, we were able to examine his serum for the expression of these idiotypic determinants over time. The results shown in Fig. 1 summarize these studies. The anti-insulin response remained constant throughout the period of study, and fell only after insulin was discontinued. Interestingly, when two monocloncal anti-idiotypes were employed a different pattern was seen. As shown in the figure, these idiotopes are variably expressed during insulin treatment and eventually are lost. Our findings suggest that anti-insulin antibodies are in dynamic flux during continuous insulin therapy, and these idiotypes expressed on anti-insulin antibodies may be the target for immune regulation.

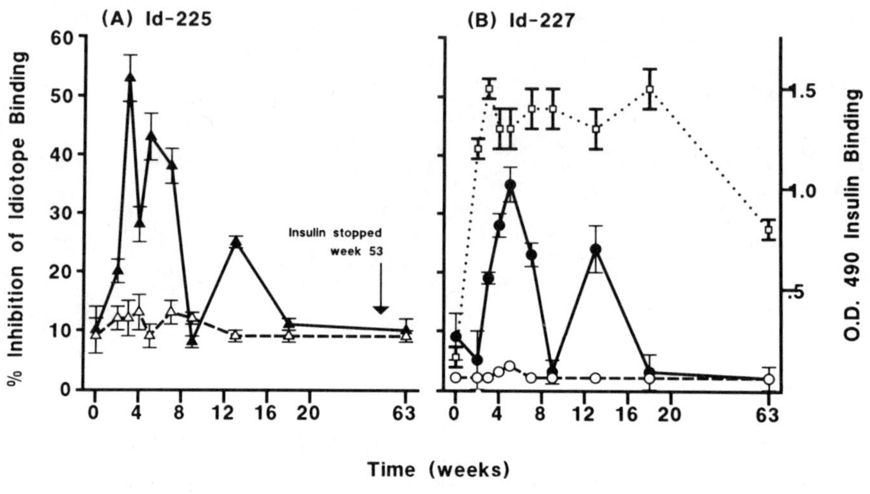

FIGURE 1. The expression of two idiotypic determinants (Id225 and Id227) on anti-insulin antibodies are examined over time in relationship to total anti-insulin antibody (open squares in panel B).

DISCUSSION

Therapeutic administration of the protein hormone insulin results in the generation of antibodies and T lymphocytes that recognize the native molecule. Both the antibody and T cell responses of a given individual are heterogeneous and extensive crossreactivity is observed at both the antibody and T cell levels. Responses at the cellular level are regulated by class II MHC antigens, and we show that HLA/DR molecules appear to be required for the recognition of insulin by immune T cells. Interestingly, an individual may employ multiple restriction elements for the recognition of the same epitope. The data shown here demonstrate that the A loop of beef insulin is recognized in association with either HLA/DR1 or HLA/DRw6 determinants. Neither of these HLA/DR antigens has been associated with insulin immunity in the previously reported population studies of anti-insulin antibody (16) or T cell proliferation (8). Thus, the overall magnitude of the immune response reflected in population studies appears to determined by additional immunoregulatory elements. One potential important regulatory element that we have examined is the expression of unique or idiotypic determinants on anti-insulin antibodies. Two monoclonal anti-idiotopes recognize determinants on anti-insulin antibodies present shortly after the institution of insulin therapy. The expression of these idiotopes varies during 9 months of continuous insulin therapy. Despite the variable expression of the two idiotypic determinants, the magnitude of the anti-insulin antibody response was constant during this same time period. Two interpretations are possible: first somatic mutation events may result in loss of the idiotypic determinants recognized by our antibody; or second, auto-anti-idiotypic antibodies, may be regulating the response of select clones. The two cycles of disappearance and reappearance observed with one anti-idiotope is most consistent with a regulatory network. Studies are in progress to clarify these two possibilities.

The data obtained from studies of T cell and antibody responses to a "simple" protein demonstrate the multiple complexities present in the outbred human population. In contrast to inbred animals most individuals are capable of generating an immune response to insulin. The induction of this immune respones requires T cell recognition of insulin in association with class II HLA antigens. Continuous administration of insulin may establish a dynamic state of idiotypic anti-idiotypic interactions that maintain the overall antibody response at a constant level. Additional regulatory elements that include T suppressor cells and tolerance to self proteins have been described in animal studies and undoubtedly will also play a role in human immune responses to insulin. As more diabetics receive therapy with autologous insulin, these additional regulatory factors may be identified in man.

ACKNOWLEDGMENTS

We wish to thank Dr. Marilyn Pollack for carrying out tissue typing on our patients. Figure 1 was used with the permission of Dr. Joseph Feldman, Journal of Immunology. The skillful and enthusiastic assistance of Mary Guerrero in the preparation of this manuscript is greatly appreciated.

REFERENCES

1. Deckert T (1985). The immunogenicity of new insulins. Diabetes 34 (suppl 2):94.
2. Thomas JW, Virta VJ, Nell LJ (1985). Heterogeneity and specificity of human anti-insulin antibodies determined by isoelectric focusing. J Immunol 134:1048.
3. Palmer JP, Asplin CM, Clemons P (1983). Insulin antibodies in insulin-dependent diabetics before insulin treatment. Science 222:1337.
4. Srikanta S, Ricker AT, McCulloch DK, Soelder JS, Eisenbarth GS, Palmer JP (1986). Autoimmunity to insulin, beta cell dysfunction and development of insulin-dependent diabetes mellitus. Diabetes 35:139.
5. Keck K (1975). Ir-gene control of immunogenicity of insulin and A-chain loop as a carrier determinant. Nature 254:78.
6. Barcinski MA, Rosenthal AS (1977). Immune response gene control of determinant selection. I. Intramolecular mapping of the immunogenic sites on insulin recognized by guinea pig T and B cells. J Exp Med 145:726.
7. Schwartz, RH (1986). Immune response (Ir) genes of the murine major histocompatibility complex. Advances in Immunol 38:31.
8. Mann DL, Mendell N, Kahan CR, Johnson AH, Rosenthal A (1983). In vitro lymphocyte proliferation response to therapeutic insulin components. J Clin Invest 72: 1130.
9. Nell LJ, Virta VJ, Thomas JW (1985). Recognition of human insulin in vitro by T cells from subjects treated with animal insulins. J Clin Invest 76:2070.
10. Kurnick JT, Ostberg L, Stegagno M, Kimira AK, Orn A, Sjoberg O (1979). A rapid method for the separation of functional lymphoid cell populations of human and animal origin on PVP silica (Percoll) density gradients. Scand J Immunol 10:563.
11. Lamb JR, Eckels DE, Lake P, Johnson AH, Hartzman RJ, Woody JN (1982). Antigen-specific human T lymphocyte clones:induction, antigen specificity, and MHC restriction of influenza virus-immune clones. J Immunol 128:233.

12. Shaw S, Pollack MS, Payne SM, Johnson AH (1980). HLA-linked B. cell alloantigens of a new segregant series. Population and family studies of the SB antigens. Human Immunol 1:177.
13. Nell LJ, Virta VJ, Thomas JW (1985). Application of a rapid enzyme-linked immunosorbent microassay (ELISA) to study human anti-insulin antibody. Diabetes 34:60.
14. Thomas JW, Virta VJ, Nell LJ (1986). Idiotypic determinants on human anti-insulin antibodies are cyclically expressed. J Immunol 137:1610.
15. Jerne NK (1974). Towards a network theory of the immune system. Ann Immunol (Paris) 125:373.
16. Bertrams J, Jansen FK, Gruneklee D, Reis HE, Drost H Beyer J, Gries FA, Kuwert E (1976). HLA antigens and immunoresponsiveness to insulin in insulin dependent diabetes mellitus. Tissue Antigens 8:13.

The Pharmacology and Toxicology of Proteins, pages 199–204

RESPONSE OF RHESUS MONKEYS TO HUMAN MONOCLONAL ANTIBODIES: EFFECT OF SUBSTITUTING THE ANTIBODY

Paul H. Ehrlich, K. Elisabeth Harfeldt,
Zeinab A. Moustafa, James C. Justice, and Lars Östberg

Monoclonal Antibody Department, Sandoz Research Institute,
Sandoz Pharmaceuticals Corp., E. Hanover, N.J. 07936.

ABSTRACT We have shown previously that rhesus monkeys are, in general, tolerant to injections of human monoclonal antibodies. Multiple injections of one antibody, EV2-7, over 200 days resulted in no immune response in three monkeys. The same result was obtained with one monkey injected with a different monoclonal antibody, EV1-15. A second rhesus monkey injected with EV1-15 developed an anti-idiotypic response after the second injection. We reasoned that substituting EV2-7 for EV1-15 in this monkey would be possible since anti-idiotypic antisera against each antibody are not cross-reactive with the other antibody. This could be a model for switching to equivalent monoclonal antibodies in patients once an immune response has developed to an initial monoclonal antibody. However, soon after the initial injection of EV2-7 into the monkey with the anti-EV1-15 response, the EV2-7 was eliminated from the circulation. Therefore, substitution of monoclonal antibodies with non-cross-reacting idiotypes may not be a viable therapeutic strategy.

INTRODUCTION

Monoclonal antibodies have great potential for therapy due to their specificity and lack of toxicity. One of the major drawbacks to monoclonal antibody therapy is the ability of the patient to mount an immune response to these large molecules. It can be expected that human monoclonal antibodies can perform much better in this regard than mouse monoclonal antibodies, since human monoclonal antibodies have greatly decreased immunogenicity in rhesus monkeys (1). However, anti-idiotypic or anti-allotypic responses may be expected in some cases (1). In addition, mouse monoclonal antibodies may still be used when no acceptable human antibody is available. It has been proposed (2) that switching from one mouse monoclonal antibody to a different antibody with similar specificity but a non-cross reacting idiotype could alleviate, at least temporarily, the effects of the anti-monoclonal antibody. This proposal was based on data from a patient who received injections of a mouse monoclonal antibody (3). It was shown that the presence of anti-isotypic antibodies did not interfere with the function of the mouse monoclonal antibody. However, the mouse monoclonal antibody elicited both anti-isotypic and anti-idiotypic antibodies in the majority of patients (4) and rhesus monkeys (5). By switching patients to a different mouse monoclonal antibody with a non-cross-reacting idiotype after an initial anti-mouse monoclonal antibody response appeared, anti-idiotypic antibodies directed against the first monoclonal antibody would be rendered harmless and an immunological state similar to the patient with only anti-isotypic antibodies would be engineered. We report here that experiments with a rhesus monkey injected with human monoclonal antibodies indicate that such a strategy is unlikely to work since each change of monoclonal antibody idiotype will at best give a small respite from the immune response.

MATERIALS AND METHODS

Human monoclonal antibodies EV2-7 and EV1-15 were isolated from hybridomas constructed by fusing human spleen cells to a mouse heteromyeloma (6) and have been previously described (1). Both antibodies are of the IgG1, κ type and neutralize human cytomegalovirus (CMV). Injections of rhesus monkeys and immunoassay methods for measuring serum levels of monoclonal antibodies and characterizing an anti-antibody response have also been previously described (1).

RESULTS

Further Characterization of Monoclonal Antibodies EV2-7 and EV1-15

The human monoclonal antibodies were further characterized in order to confirm that they are not related. A sandwich ELISA was set up in which goat anti-EV2-7 idiotype was coated on the solid phase and any solid phase-bound immunoglobulin was detected with rabbit anti-EV2-7 idiotype and peroxidase-conjugated goat anti-rabbit IgG. In this assay, EV1-15 was less than 2×10^6 fold less active than EV2-7. In addition, preliminary evidence indicates that the antibodies bind antigens of different molecular weight.

Rhesus Monkey Responses to Antibody EV1-15

One rhesus monkey (813) injected with monoclonal antibody EV1-15 had no apparent immune response to this antibody as shown in Figure 1. After each injection there is a fast decline in serum concentration of the monoclonal antibody due to distribution to additional compartments, then a slow decay typical of metabolism of the immunoglobulin. Monkey 817 did not respond in this way (1). After the second injection, all EV1-15 was removed from the circulation and anti-idiotypic antibodies appeared

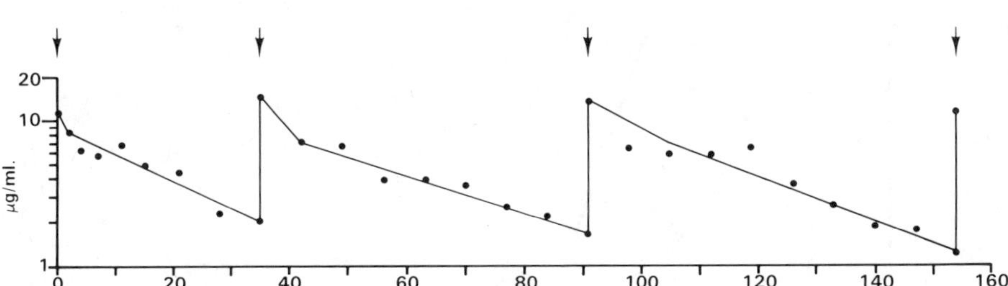

Figure 1. Serum concentration of monoclonal antibody EV1-15 in Rhesus Monkey 813 vs. time. Arrows indicate time of injection.

in the serum (1). Within the limits of our assays, the anti-EV1-15 immune response was totally anti-idiotypic, i.e., EV1-15 was more than 8,000 times more effective than pooled normal human IgG in inhibiting the Monkey 817 anti-EV1-15 (in a different assay system, Monkey 817 anti-EV1-15 was ineffective in inhibiting the interaction of ^{125}I-EV2-7 with goat anti-EV2-7 idiotype). A third injection of EV1-15 increased the anti-EV1-15 idiotype titer.

Rhesus Monkey 817 Response to Antibody EV2-7

After a rest period of five months, Monkey 817 was injected with 3 mg of human monoclonal antibody EV2-7. An immune response developed quickly and this antibody was removed from the circulation within two weeks as shown in Figure 2. In contrast to the Monkey 817 anti-EV1-15 response, the anti-EV2-7 response was mostly, if not totally, directed against pooled human IgG as shown in Table 1.

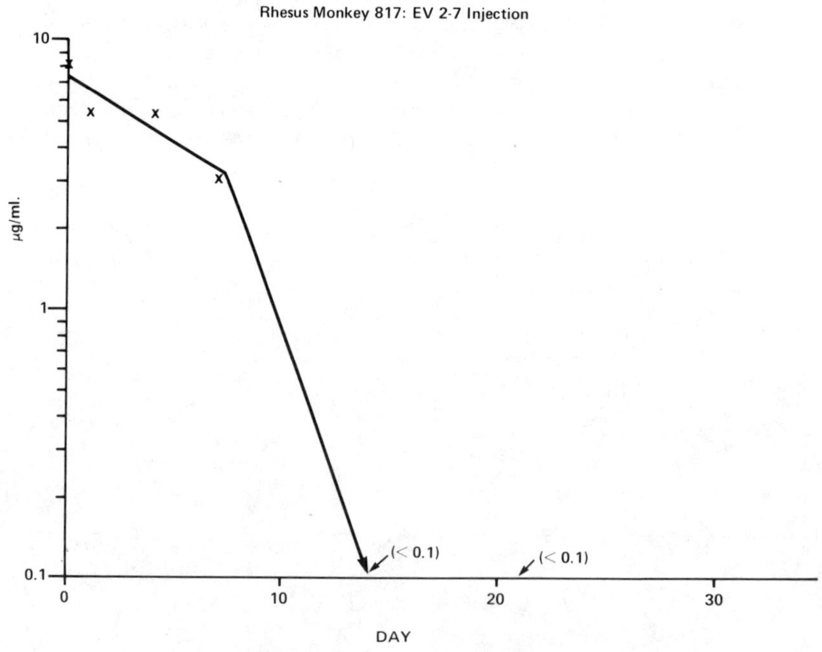

Figure 2. Serum concentration of monoclonal antibody EV2-7 in Rhesus Monkey 817 vs. time after injection.

TABLE 1
ANALYSIS OF MONKEY 817 SERUM AFTER INJECTION OF MONOCLONAL ANTIBODY EV2-7*

Sample	% Inhibition
0.025 μg/ml EV2-7 + Normal Monkey Serum	0
0.025 μg/ml EV2-7 + Monkey 817 Serum	100
0.025 μg/ml EV2-7 + Monkey 817 Serum	
+ 0.02 μg/ml pooled human IgG	97
+ 0.20 μg/ml pooled human IgG	93
+ 2.0 μg/ml pooled human IgG	73
+ 20 μg/ml pooled human IgG	35

*Results from an ELISA performed as follows: ELISA wells were coated with affinity purified goat anti-EV2-7 idiotype antibodies. Samples were pre-incubated for 1 hr at 37°C and then transferred to the coated wells. After incubation with the samples, the wells were incubated with a rabbit anti-EV2-7 idiotypic antiserum and then with horseradish peroxidase conjugated goat anti-rabbit IgG.

Monkey 817 serum could inhibit binding of antibody EV2-7 to goat anti-EV2-7 idiotypic antibodies. However, this inhibition was inhibited by pooled normal human IgG indicating that the epitope recognized by the Monkey 817 antibodies is also present on normal human IgG. In a separate experiment, it was shown that the pooled human IgG could completely inhibit the inhibition caused by the Monkey 817 serum, indicating that there was little, if any, anti-EV2-7 idiotypic reactivity.

DISCUSSION

Human monoclonal antibodies EV2-7 and EV1-15 are distinct antibodies with non-cross-reactive idiotypes. The above results indicate that substituting one monoclonal antibody with a non-cross-reactive idiotype for another monoclonal antibody that has elicited an immune response is not a viable therapeutic strategy. An exception can be made in those cases in which therapy would be complete in less than two weeks (however, the initial antibody should be sufficient in those circumstances). What is especially noteworthy is that the rapid immune response after injection of antibody EV2-7 was preceded by an anti-EV1-15 response that was apparently totally anti-idiotypic. In retrospect, it seems likely that there was a very small component of anti-human IgG and memory cells were formed that were greatly expanded by the injection of a human IgG without the same idiotype. These results are in contrast to those reported previously (2,3). Although rhesus monkey 817 did not produce anti-EV2-7 idiotypic antibodies, the EV2-7 antibody was cleared from the circulation. Thus, unlike the patient injected with OKT3 mouse monoclonal antibody, antibodies to the constant domains were sufficient to render the monoclonal antibody non-functional. It is therefore probable that what Baudrihaye et al (3) call an unusual response for their patient may not, in fact, be typical of anti-constant region responses in monkeys (and possibly in humans). Therefore, it may be necessary in most cases to avoid an immune response to both constant and variable domains.

REFERENCES

1. Ehrlich PH, Harfeldt KE, Justice JC, Moustafa ZA, Östberg L (1987). Rhesus monkey responses to multiple injections of human monoclonal antibodies. Hybridoma. (In press)
2. Chatenoud L, Jonker M, Villemain F, Goldstein G, Bach J-F (1986). The human immune response to the OKT3 monoclonal antibody is oligoclonal. Science 232:1406.
3. Baudrihaye M-F, Chatenoud L, Kreis H, Goldstein G, Bach J-F (1984). Unusually restricted anti-isotype human immune response to OKT3 monoclonal antibody. Eur J Immunol 14:686.
4. Jaffers, GJ, Colvin RB, Cosimi AB, Giorgi JV, Goldstein G, Fuller TC, Kurnick JT, Lillehei C, Russell PS (1983). The human immune response to murine OKT3 monoclonal antibody. Transpl Proc XV:646.
5. Villemain F, Jonker M, Bach J-F, Chatenoud L (1986). Fine specificity of antibodies produced in rhesus monkeys following in vivo treatment with anti-T cell murine monoclonal antibodies. Eur J Immunol 16:945.
6. Östberg L, Pursch E (1983). Human x (Mouse x Human) hybridomas stably producing human antibodies. Hybridoma 2:361.

The Pharmacology and Toxicology of Proteins, pages 205–219

CONVERSION OF ANTIGENS TO TOLEROGENIC DERIVATIVES BY CONJUGATION WITH MONOMETHOXYPOLYETHYLENE GLYCOL.[1]

A.H. Sehon, C-J.C. Jackson, V. Holford-Strevens, I. Wilkinson, P. K. Maiti and G. Lang

MRC Group for Allergy Research, Department of Immunology, Faculty of Medicine, The University of Manitoba, Winnipeg, MB, Canada, R3E 0W3

ABSTRACT Murine antibody responses to antigenic proteins were specifically suppressed by tolerogenic conjugates of the corresponding proteins and an appropriate number of monomethoxypolyethylene glycol chains. Thus, treatment of mice with conjugates of ovalbumin prior to sensitization with ovalbumin by a protocol leading to IgE and IgG antibody formation resulted in suppression of both antibody responses, which was mediated by carrier specific suppressor T cells. The degree of suppression depended on the number of polymer chains in the conjugate, the molecular weight of the polymer and the dose used. Treatment of presensitized mice with appropriate conjugates resulted in significant downregulation of the IgE response and the residual IgE antibodies had lower affinities for ovalbumin. As a model for establishing the tolerogenicity of similar conjugates of xenogeneic monoclonal antibodies in man, the reciprocal system, consisting of human myeloma IgG and the corresponding monomethoxypolyethylene glycol conjugates for immunization and tolerance induction, respectively, was tested in mice. These conjugates induced specific tolerance which was transferable to normal mice by spleen cells of tolerized mice possessing the properties of suppressor cells. This procedure was also successful in producing tolerogenic conjugates of potentially antigenic,

[1]Supported by grants from the Medical Research Council of Canada and the National Institute of Allergy and Infectious Diseases (AI 14526), National Institutes of Health, Bethesda, MD, USA.

biologically active factors synthesized by r-DNA technology, such as recombinant human growth hormone. The results of all these studies hold the promise that tolerogenic monomethoxypolyethylene glycol conjugates of allergens and of diverse xenogeneic proteins may prove therapeutically useful in man in relation to (i) allergies of the immediate type, (ii) organ transplantation, localization of tumors by immuno-imaging and tumor destruction by immunotoxins, and (iii) the use of antigenic factors produced by rDNA technology.

INTRODUCTION

Since traditional hyposensitization therapy with allergenic extracts is associated with the risk of systemic anaphylactic reactions, the doses of allergens administered are, by necessity, extremely small. Therefore, generally, the course of the classical therapeutic regimen is extended over a number of years. To circumvent these disadvantages, many attempts have been made to decrease the allergenicity of the therapeutic preparations by chemical modification, e.g., urea denaturation and polymerization with glutaraldehyde (reviewed in 1). On the basis of (i) reports by Abuchowski and his colleagues (2,3) that certain proteins lost their immunogenicity after conjugation with monomethoxypolyethylene glycol (mPEG) and that the clearance rates of the resulting conjugates were significantly lower than those of the unmodified proteins, and (ii) the established fact that the clearance time of some antigens which had been rendered tolerogenic by deaggregation [e.g., xenogeneic immunoglobulins (XIg)] was longer than that of the same antigens in their original form (4), it was hypothesized and confirmed that mPEG conjugates of various antigens would be tolerogenic (5). Thus, conjugates of ovalbumin (OA) and of other proteins with mPEG preparations (of molecular weights in excess of 3,000 Da) were shown to abrogate specifically the primary IgE and IgG responses in mice and rats to the corresponding unmodified antigens (6-11). Moreover, as a result of the masking of some of the epitopes of allergens, tolerogenic mPEG conjugates of allergens were essentially devoid of allergenicity.

Therapeutic procedures involving injections into humans of XIg are becoming increasingly prevalent (12), e.g., administration of (i) mouse monoclonal antibodies to human T cell subsets in organ transplantation (13), (ii) "magic bullets", i.e., immunotoxins for the selective destruction of tumors (14), and (iii) radio-labeled anti-tumor immunoconjugates for diagnostic imaging of tumors (15,16). However, it has become apparent that patients produce antibodies to the injected murine monoclonal antibodies used for these purposes (15-21). Hence, the therapeutic or diagnostic effectiveness of these immunoconjugates may be counteracted by their inherent immunogenicity; the consequent immune responses may lead -- following re-administration of the monoclonal antibody preparations -- to anaphylactic symptoms such as broncho-spasm, dyspnoea and hypotension (18,19-21). Moreover, as recently demonstrated in patients receiving antibodies labeled with different radioactive elements for diagnostic imaging, the resulting immune complexes consisting of the patients' antibodies and XIg were deposited primarily in the liver (15). Therefore, it is imperative to devise strategies to render the patients tolerant to these foreign proteins prior to initiation of the treatment *per se*, i.e., to activate pathways in the regulation of the immune response leading to its specific suppression to XIg or to the corresponding immunotoxins.

On the basis of the studies with $OA(mPEG)_n$ conjugates referred to above (1,22), we have extended our investigations to the production of tolerogenic mPEG derivatives of XIg; for this purpose a murine model system utilizing human myeloma IgG (HIgG) as a source of XIg was used. We have thus induced suppression of the immune response in mice to these XIg (23,24). The proposed extension of these findings could, therefore, be of relevance to the treatment of patients with mouse or rat monoclonal antibodies. Similar applications may be also envisaged in relation to treatment with potentially immunogenic hormones [e.g., as demonstrated below for the synthetic human growth hormone (r-hGH)] or other biologically active agents produced by rDNA technology.

RESULTS AND DISCUSSION

$OA(mPEG)_n$ Conjugates.

Dependency of tolerogenicity on molecular composition of mPEG conjugates. It was demonstrated that there was a stringent dependency of the degree of tolerogenicity of conjugates of a given antigen on the molecular composition of its corresponding mPEG conjugates. Thus, as illustrated in Table 1, it is evident that (i) administration of a dose of 50 μg per mouse of a conjugate containing on the average 8 mPEG chains per OA molecule (i.e., possessing an average degree of conjugation, n, of 8), one day prior to immunization, induced essentially 100% suppression, and (ii) mPEG derivatives with a higher or lower degree of conjugation had a diminished suppressogenic activity. It is to be noted also that at the same dose the unmodified OA was markedly less suppressogenic than any of the conjugates. Moreover, as is evident from the data in this Table, the dose required for suppression of IgG1 antibody (which represents the major isotype class of mouse immunoglobulins elicited by the immunization procedure used) was higher than that required for the downregulation of the IgE response.

Mechanism of the tolerance induced by $OA(mPEG)_n$ conjugates. With respect to the mechanism underlying the observed immunologically specific suppression of IgE antibodies by $OA(mPEG)_n$ conjugates, it had been originally shown that this effect was due, at least in part, to suppressor T (Ts) cells (7); this mechanism was confirmed also by others (10). More recently, the possibility that treatment with $OA(mPEG)_n$ may also result in the inactivation of B cells was proven in adoptive transfer

TABLE 1

THE EFFECT OF $OA(mPEG)_n$ ON ANTI-OA IgG1 AND IgE RESPONSES

Compound used for treatment	IgG1 titers[a] 50 μg[b]	IgG1 titers[a] 150 μg	IgE titers[a] 50 μg	IgE titers[a] 150 μg
OA	6,400	3,800	960	80
$OA(mPEG)_3$	1,600	5,000	4	<4
$OA(mPEG)_8$	600	<250	<4	<4
$OA(mPEG)_{10}$	2,000	340	160	<4
$OA(mPEG)_{12}$	2,300	ND[c]	640	12
PBS[d]	5,000	5,000	5,120	5,120

[a]IgE and IgG1 anti-OA determined by passive cutaneous anaphylaxis (PCA) and enzyme linked immunosorbent assay (ELISA), respectively.

[b]Treatment dose.

[c]Not determined

[d]Phosphate buffered saline, pH 7.4.

experiments involving the injection of surface Ig^+ splenic cells (85-90% B cells) from murine donors, which had been treated with $OA(mPEG)_n$ prior to transfer into irradiated (700R) recipients possessing helper T (Th) cells to OA (for details see reference 25). It was thus shown that by comparison with appropriate controls, the B cells of animals pretreated with $OA(mPEG)_n$ remained essentially immunosilent in spite of the presence of Th cells for the carrier epitopes of OA.

From the data in Table 2, it is evident that the degree of suppression of IgE antibodies in presensitized mice depended on the number of weekly injections of the tolerogen and that by day 42 the circulating IgE, as measured by PCA, was almost totally abrogated by 4 injections (i.e., the suppression was of the order of 99%). It is also obvious from the data listed in the last column of Table 2 that the minimum dose of unmodified OA which was necessary for eliciting a Prausnitz-Küstner (P-K) reaction in skin sites of rats sensitized with sera from successive

TABLE 2
EFFECT OF NUMBER OF INJECTIONS OF $OA(mPEG)_6$ ON PCA AND P-K REACTIONS

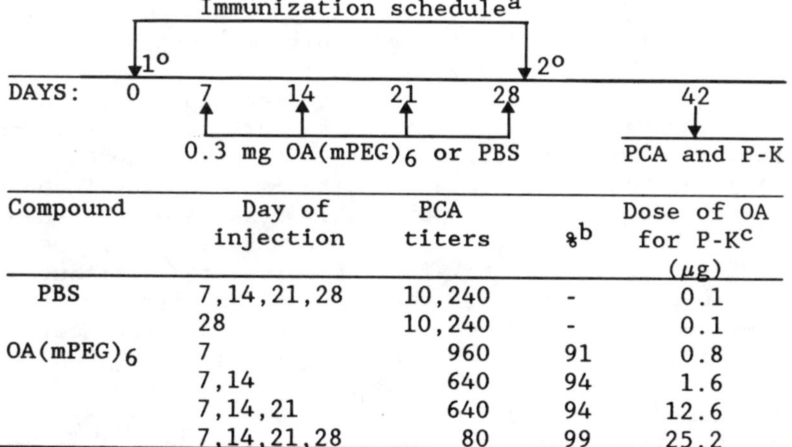

Compound	Day of injection	PCA titers	%[b]	Dose of OA for P-K[c] (µg)
PBS	7,14,21,28	10,240	-	0.1
	28	10,240	-	0.1
$OA(mPEG)_6$	7	960	91	0.8
	7,14	640	94	1.6
	7,14,21	640	94	12.6
	7,14,21,28	80	99	25.2

[a] The immunization schedule consisted of two i.p. injections of 1µg of OA plus 1mg $Al(OH)_3$ on days 0 and 29.

[b] Suppression of IgE titers expressed as percentages with respect to PBS controls.

[c] Minimum dose for elicitation of a P-K reaction.

bleedings was greater than the dose required for the serum of nontolerized animals, i.e., control animals which had been given PBS in lieu of the conjugate. Thus, the challenge dose of OA required for the manifestation of the P-K reaction with the serum from animals that had received 4 injections of the tolerogen was 250-fold higher than that required with control sera. In accordance with the law of mass action these results lead to the interpretation that the residual serum IgE, after treatment of mice with $OA(mPEG)_n$, had a lower average affinity for OA than the serum of control mice. Furthermore, these results support the view that the B_ϵ cells susceptible to tolerance by treatment with mPEG conjugates possess surface IgE of higher affinity and would be, therefore, inactivated more readily by interaction with the tolerogen in their early phase of differentiation and expansion.

Since $protein(mPEG)_n$ conjugates persist longer in circulation than the unmodified antigens (26,27), it was considered that the mPEG chains might interfere with the uptake and subsequent clearance of these conjugates by cells of the reticuloendothelial system. Hence, it may be visualized that macrophages (viz., thioglycollate induced peritoneal exudate cells) might show a reduced capacity to present OA determinants of $OA(mPEG)_n$ conjugates to Th cells. It was indeed found that the minimum concentration of $OA(mPEG)_n$ required to associate *in vitro* with adherent cells to induce priming was higher than the minimum concentration of OA required (28); the concentrations of both OA and $OA(mPEG)_n$ were expressed in terms of the corresponding protein contents of the solutions. Moreover, in exploratory studies with OA and an $OA(mPEG)_{10}$ conjugate, both trace labelled with ^{125}I, it was found that adherent cells took up appreciably less $OA(mPEG)_n$ than OA when exposed to the same concentrations of OA and $OA(mPEG)_n$. It is, therefore, possible to attribute the differences observed in the pulsing experiments to the lesser uptake of mPEG conjugates by the adherent cells or to their diminished ability to process the conjugated antigens, or to both. Obviously, it remains to be established if pulsed macrophages (Mϕ), capable of priming for Th cells, would be generated *in vivo* by injection of $OA(mPEG)_n$ into naive mice, or if under *in vivo* conditions Mϕ would play a role in the development of suppression of humoral responses.

HIgG(mPEG)$_n$ Conjugates

In one set of experiments, three HIgG(mPEG)$_n$ conjugates were prepared with n values of 5, 13 and 20 (23). To determine the tolerogenic capacity of these conjugates, test groups (three mice per group) were injected with varying doses of HIgG(mPEG)$_n$; control groups received "sham-PEGylated" HIgG (isolated from a mixture of HIgG and inactive mPEG intermediate), or PBS. Six days later all mice were immunized i.p. with 10 μg of heat aggregated HIgG (ha-HIgG) and their IgG1 serum antibody titers to HIgG were determined after a further 14 days. For this purpose, sera of mice in a given group were pooled for determination of antibodies by ELISA. As is evident from the values in Figure 1, treatment of the mice with HIgG(mPEG)$_n$ conjugates resulted in marked suppression of the immune response. Moreover, this effect was related to the dose of conjugate administered and the degree of conjugation, viz., maximum suppression (95%) was observed in mice treated with 100 μg of HIgG(mPEG)$_{20}$, i.e., with this most tolerogenic conjugate at the highest dose tested (23).

The tolerance induced by HIgG(mPEG)$_n$ could be transferred to normal mice by whole spleen cells or by a surface Ig$^-$ subpopulation of spleen cells from tolerized donors, a finding which is consistent with the participation of Ts cells in the tolerized mice (24). Indeed, this interpretation is supported by the results shown in Table 3, which demonstrate that injection of unfractionated spleen cells or sIg$^-$ spleen cells from mice which had been treated with HIgG(mPEG)$_{20}$ resulted in 82-89% suppression of the subsequent IgG1 response in the cell recipients to an immunogenic dose of ha-HIgG. This suppression was antigen specific, since the antibody response to an unrelated antigen, such as OA, was not suppressed (24). Moreover, as shown in Table 4, the suppression could also be transferred by a soluble extract prepared by 3 cycles of freezing and thawing of sIg$^-$ spleen cells (F/T extract) of the tolerized donors but not of normal donors. Similar results were also obtained using an HIgG(mPEG)$_{17}$ conjugate prepared from a different myeloma IgG derived from another patient (24).

Transferable suppression could also be demonstrated with these conjugates 14 days after treatment with the tolerogen, but not 28 or 42 days after treatment (24). However, it is important to note that the tolerance in the treated mice persisted for much longer(i.e., up to at least

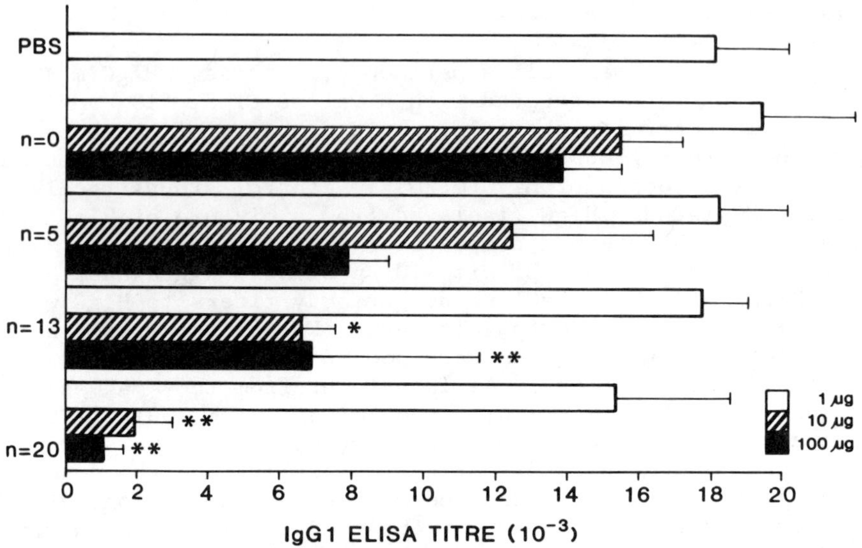

FIGURE 1. Downregulation of IgG1 antibody responses to HIgG by HIgG(mPEG)$_n$ conjugates. Mice received, i.v., PBS or 1µg, 10µg or 100µg of HIgG(mPEG)$_n$, and were immunized 6 days later with ha-HIgG and bled after a further 14 days for assay of antibodies to HIgG. The means of titers obtained in three separate experiments are illustrated. The means significantly different from the mean titer for the group treated with PBS are indicated by one ($p < 0.05$) or two ($p < 0.01$) asterisks.

57 days) than the period during which transferable suppression could be induced. This finding is similar to that reported for the tolerance induced by deaggregated HIgG (29) and may reflect multiple pathways of tolerance. In a subsequent experiment, using a slightly different protocol, an HIgG(mPEG)$_{18}$ conjugate prepared from a third myeloma HIgG was also shown to induce transferable suppression. In this experiment (Table 5), mice receiving spleen cells from donors treated with HIgG(mPEG)$_{18}$ showed 67% to 87% suppression of their anti-HIgG responses in relation to the responses of mice which had received spleen cells of control animals treated with PBS.

In another experiment, illustrated by the protocol and the results given in Table 6, the effect of the administration of multiple injections of a tolerogenic HIgG(mPEG)$_{33}$ conjugate was shown to be nonimmunogenic. In this experi-

TABLE 3
TRANSFER OF TOLERANCE BY SPLEEN CELLS FROM TOLERIZED MICE[a]

Treatment of cell donors (Day -6)	Spleen cells transferred (Day 0)	IgG1 anti-HIgG titers of recipients[b]	
		Day 14	Day 21
PBS	5 X 10^7 SC	18,000 (1.20)	18,000 (1.35)
HIgG(mPEG)$_{20}$	5 X 10^7 SC	3,100*(1.19)	3,200*(1.38)
HIgG(mPEG)$_{20}$	2 X 10^7 sIg$^-$	2,000*(1.09)	2,200*(2.00)
HIgG(mPEG)$_{20}$	3 X 10^7 sIg$^+$	13,000*(1.32)	15,000 (1.30)

[a]Cell donors received 100 μg HIgG(mPEG)$_{20}$ or PBS on day-6; unfractionated spleen cells (SC) or sIg$^-$ or sIg$^+$ spleen cells were transferred on day 0. Cell recipients were immunized with 10 μg ha-HIgG one day after cell transfer and bled 14 and 21 days later.

[b]The titers are expressed as geometric means of 4 to 12 experiments; the standard deviations are indicated in brackets and the means significantly different ($p<0.01$) from those of the PBS control groups are identified by asterisks.

TABLE 4
TRANSFER OF TOLERANCE BY F/T EXTRACT OF SPLEEN CELLS FROM TOLERIZED MICE[a]

Treatment of cell donors (Day -6)	IgG1 anti-HIgG titers of recipients[b]	
	Day 14	Day 21
PBS	13,000 (1.42)	16,000 (1.34)
HIgG(mPEG)$_{20}$	2,400*(2.71)	2,700*(2.13)

[a] Cell donors received 100 μg HIgG(mPEG)$_{20}$ or PBS on day -6; an extract of spleen cells was prepared and transferred on day 0. The recipient mice were immunized with 10 μg ha-HIgG one day after i.v. injection of the F/T extract and bled 14 and 21 days later.

[b] The titers are expressed as geometric means of six experiments; the standard deviations are indicated in brackets and the means significantly different ($p<0.01$) from those of the control groups are identified by asterisks.

TABLE 5

TRANSFER OF TOLERANCE BY SPLEEN CELLS FROM TOLERIZED MICE[a]

Treatment of cell donors	IgG1 anti-HIgG titers of recipients[b] Day 16	Day 23	Day 30
PBS	42,000	32,000	20,500
$HIgG(mPEG)_{18}$	14,000	5,800	3,300
Suppression	67%	82%	87%

[a] In this experiment the cell donors received 100 μg $HIgG(mPEG)_{18}$ or PBS on day -17, 20 μg ha-HIgG on day -10, and the spleen cells were transferred on day 0 into naive recipients which were immunized with 20 μg ha-HIgG one day later.

[b] The titers are given for bleeding at days 16,23, and 30 after transfer of donor cells.

ment the conjugate was prepared from the same myeloma HIgG as used for the preparation of the $HIgG(mPEG)_{18}$ conjugate in the experiment illustrated in Table 5. Thus, whereas one immunizing injection of this HIgG resulted in an IgG1 antibody titer of 90,000, mice given $HIgG(mPEG)_{33}$ six days prior to and on the day of immunization had a circulating anti-HIgG antibody titer of only 5,000 (i.e., 94% suppression was achieved by this protocol).

Taking an overview of all the results, it is suggested that mPEG conjugates of XIg may be of use in eliminating or minimizing the observed undesirable immune responses of patients to the administered XIg. If mPEG modified XIg are proven to suppress in patients the immune responses to the corresponding unmodified XIg, $XIg(mPEG)_n$ conjugates could be routinely used to tolerize patients prior to their being treated with the XIgG. It is important to point out that the possible therapeutic success of this regimen is predicated on the use of the recommended two-step procedure, involving first the immunological suppression of the patient with $XIg(mPEG)_n$ and, secondly, administration of the therapeutically beneficial XIg. This dual step treatment is essential since coupling of the mPEG chains to antibodies was shown to result in marked reduction or total loss of their antigen binding capacity(30). Moreover, to ensure maintenance of specific immunological suppression during prolonged regimens involving multiple injections of XIg, it may also be necessary to administer repeated injections at short intervals of the corresponding tolerogenic mPEG derivatives.

TABLE 6
NON-IMMUNOGENIC AND TOLEROGENIC EFFECTS OF MULTIPLE INJECTIONS OF HIgG(mPEG)$_{33}$[a]

	Treatment of mice		
Day -6 (i.v.)	Day 0 (i.p.)	Day 0 to day 12 (i.p.)	IgG1 titers on day 19
HIgG(mPEG)$_{33}$	PBS	HIgG(mPEG) on days 0,2,4,6,8,10,12	<100
HIgG(mPEG)$_{33}$	PBS	HIgG(mPEG) on days 0,4,8,12	<100
HIgG(mPEG)$_{33}$	PBS	HIgG(mPEG) on days 0,6,12	<100
HIgG(mPEG)$_{33}$	HIgG	HIgG(mPEG) on day 0	5,000
PBS	HIgG	PBS on days 0,2,4,6,8,10,12	90,000
PBS	PBS	PBS on days 0,2,4,6,8,10,12	<100

[a]The injections consisted of 40 μg of HIgG or of 100μg of HIgG(mPEG)$_{33}$.

r-hGH(mPEG)$_n$ conjugates

In the exploratory experiments illustrated in Figure 2 for the induction of tolerance in mice to r-hGH, which had been shown to be immunogenic in man (31), it was established that i.p. pretreatment of mice with r-hGH(mPEG)$_2$ 6 days prior to immunization with 10 μg of r-hGH (in the presence of 1 mg Al(OH)$_3$ as adjuvant) resulted in at least 90% suppression of the IgG1 antibody response to r-hGH. By contrast, pretreatment of mice with the unmodified r-hGH -- in lieu of r-hGH(mPEG)$_2$ -- enhanced the IgG1 antibody titers at all doses tested, in relation to the antibody responses of control groups of animals receiving PBS in lieu of the conjugate.

CONCLUSIONS

It has been shown in murine model systems that conjugates of antigenic proteins and optimal numbers of mPEG chains are capable of inducing specific suppression of IgE and IgG1 antibodies in mice to immunization with the

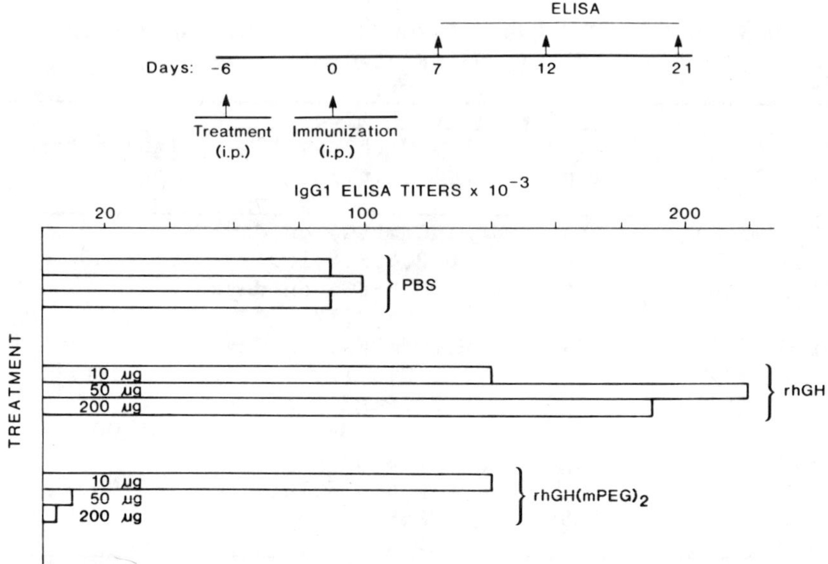

Figure 2. Suppression of the primary murine antibody response to r-hGH by treatment with r-hGH(mPEG)$_2$. Mice received PBS or various doses of r-hGH or of r-hGH(mPEG)$_2$ and were immunized 6 days later with 10μg r-hGH mixed with Al(OH)$_3$ as adjuvant. They were bled after a further 14 days for determination of the anti—r-hGH IgG1 antibody titers by ELISA. Each treatment group consisted of 3 mice. The doses indicated are expressed in terms of the protein content of both the unmodified and the conjugated r-hGH.

corresponding unmodified proteins. These results hold, therefore, the promise that tolerogenic mPEG conjugates of allergens, xenogeneic immunoglobulins and of antigenic biologically active factors may prove to be valuable therapeutic agents in man.

ACKNOWLEDGEMENTS

The myeloma sera were provided by Drs. G. Delespesse and F. Paraskevas. We acknowledge the assistance of Ms. M. Cheang for statistical analysis of the data, Mr. B. Wakefield for technical help and Ms. G. Falkenberg for her expert secretarial assistance.

REFERENCES

1. Sehon AH (1982). Suppression of IgE antibody responses with tolerogenic conjugates of allergens and haptens. Progr Allergy 32: 161.
2. Abuchowski A, van Es T, Palczuk NC, Davis FF (1977). Alteration of immunological properties of bovine serum albumin by covalent attachment of polyethylene glycol. J Biol Chem 252: 3578.
3. Abuchowski A, McCoy JR, Palczuk NC, van Es T, Davis FF (1977). Effect of covalent attachment of polyethylene glycol on immunogenicity and circulating life of bovine liver catalase. J Biol Chem 252: 3582.
4. Weigle WO (1973). Immunological unresponsiveness. Adv Immunol 16: 61.
5. Lee WY, Sehon AH (1977). Abrogation of reaginic antibodies with modified allergens. Nature 267: 618.
6. Lee WY, Sehon AH (1978). Suppression of reaginic antibodies with modified allergens. II Abrogation of reaginic antibodies with allergens conjugated to polyethylene glycol. Int Archs Allergy Appl Immunol 56: 193.
7. Lee WY, Sehon AH, Akerblom E (1981). Suppression of reaginic antibodies with modified allergens. IV Induction of suppressor T cells by conjugates of polyethylene glycol (PEG) and monomethoxyPEG with ovalbumin. Int Archs Allergy Appl Immunol 64: 100.
8. King TP, Kochoumian L, Lichtenstein LM (1977). Preparation and immunochemical properties of methoxy-polyethylene glycol-coupled and N-carboxymethylated derivatives of ragweed pollen allergen, Antigen E. Arch Biochem Biophys 178: 442.
9. King TP, Kochoumian L, Chiorazzi N (1979). Immunological properties of conjugates of ragweed pollen antigen E with methoxypolyethylene glycol or a copolymer of D-glutamic acid and D-lysine. J Exp Med 149: 424.
10. Savoca KV, Davis FF, Palczuk NC (1984). Induction of tolerance in mice by uricase and monomethoxypolyethylene glycol-modified uricase. Int Archs Allergy Appl Immunol 75: 58.
11. Kawamura K, Igarashi T, Fujii T, Kamisaki T, Wada H, Kishimoto S (1985). Immune responses to polyethylene glycol modified L-asparaginase in mice. Int Archs Allergy Appl Immunol 76: 324.
12. Dillman RO (1985). Monoclonal antibodies in the treatment of cancer. CRC Crit Rev Oncol Hematol 1: 357.

13. Ortho Multicenter Transplant Study Group (1985). A randomized clinical trial of OKT3 monoclonal antibody for acute rejection of cadaveric renal transplants. New Engl J Med 313: 337.
14. Frankel AE, Houston LL, Issell BF, Fathman G (1986). Prospects for immunotoxin therapy in cancer. Ann Rev Med 37: 125.
15. Larson SM, Brown JP, Wright PW, Carrasquillo JA, Hellstrom I, Hellstrom KE (1983). Imaging of melanoma with I-131-labelled monoclonal antibodies. J Nucl Med 24: 123.
16. Pimm MV, Perkins AC, Armitage NC, Baldwin RW (1985). The characteristics of blood-bone radiolabels and the effect of anti-mouse IgG antibodies on localization of radiolabeled monoclonal antibody in cancer patients. J Nucl Med 26: 1011.
17. Miller RA, Levy R (1981). Response of cutaneous T cell lymphoma to therapy with hybridoma monoclonal antibody. Lancet ii: 226.
18. Dillman RO, Shawler DL, Collins HA, Beauregard JC, Wormsley SB, Royston I (1982). Murine monoclonal antibody therapy in two patients with chronic lymphocytic leukemia. Blood 59: 1036.
19. Dillman RO, Beauregard JC, Shawler DL, Sobol RE, Royston I (1983). Results of early trials using murine monoclonal antibodies as anti-cancer therapy. In Peeters H (ed): "Protides of the biological fluids, Proceedings of the 30th colloquium" New York: Pergamon, p. 353.
20. Shawler DL, Bartholomew RM, Smith LM, Dillman RO (1985). Human immune response to multiple injections of murine monoclonal IgG. J Immunol 135: 1530.
21. Sears HF, Alkinson B, Mattis J, Ernst C, Herlyn D, Hayry P, Steplewski Z, Koprowski H (1982). Phase-I clinical trial of monoclonal antibody of gastrointestinal tumours. Lancet i: 762.
22. Sehon AH, Lang G, Carter BG (1983). Downregulation of IgE antibodies by suppressogenic conjugates. In Tada T and Yamamura Y (ed): "Progress in Immunology V" New York: Academic Press, p. 483.
23. Wilkinson I, Jackson C-J.C, Lang GM, Holford-Strevens V, Sehon AH (1987). Tolerogenic polyethylene glycol derivatives of xenogeneic monoclonal immunoglobulins. Immunol Letters (in press).
24. Wilkinson I, Jackson C-JC, Lang GM, Holford-Strevens V, Sehon AH (1987). Tolerance induction in mice by

conjugates of monoclonal immunoglobulins and monomethoxypolyethylene glycol. Transfer of tolerance by T cells and by T cell extracts. J Immunol (in press).
25. Sehon AH, Lang GM (1986). The use of nonionic, water soluble polymers for the synthesis of tolerogenic conjugates of antigens. In Singhal SK, Delovich TL (eds): "Mediators of Immune Regulation and Immunotherapy", New York: Elsevier, p. 190.
26. Davis FF, Abuchowski A, van Es T, Palczuk NC, Savoca K, Chen RH-L, Pyatak P (1980). Soluble, nonantigenic polyethylene glycol-bound enzymes. In Goldberg EP, Nakajima A (eds): "Biomedical Polymers. Polymeric materials and pharmaceuticals for biomedical use", New York: Academic Press, p 441.
27. Lee WY, Sehon AH (1978). Suppression of reaginic antibodies with modified allergens. I Reduction in allergenicity of protein allergens by conjugation to polyethylene glycol. Int Archs Allergy appl Immunol 56: 159.
28. Holford-Strevens V, Jackson C-JC, Charlton J, Akiyama KA, Lang GM, Carter BG, Sehon AH (1987). Induction of *in vivo* helper activity for murine antibody responses by macrophages pulsed with ovalbumin-monomethoxypolyethylene glycol (OA-mPEG) conjugates. Cell Immunol 104: 245.
29. Doyle MV, Parks DE, Weigle WO (1976). Specific suppression of the immune response by HGG tolerant spleen cells I. Parameters affecting the level of suppression. J Immunol 116: 1640.
30. Suzuki T, Kanbara N, Tomono T, Hayashi N, Shinohara I (1984). Physicochemical and biological properties of poly(ethylene glycol)-coupled immunoglobulin G. Biochim Biophys Acta 788: 248.
31. Product monograph. 1986. Genentech Inc. South San Francisco CA. USA.

The Pharmacology and Toxicology of Proteins, pages 221–233

Immunosuppression by Alpha-2 Macroglobulin: The Role of the Macrophage[1]

William J. Hubbard[2], Gregory T. Stelzer[3], John R. Yannelli and Bruce D. Gitter[2,4]

Biotherapeutics, Inc.; Franklin, TN

ABSTRACT Previous research has indicated that the proteinase inhibitor Alpha-2 macroglobulin (α2M) was immunosuppressive in its proteinase-complexed form and that its effects appeared to be exerted upon adherent cells (e.g., macrophages). This study examines the effects of α2M on peripheral blood monocytes and the monocytic cell line U-937. At the functional level it was found that α2M effectively blocked antigen presentation and promoted the synthesis of prostaglandin E_2, but that the suppression was not attributable to prostaglandin. Flow cytometric analysis with monocytes and U-937 demonstrated that α2M somehow interfered with the signal which promoted the expression of DR antigens and complement receptors, as well as blocking the expression of basal

[1]This work was supported by the Internal Research Program of Biotherapeutics, Inc.
[2]Previous address: Department of Microbiology and Immunology, Duke University Medical Center, Durham, NC
[3]Address: Center for Clinical Studies, International Clinical Laboratories, Nashville, TN
[4]Address: EIi Lilly and Co., Indianapolis, IN

levels of DR. In addition α2M caused a cell cycle arrest in U-937 at G_2/M which was distinct from that of the activating agents "differentiation"-associated arrest at G_1 induced by IFNγ. The patterns seen with cytochemical staining for U-937 indicate that the treatments did not induce markedly different patterns in expression for the enzymes surveyed. One exception was found with acid phospatase which was expressed on a fraction of the treated cells but not on controls. These findings give insight into how α2M blocks the early events in immune reactions and suggests a role in evasion of the immune response by cancer. The blockade of cell division is an additional area for consideration for α2M as a regulatory molecule.

INTRODUCTION

There has been a longstanding interest in the immuno-regulatory properties of alpha globulins (1-3). Alpha-2 Macroglobulin (α2M) is one of these proteins that has been investigated principally as an immunosuppressant (4,5) with some particular emphasis in the areas of cancer (6,7), transplantation (8,9) and antibody production (10,11). The physiological and biochemical basis for this regulation of cellular activity has been proposed to relate to the change of state undergone by α2M when it binds and inactivates a proteolytic enzyme (12); a change that confers rapid receptor engagement leading to clearance and degradation of the α2M-proteinase complexes (13-15). The receptors for α2M have not been broadly surveyed for their tissue distributions, but they are well characterized on macrophage or reticuloendothelial system elements and fibroblasts (13-16). In addition these cell types, tumor cells (17) synthesize α2M *in vitro* but do not uniformly express α2M receptors (14). It should be noted that the receptors will recognize only the altered form (electrophoretically "fast") of α2M and not the native (electrophoretically "slow") form. Only the "fast" form of α2M has been used in this study to produce the suppressive effects and in fact it is problematic to retain the slow form in culture due to the release of proteinases from the cells.

Previous *in vitro* characterization of immunosuppression by α2M has indicated that the monocyte/macrophage (Mϕ) is the cellular target, (18) e.g. its loss of "accessory cell" collaboration would preclude the sensitization and clonal expansion of otherwise appropriately stimulated lymphocytes. In this study we will expand on this observation with an examination of the changes induced in the Mϕ by α2M. In addition to peripheral blood monocytes we have utilized the promonocytic cell line U-937 (19) which can be driven to differentiate with signals from biological response modifiers such as gamma interferon (IFNγ) and phorbol esters. This obviates many of the problems associated with removal of the tightly adherent Mϕ from its plastic substrate and additionally, since the U-937 is a dividing cell, allows for examination of cell cycle parameters by flow cytometry.

MATERIALS AND METHODS

Purification of α2M.

The methods for purification of the protein and its conversion to the proteinase-complexed form have been described previously (18,20). The outdated human plasma used as the starting material was a generous gift of the Tennessee Valley Red Cross, Nashville, TN.

Cellular Methods.

Methods for monocyte purification have been described by Fischer and her associates (21). The function of antigen presentation by human Mϕ was measured using tetanus toxoid antigen according to the method of Alpert, et al. (22). The cell line U-937 was obtained from the American Type Culture Collection (Washington, DC). Activation of Mϕ or U-937 was accomplished with exposure to recombinant γ-interferon (IFNγ; Biogen, Cambridge, MA) or IFNγ plus phorbol myristate acetate (PMA, Sigma), respectively.

Prostaglandin Measurement.

Kits for the determination of prostaglandin E_2 (PGE_2) were obtained from Seragen (Boston, MA) and were used according to the manufacturer's instructions.

Flow Cytometry.

Flow cytometric analyses were performed on either an Ortho Cytofluorograph IIs or a Coulter Epics 5 instrument. Class II antigen or complement receptor (as antigen) were detected with monoclonal antibodies versus HLA-DR or CD-11 respectively (obtained from Becton Dickenson, Ortho or Coulter). The propidium iodide method was employed for analysis of the cell cycle vis-a-vis DNA content (23).

Cytochemical Staining.

Reagents for the detection of nonspecific esterase, acid phospatase, myeloperoxidase and β-glucuronidase were used according to the suppliers directions (Sigma).

RESULTS

Antigen Presentation.

The presentation of Tetanus Toxoid (TT) antigen by autologous Mϕ to T cells from sensitized donors was almost completely blocked by 1 mg/ml α2M. Taking the level of stimulation of the 18 hr TT-pulsed Mϕ (which were subsequently cultured with the T cells) as 100%

(≅30,000 cpm as ^{3}H-thymidine incorporation; pulsed for last 18 hr of 7 day culture), the α2M treatment produced only 16% incorporation of ^{3}H-thymidine which was identical to the controls using Mϕ not exposed to TT. T cell controls exposed to TT (but not α2M) ostensibly without Mϕ present, were stimulated to incorporate ^{3}H-thymidine at a 36% of control level.

Prostaglandin E_2 Synthesis.

In order to determine if PGE_2 was produced by α2M treated Mϕ, 7.5×10^5 cells were exposed to α2M and followed for PGE_2 synthesis and liberation into the culture medium (0.5 ml). Periodic sampling revealed that, at a dose of 1 mg/ml of α2M, PGE_2 synthesis proceeded in an approximately linear fashion during the 18 hr test period and reached a level of synthesis (80%) of the poly I:C control dose (100 μg/ml). Lower α2M doses evoked proportionally lower levels of PGE_2 synthesis (100 μg α2M promoted synthesis at 36% of poly I:C control; 10 μg- 14%; 1 μg-8%; 0.1 μg-no effect). The range of response to the poly I:C stimulus for normal Mϕ donors (n=6) ranged from 2-7 ng/ml PGE_2 in 18 hr,with 3 ng/ml being average. The stimulation by either α2M or poly I:C of PGE_2 synthesis was completely prevented by the presence in the cultures of 10^{-6}M indomethacin (10^{-7}M produced a 90% blockade, 10^{-8}M-60%). However, it was found that the immunosuppression of the mixed lymphocyte response by α2M was unaffected by the addition of 10^{-6}M indomethacin (data not shown). Thus it is dubious that α2M is acting to suppress the immune response solely by generating PGE_2 (and possibly other aracadonic acid pathway products that might be synthesized).

α2M and Soluble Mϕ Product Immunosuppression.

It was of interest to assess whether the cultures contained other soluble immunosuppressive products, generated as a result of exposure to α2M. Possible candidates could be degradation fragments of the α2M (and its associated proteinase) or other pharmacologically active substances released into the culture medium as promoted by α2M. For this purpose 10^6 purified monocytes were plated on 6 cm petri dishes and treated with varying doses of α2M (0.1, 0.25, 0.5, 1.0 mg/ml and none in 3 ml medium). Supernatants were harvested and centrifuged after 24 hr. These "conditioned media" were then applied to mixed lymphocyte cultures (as described in reference 18) at different proportions (ratio of α2M-conditioned to normal medium 0:100, 10:90, 25:75, 50:50, 75:25). In no instance was there any diminution (or enhancement) in the response as measured by ^{3}H-thymidine incorporation.

Flow Cytometric Analysis of Mϕ and U-937.

Monocytes ($3x10^6$ per T_{25} flask containing 3 ml medium) were stimulated for 2-3 days with $1x10^3$ U/ml IFNγ and removed from the flasks by treatment with versene and scraping. In a similar fashion, U-937 were treated with like doses of IFNγ plus 2 μg/ml PMA to enhance activation (Harold Miller, personal communication). Both the Mϕ and U-937 were stimulated to express much higher levels of DR antigen and complement receptor (CD-11). However, when treated with 1 mg/ml α2M in addition to the IFNγ and PMA signal, the increased expression of DR and CD-11 was substantially reduced in a similar fashion. For example, with a DR determination using U-937 the logarithmic scale peak mean channels were: control=438±82, IFNγ+PMA=571±78 and IFNγ+PMA+α2M=481±102. The complement receptor pattern of expression with identically-treated U-937 was virtually the same as those observed for DR expression. A similar blockade of the smaller "basal" increase in DR expression was seen in unstimulated (ie, fresh from the peripheral blood) monocytes whose DR expression increased after 3 days of culture. Treatment of Mϕ or U-937 with α2M alone was found to increase the wide angle scatter pattern, indicative of increased cellular vacuolization or granularity. Using propidium iodide, the patterns of cell cycle distribution was examined for U-937. The control cells displayed a typical DNA content of a dividing but asynchronous cell population. The IFNγ+PMA treatment gave cells that were skewed toward the G_1 distribution with fewer cells in S phase than controls; a pattern that is typical for cells that are driven toward differentiation. In this instance only a portion of the cells are seen microscopically to adhere or form clumps. The larger number of cells remaining were present as solitary suspension cells in the culture medium indistinguishable from the control cells. With the IFNγ+PMA+α2M or α2M (alone) treatment the cells were shifted markedly to the G_2/M compartment with a loss of S compartment cells. These cultures were visually similar to the activated cells but with fewer cells adhering and clumping. The Mφ exposed to α2M were seen to take on a more rounded appearance (with or without IFNγ) as compared to the "fried egg" appearance of the control and "activated" cells. There was evidence of cell lysis in cultures treated with the PMA+IFNγ and/or α2M.

Cytochemical Studies.

For this preliminary examination of the enzyme expression in U-937, substrates which identify nonspecific esterase, myeloperoxidase, acid phospatase and β-glucuronidase were selected to detect some "typical" Mφ enzymes. Of these myeloperoxidase and β-glucuronidase were undetectable and presumed not to be present while nonspecific esterase was conversely found to be expressed in all 4 treatment groups. These were 1)untreated controls, 2) α2M only (1mg/ml), 3) PMA+IFNγ and 4) PMA+IFNγ+α2M at culture initiation. The only deviation from the all or none pattern was with acid phospatase which showed a lack of expression in controls with a 6% expression (e.g., 6/100 of the cells expressed the trait) in groups 2 and 3 and 14% expression in group 4. This particular pattern of enzyme expression roughly correlates with the disparate microscopic morphological appearances for U-937 given IFNγ+PMA and/or α2M treatment as described in the preceding section. We were unable to ascertain if the adherent or clumping cells actually were the acid phospatase positive cells because the cells were transferred for "cytospin" preparation. Parenthetically, while the α2M (only) treated cultures were apparently healthy, the presence of a 1 mg/ml pulse at culture initiation resulted in a 32% reduction in cell numbers after 5 days as compared to untreated controls. This would seem to correlate with the apparent G_2/M arrest for these cells seen in flow cytometry.

DISCUSSION

The exposure of mononuclear phagocytic cells to α2M appears to induce marked physiological, morphological and biochemical changes. It would not seem that the α2M mediates these changes through a generalized "shutdown" of cellular functions, but rather by selective metabolic activation and suppression. For example, the activation of the prostaglandin synthesis and the increased light scatter in combination with the blockade of DR and CD-11 expression indicate that a possible preemption of the potent activation signals has taken place to facilitate the uptake and catabolic degradation of the α2M-proteinase complexes. It is known that receptor mediated clearance of the α2M is rapid and that the occupied receptors move to coated pits for interiorization into phagolysosomes (16). Further experimentation will be required to ascertain if there is indeed a hierarchy of signalling for the Mϕ vis-a-vis the wide variety of pharmacologic and physiological agents which can manipulate this cell type. However, given the potentially destructive nature of "free" proteinases it is not difficult to accept a (teleological) speculation that inhibition and removal of the enzyme (as enzyme-inhibitor complexes) would have a high biological priority. Likewise, the concomitant "sensing" of this processing would represent the feedback regulation which would down-regulate the release of proteinases. This is supported by the report of Johnson and his colleagues demonstrating that α2M-proteinase complexes are capable of downregulating Mϕ proteinase secretion, including a cytotoxic proteinase which is tumoricidal *in vitro* (24).

The failure to express (increased levels of) DR antigen would be the best candidate to explain immunosuppression insofar as the Mϕ's role in stimulating lymphocytes is concerned. It is obligatory that the Mϕ presents antigen in a context of class II cell-surface antigens identical to those of the recipient lymphocyte's phenotype. Even with appropriately processed antigen plus lymphokine signals, a lack of DR signal would preclude lymphocyte stimulation (25). A loss of cell motility could likewise be contributory to immunosuppression. It was found that a partial restoration of the α2M-mediated loss of mixed lymphocyte allogeneic reactivity could be accomplished through the use of V-bottom rather than flat bottom microtiter plates (26), which forces the cells into intimate contact. Finally, the failure of the Mϕ to express complement receptors further strengthens the case for α2M-induced anaergy and suggests the areas of antibody-mediated and nonspecific host defenses for investigation.

A lack of suppression mediated by soluble products in supernatants of α2M treated Mϕ was quite clear in these experiments. However, this approach might be an inadequate means to assess the existence or activity of these putative mediator substances from Mϕ catabolism or synthesis. The *in vitro* culture is a stagnant environment in terms of cell trafficking and the flow of physiological fluids. Furthermore, the presence of large numbers of highly active cells might mitigate for inactivation of the potential mediators during the long period of culture (18-24 hr). If not systemic, soluble products of synthetic cascades or degradative pathways could be of local importance. Indeed, such compounds *in vivo* may account for short (or long) acting signals for cell recruitment and changes in vascular permeability for a focus of Mϕ activity. In the case of cancer it could be conceived that the invasive process and necrotic centers of a tumor would be cause for the release of substantial quantities of proteolytic enzymes which ultimately would be found in an irreversible, highly reactive complex with α2M. This, in combination with the inability to present antigen, might explain at least in part the paradoxical situation of tumors being infiltrated by Mϕ and lymphocytes while apparently unaffected by their presence.

A systemic accumulation of α2M might indicate a generalized failure of the reticulioendithelial system since the circulating "fast" form has a half-life of roughly 5 min in the mouse and man (14,15). We have found substantial accumulations of proteinase complexed α2M in the plasma of melanoma patients (67% of greater than 400 individuals) with the presence of similar complexes being undetectable in 50 normal controls (W. J. Hubbard and H. S. Siegler, unpublished observations). A similar contribution to anaergy could be made in situations where immunologic unresponsiveness is observed in trauma or infectious disease. On the positive side, Proud and his associates have attributed enhanced allograft survival to the presence of circulating "suppressive" α2M (27).

An additional mechanism of immunosuppression for α2M has been attributed to the degradation of Interluken 2 (IL2) as observed experimentally *in vitro* with α2M-trypsin complexes (28,29). With the caveat that trypsin as a digestive enzyme is probably not representative of the proteinases that might be found in the periphery or lymph, these observations nonetheless point toward possible interference with the *in vivo* biological activity of IL2. Our finding of significant accumulations of α2M-proteinase complexes in the circulation of cancer patients indicates the need for preliminary studies to measure the effects of the α2M complexed enzyme on systemically administered therapeutic IL2.

The cell cycle arrest seen at G_2/M in the U-937 is another observation that may point to a possible regulatory role for α2M. It has been speculated that the release of proteinases by actively dividing cells would cause a feedback dampening on cell division (12). Indeed, we found that a pulse of α2M at culture initiation for U-937 would cause a reduction by ½ in cell numbers in 5 days. With the knowledge that α2M receptors are expressed on fibroblasts, one could reason that α2M might have a role in regulating cell division. As such, examining the sensitivity of tumor cell division to α2M (in comparison to normal cells) might add a new dimension the "escape" of cancer cells from physiological growth regulation. In addition, without receptor mediated uptake of α2M, a tumor cell would avoid the energy expenditure for catabolic degradation of the α2M-enzyme complexes incurred by normal cells. If α2M-proteinase complexes acted to the detriment of the immune response but did not affect the division of tumor cells, intervention to remove the complexes might help tip the balance in favor of the host.

ACKNOWLEDGEMENTS

There is great appreciation for scientific discussion and suggestions from Drs. Bernard Amos, Sal Pizzo, Barry Anderson and Gary Thurman as well as for support and review of this manuscript by Dr. Bob Oldham. Thanks also go to Patti Ladd and Keith Shults for expert technical assistance and Sherri Gossett for help in preparation of this manuscript.

REFERENCES

1. Kamrin BB (1959). Proc Soc Exp Biol Med **100**:58.

2. Mowbray J, SR Cooperbrand (1974) *in* Progress in Immunology II L Brent and J Holborow (eds) 383-6 North Holland Pub Co NY NY.

3. Cooperbrand SR, R Nimberg, K Schmid and JA Mannick (1976). Transplant Proc **8**:225.

4. Goutner S, IT Simmler and C Rosenfeld (1976). Differentiation **5**:171.

5. Streilen JS, DA Hart (1976).Fed Proc **36**:1270.

6. Ford WH, EA Caspary and B Shenton (1973). Clin Exp Immunol **15**:169.

7. von Ardenne M and RA Chaplain (1972). Experientia **29**:1271.

8. Shenton BK, G Proud, BM Smith, RMR Taylor (1979). Transplant Proc **11**:171.

9. Donnelly PK et al (1983). Br J Surgery **70**:614.

10. James K (1980). Trends in Biochem Sci **5**:43.

11. Teodrescu M et al (1984). Arth Rheum **27**:1122.

12. Hubbard WJ (1978). Cell Immunol **39**:388.

13. Ahlsson K, C-B Laurell (1976). Clin Sci Mol Med **51**:87.

14. Van Leuven FV, JJ Cassiman, H Van Den Berghe (1979). J Biol Chem **254**:5155.

15. Imber MJ, SV Pizzo (1981). J Biol Chem **256**:8134.

16. Kaplan J, FA Ray, EA Keough (1981). J Biol Chem **256**:7705.

17. Bizik J et al. (1986). Int J Cancer **37**:81.

18. Hubbard WJ, AD Hess, S Hsia, DB Amos (1981). J Immunol **126**:292.

19. Sundstrom C, K Nilsson (1976). Int J Cancer **17**:565.

20. Hubbard WJ et al. (1983). Ann NY Acad Sci **421**:332.

21. Fischer DG, WJ Hubbard, HS Koren (1981). Cell Immunol **58**:426.

22. Alpert SD, ME Jonsen, MD Broff, E Schneeberger, RS Geha (1981). J Clin Immunol **1**:21.

23. Krisan A, (1975). J Cell Biol **66**:188.

24. Johnson WJ, SV Pizzo, MJ Imber, DO Adams (1982). Science **218**:574.

25. Unanue ER (1981). Adv Immunol **31**:1.

26. Anderson BA (1983). Doctoral Thesis Duke University Medical Center.

27. Proud G, BK Shenton, BM Smith (1979). Br J Surgery **66**:678.

28. Mannhalter JW, W Borth, MA Eibel (1986). J Immunol **136**:2792.

29. Borth W, M Teodorescu (1986).Immunology **57**:367

SECTION VI: CLINICAL PHARMACOLOGY AND THERAPEUTICS

B. Issell, Cetus Corporation, initiated this session with a summary of the issues and barriers to cancer therapy with monoclonal antibodies (Mab). Assuming that adequate tumor associated epitopes for these Mabs exist, delivery of effective amounts must overcome the following potential barriers: 1) containment within the vascular compartment; 2) non-specific uptake by the reticuloendothelial system; 3) blockade by shed antigen; 4) host antibodies against the therapy; 5) removal by proximal high affinity sites and; 6) antigen heterogeniety and modulation. Additional problems are the need for high local concentrations of effector cells and complement for the cytotoxicity of the antibodies. Conjugates of immunotoxins with antibodies may alter the half-life or be stripped from the antibody. Approaches to overcome these problems include use of combination of monoclonal antibodies directed at various epitopes, antibody fragments, alteration of tissue permeability by other pharmacologic agents, coadministration of ligands to block nonspecific uptake, suppression of antigenicity of the Mabs by PEG and local delivery to lymphatics. Two papers reported uses of Mabs: M. Pollack, Uniformed Services University, Mab therapy of infectious diseases and J. Larrick, Cetus Corporation, anticomplement C5a to evaluate the role of this component in sepsis and adult respiratory distress syndrome.

E. Borden, University of Wisconsin, defined the therapeutic and immunomodulatory effects of interferons alpha, beta and gamma. Collaborative studies have identified clinical activity in multiple myeloma, hairy cell leukemia, renal carcinoma, breast cancer and malignant melanoma. Challenges include definition of the optimal dosage, identification of active analogues and determination of effectiveness of combination therapy with other biologic and pharmacoloic agents.

J. Greiner, National Cancer Institute, showed that human recombinant interferon alpha is a multigene family with marked differences in the ability of different species to augment expression of surface antigens on human breast carcinoma cells.

E. Bradley, Cetus Corportion, reviewed their phase II trials with recombinant interleukin-2 (IL-2) in patients with metastatic cancer. Clinical responses have been seen in varous malignancies with and without infusion of lymphocytes. Toxicity and clinical response are related to dose and schedule. Optimal dosages for _in vitro_ lymphokine-activated killer cell (LAK) activation correlate with _in vivo_ LAK activation and clinical response. Antibodies can be detected to IL-2 both before and after treatment but allergic reactions have not been reported. Combination therapy with interferons, monoclonal antibodies, tumor necrosis factor, tumor vaccines, surgery and conventional chemotherapy are under investigation.

E. Grossbard, Genentech Incorporated, presented the current results of clinical trials with tissue type recombinant plasminogen activator. A single and double chain preparation have different pharmacokinetic properties but similar final incidence of recanalization of coronary vessels. Since this plasminogen activator is more specific than urokinase or streptokinase, it may have much greater clinical appreciability. Plasminogen activator is one of the factors that tumor cells secrete to locally aid in their invasion. Protease inhibitors isolated from fibroblasts and vascular smooth muscle cells have the potential to block the destruction of extracellular matrix and tumor invasion. This work was reviewed by J. Baker, University of Kansas, and W. Laug, Childrens Hospital of Los Angeles.

Recombinant human growth factor was discussed by B. Sherman, Genentech Incorporated. Their product is a 191 aminoacid peptide with a N-terminal methionine. Material lacking this methionine has similar pharmacokinetics properties and biologic effects. Clinical trials have shown significant growth enhancement in children with growth hormone deficiencies. Variation in individual responses may be related to basal somatomedin C concentration, adiposity, height and to maternal height. Low titer antibodies have been found but no clinical allergic reactions. A. Tonelli, American Cyanamid Company, presented a poster on a luteinizing hormone releasing hormone encapsulated in a biodegradable polymer. Release from the microcapsule correlated with suppression of testosterone in dogs. Clinical trials have shown that patients with prostate cancer can be treated with monthly intramuscular injections.

P. Needleman, Washington University reported his discovery and characterization of a vasoactive, natriuretic-diuretic peptide from rat atria, atriopeptin. It is a 28 amino acid peptide that is stored as a 126 amino acid prohormone and released in response to stretch. Antisera, cDNA probes and synthetic peptides have been used to show its localization in the heart and brain, its receptors in the kidney and its rapid degradation in the kidneys. Therapeutic uses are being investigated.

Finally, N. Bang, Eli Lilly Company, summarized their work with human protein C from plasma and recombinant protein C from cloned human liver. When activated this 461 amino acid protein specifically inhibits activated clotting factors V and VIII. Preclinical trials show antithrombotic activity without increased blood loss during surgery.

The Pharmacology and Toxicology of Proteins, pages 239–254

THE PROSPECTS FOR MONOCLONAL ANTIBODY THERAPY IN INFECTIOUS DISEASES[1]

Matthew Pollack

Department of Medicine, Uniformed Services University School of Medicine, Bethesda, Maryland 20814

Introduction

No monoclonal antibodies have undergone clinical evaluation in human infections to date. Much is known, however, concerning the biologic properties of antibodies in general, and there is a rapidly growing body of experimental data elucidating the specificity and function of monoclonal antibodies directed against various constituents of a wide variety of human pathogens. Thus, existing information on both polyclonal and monoclonal antibodies permit relatively informed predictions regarding future clinical applications of monoclonal antibodies in infectious diseases. The purpose of this brief review is to characterize antibodies against infectious agents and to illuminate specific issues concerning the immunoprophylactic or therapeutic use of antibodies, particularly monoclonal antibodies, in infectious diseases.

Intracellular vs. Extracellular Parasites

It is useful to think of pathogens as belonging to one of two categories -- obligate intracellular parasites or extracellular parasites. In the former case, bacteria like *Mycobacterium tuberculosis* and *Listeria monocytogenes*, or viruses like *herpes simplex* and HIV-III have evolved mechanisms for multiplying within cells of the infected host. In the latter case, pathogens like encapsulated and/or Gram-negative bacteria, as well as many multicellular parasites, live and proliferate outside of cells. The consequences of phagocytosis are different

[1]This work was supported by grant AI-22706 from the National Institute of Allergy and Infectious Diseases.

for intracellular and extracellular parasites. For the intracellular parasite, phagocytosis represents an integral part of the life-cycle of the organism and is actually required for the establishment or maintenance of an infection. But for the extracellular parasite, ill-equipped for intracellular survival, phagocytosis by host white blood cells usually results in the organism's demise (1).

Intracellular parasites, sequestered within infected cells, are relatively inaccessible to antibodies. Extracellular parasites, on the other hand, have evolved sophisticated antiphagocytic mechanisms to sustain their survival outside of cells. For most extracellular bacteria, these mechanisms involve capsules or other surface molecules like lipopolysaccharide (LPS), which protect the bacteria from the bactericidal and opsonic activities of complement. Antibodies to these antiphagocytic moieties overcome this barrier effect by fixing complement on the microbial surface. Attached complement is then recognized by C3 receptors on host phagocytic cells, leading to attachment, ingestion, and intracellular killing of the organism.

The distinction between obligate intracellular and extracellular parasites, and the very different role played by antibodies in host defenses against these two groups of pathogens, holds important implications for passive immunoprophylaxis or therapy, whether accomplished with polyclonal or monoclonal antibodies. On the other hand, this sharp dichotomy is somewhat overdrawn. Viruses, for example, which are primarily intracellular parasites, may manifest an extracellular phase during which they are susceptible to neutralizing antibodies. In addition, it is obvious that antibodies operate in conjunction with cellular immune mechanisms, which are particularly relevant in host defense against obligate intracellular parasites. Some antibodies, for example, drive the process of antibody-dependent cell-mediated cytotoxicity, which results in the lysis of virus-infected cells. In addition, cytophilic antibodies serve a critical function in the communication which occurs between monocytes and lymphocytes in response to specific antigenic stimuli.

Active vs. Passive Immunization

The distinctions between active and passive immunization are also important in evaluating the possible practical applications of antibody-mediated immunity. Active immunization has the potential advantage of recruiting diverse cellular and humoral elements of the vaccinee's immune system. In addition, active immunization may induce a polyclonal response (2) that embraces antibodies of diverse specificity, isotype, and anatomical distribution -- factors that may enhance protective immunity. Active immunization, on the other hand, must be accomplished in a well-defined high risk group prior to an infectious exposure in order to allow adequate time for a protective antibody response. This might be impractical if the attack rate of the infection is low and the susceptible population large and/or poorly defined. Moreover, effective immunization requires an immunogenic and safe vaccine. These requirements may be difficult to satisfy, particularly if susceptible individuals are at the extremes of age or are immunocompromised (3).

Prophylaxis vs. Treatment

Another important issue relating to immunization against infection involves the choice between prophylaxis and therapy. Active immunization necessarily involves prophylaxis rather than treatment, even if this represents post-exposure prophylaxis (e.g., immunization against rabies following the bite of a rabid animal). Passive immunization, on the other hand, may involve either prophylaxis or treatment. Patients with congenital or acquired hypogammaglobulinemia, for example, receive regular immunoglobulin injections to prevent bacterial and viral infections. Specific antibody (serum) therapy, on the other hand, has been employed in the treatment of already established infections such as pneumococcal pneumonia, bacterial meningitis, *Pseudomonas* burn wound sepsis, and hemorrhagic fevers. Many currently practiced examples of passive prophylaxis actually involve the post-exposure period (e.g., hepatitis B immune globulin in neonates born of HbSAg-positive mothers, zoster immune globulin in immunosuppressed children exposed to chicken pox, CMV hyperimmune intravenous immune globulin (IGIV) in bone marrow transplant patients). Effective passive prophylaxis imposes a number of stringent requirements. As in active immunization, a high risk group must be identified prospectively and immunized expectantly. Unlike active immunization, however, passive prophylaxis must be carried out in the immediate pre- or post-exposure periods in order to insure adequate specific antibody levels in the face of a relatively short immunoglobulin half-life. Monoclonal antibodies may accentuate certain practical limitations of passive prophylaxis because of their exquisite specificity (i.e., narrow spectrum) and, particularly in the case of murine antibodies and IgM, short half-life (4).

Passive therapy circumvents some of the limitations of passive prophylaxis. Therapy focuses on established disease, thus obviating the need for identifying and administering antibody to an entire group at risk in order to prevent a relatively small number of cases. Moreover, treatment, while possibly more intense than prophylaxis, is likely to be of shorter duration. Therapy thus represents a more efficient approach to antibody utilization. A further argument in favor of the more selective use and shorter duration of antibody therapy, in contrast to prophylaxis, is based on the likelihood that the repeated use of monoclonal antibodies derived from non-human germ lines will induce anti-immunoglobulin antibodies which limit antibody efficacy (5).

Notwithstanding the above, antibody prophylaxis does hold certain potential advantages over therapy. For example, certain antibodies which recognize microbial structures involved in adherence to epithelial surfaces, may be effective in preventing colonization but ineffective in treating an established infection. Other antibodies may promote microbial opsonophagocytosis or lysis, but be required in much larger amounts to treat an infection, once established, than to prevent disease at an earlier stage. Moreover, the pathogenesis of systemic infections involves the complex interplay of microbial and host factors. Once underway, certain elements of the resulting disease process, particularly those involving host responses to the infectious agent, may be irreversible, even in the presence of antibodies directed against bacterial virulence factors critical to the initiation of that process. Once again, immunoprophylaxis may be more effective than later therapeutic intervention.

Polyclonal vs. Monoclonal Antibodies

The practicality of passive prophylaxis and therapy has been facilitated by the development of human immunoglobulin for intravenous use (IGIV) (6). Most commercial IGIV's presently available contain primarily monomeric IgG in a non-reactogenic form which permits relatively safe intravenous injection of doses as high as 500 mg/kg body weight on several successive days. The greater ease of injection, more favorable pharmacokinetics, and possibility of administering large doses of IGIV, compared with previously available ISG for intramuscular use, serve as a basis for reevaluating many potential therapeutic uses of antibodies.

In an historical context, the sometimes successful treatment, during the pre-antibiotic era, of life threatening infections such as pneumococcal pneumonia with hyperimmune serum still stands as a relevant model for the therapeutic use of immunoglobulins in infectious diseases. A critical practical question, in the modern era, is whether to approach antibody-mediated prophylaxis or therapy by means of polyclonal human antibodies contained in IGIV or monoclonal antibodies generated in human, non-human, or mixed systems. The implementation of IGIV therapy in infectious diseases may be more rapidly achievable than monoclonal antibody therapy since IGIV is already in clinical use. Standard IGIV's normally contain antibodies directed against a variety of common human pathogens based on the previous antigenic experience of plasma donors (7). Moreover, specific antibody levels present in IGIV can be augmented through donor immunization or screening (8). In contrast, the development of even single monoclonal antibodies may require time-consuming and expensive research aimed at solving practical problems associated with hybridoma stability, large-scale production, and safety, as well as answering questions related to antibody efficacy.

IGIV's obviously present a significant advantage over monoclonal antibodies in so far as they provide antibodies with a broad range of specificities, avidities, and isotypes. The natural variation in Fab and Fc structure, which characterizes polyclonal antibodies, may be exploited by taking advantage of opportunities for antibody cooperation or synergy. Moreover, the presence of different immunoglobulin isotypes, and whole classes in the case of at least one experimental IGIV product which contains IgM and IgA as well as IgG, provides IGIV's with the potential for a wider variety of antibody effector functions than those residing in a single monoclonal antibody.

The clinical use of IGIV may be viewed as embracing somewhat of a shotgun approach to immunoprophylaxis or therapy, both in terms of insuring the best antibody or antibodies against a specific pathogen or broadening "coverage" to include multiple pathogens. Monoclonal antibodies, on the other hand, represent a more focused approach in which particular antibodies or combinations of antibodies are selected to provide optimal specificity (or cooperativity), avidity, and functional activity. This more focused approach demands rather detailed information on the role of specific virulence factors in the pathogenesis of particular infections, and on the functional immunochemistry of these factors. Gathering this kind of information is a time-consuming process. The payoff in the long run, however, may represent more effective clinical reagents whose therapeutic potency is better standardized and more reproducible compared with polyclonal immunoglobulins obtained from pooled human plasma.

Antimicrobial Activities of Antibodies

Immunoglobulins are bifunctional molecules with antigenic recognition and effector functions residing in two distinct structural domains commonly referred to as the Fab and Fc regions, respectively. Antibodies express many of their anti-infective properties through interactions with other elements of the humoral and cellular immune systems, the net effect of which is to mobilize these elements in a very specific manner against invading pathogens. Described below are the major mechanisms by which antibodies mediate protection against infection.

Cell Lysis

Deposition of complement on cell surfaces via either the classical or alternative pathways may result in activation of the terminal complement sequence consisting of C5-C9, which together comprise the membrane attack complex (MAC). If certain physicochemical and spatial conditions are satisfied, MAC's create pores in the plasma membrane leading to cell lysis. Complement may be fixed directly by certain molecular species found on the cell surface of some pathogens. Examples are the lipopolysaccharides (LPS) of Gram-negative bacteria and the lipoteichoic acid of *Staphylococcus aureus*. Complement fixation per se does not result in cell lysis, however. In the case of Gram-negative bacteria, the number and length of O-side chains covalently linked to the LPS core structure correlate with bacterial resistance to complement-mediated cell lysis (serum resistance) (9).

In the case of rough mutant Gram-negative bacteria, which lack O-side chains and varying portions of the core oligosaccharide, complement fixation results in cell lysis (serum sensitivity).

An alternative mechanism for complement-mediated bacterial or viral cell lysis is provided by antibodies which bind to cell surface antigens via the Fab region and fix complement through the Fc portion of the antibody molecule. This antibody function is restricted to particular IgG subclasses and IgM. The resulting antibody-mediated cytolysis plays a critical protective role, for example, in host defense against infections caused by *Neisseria species* and certain other Gram-negative bacteria. A similar mechanism is involved in the killing of some viruses.

Opsonization

Opsonization describes a process in which organisms or other particles are coated by antibodies and/or complement (or other factors) in a manner that prepares the organisms for recognition and ingestion by host phagocytes (1). Antibodies and complement serve as ligands in this process. When antibody is involved alone, it binds via the Fab region to specific antigenic determinants on the surface of an organism or other particle; the Fc moiety, in turn, binds to Fc receptors on phagocytic cell membranes. Only antibodies of certain IgG subclasses are capable of this direct linking of antigens with immunoglobulin receptors on white blood cells, leading to opsonophagocytosis (4). In the absence of antibody, complement may bind directly to the surface of an organism, as described in the preceding section, via the classical or alternative pathways. C3 is deposited on the cell surface in either case and is available to C3 receptors on phagocytic cells. The resulting attachment of organisms to phagocytes is usually followed by ingestion, and in some instances, by intracellular killing. In the presence of both antibody and complement, the Fab portion of the antibody molecule binds an antigenic determinant on the foreign particle, the Fc region fixes complement (as described above in reference to direct bactericidal reactions), and the attached C3 component binds to C3 receptors on phagocytic cells. This latter mechanism, which can involve antibodies of the IgG or IgM classes, and either the classical or alternative complement pathways, is probably the most active of the three opsonic mechanisms described (1). It is discussed in more detail below.

As previously indicated, extracellular parasites, including many bacterial pathogens, have evolved mechanisms for resisting opsonization and phagocytosis by humoral and cellular components, respectively, of the infected host's immune system. Principal among these mechanisms are cell surface structures with antiphagocytic properties. Examples are the exopolysaccharides or capsular polysaccharides of many bacteria, the LPS of Gram-negative bacteria, the M protein of *Streptococcus pyogenes* (10), and the protein A of *Staphylococcus aureus*. The virulence of many Gram-negative and Gram-positive bacteria (e.g., *Streptococcus pneumoniae*, *S. aureus*, *Escherichia coli*, *Hemophilus influenzae*, *Klebsiella pneumoniae*, *Salmonella typhi*, and *Neisseria meningitidis*) is largely attributable to the antiphagocytic properties of their polysaccharide capsules. The mechanism of this antiphagocytic effect is based on the resistance of many capsular polysaccharides to direct complement fixation, as well as the covering up of subcapsular surface structures of the organism which would otherwise constitute favorable targets for complement or antibody attachment. While unencapsulated bacteria can be opsonized by complement alone, encapsulated bacteria require specific anticapsular antibody to fix complement on the bacterial surface. When ligand, principally C3, is evenly distributed on the surface of the organism, whether fixed directly or through the interposition of complement-fixing antibody, the organism is a prime target for phagocytosis by C3 receptor-bearing polymorphonuclear leukocytes or monocytes (macrophages).

The direct implication of the above discussion is that anticapsular antibodies, on the basis of their opsonic activity, serve an important protective function in infections caused by encapsulated bacteria. This has been amply documented in a number of bacterial diseases (e.g., pneumococcal, meningococcal, and *H. influenzae* infections), but remains to be conclusively proven in others (e.g., typhoid fever) (11).

Inhibition of Microbial Adherence

Another mechanism by which antibodies may prevent disease, but to a lesser extent treat disease once established, is by blocking attachment of pathogens to epithelial surfaces. Specialized structures or organelles which mediate adherence of pathogens to mucosal surfaces have been described for a variety of organisms. Examples are the pili (fimbriae) of pathogenic gonococci (12) and *Pseudomonas* (13), "intestinal colonization factor" found in certain bacteria causing diarrhea, and M protein of *S. pyogenes* (10). Antibodies specific for some of these adherence factors appear to block bacterial adherence in *in vitro* assays and prevent colonization *in vivo*; in some instances these effects correlate with disease prevention in direct challenge experiments or clinical trials. Antibodies of the IgA class assume a central role in mucosal immunity, and appear to act, in some instances, by inhibiting microbial adherence. Locally-induced mucosal immunity, based largely on the generation of secretory IgA, may be more effective than systemically-induced or passively transferred immunity in preventing mucosal colonization and subsequent infection.

Toxin Neutralization

Specific protein toxins have been clearly implicated as major pathogenic factors in a number of bacterial diseases. Classic examples are diphtheria toxin, tetanus toxin, clostridial toxins, and cholera toxin. In all of these cases, the infections which give rise to the toxins are relatively superficial, confined mainly to mucous membranes, while the toxins themselves are largely responsible for the distinctive disease processes associated with the infections. Toxin-neutralizing antibodies can, in many instances, prevent these toxin-mediated diseases. The molecular biology and immunochemistry of individual bacterial toxins determine whether particular antibodies will mediate *in vitro* and *in vivo* toxin-neutralizing activity. Most exotoxins contain enzymatically active (A) and binding (B) domains; the A region mediates the actual toxic action of the molecule and the B region is responsible for binding to receptors on target cells. Antibodies that recognize epitopes in the A domain may neutralize a toxin's enzymatic activity in cell-free systems, while manifesting little biological activity in tissue culture or *in vivo* (14). Some evidence suggests that this results from the inability of such antibodies to remain attached to their binding site on the toxin's A subunit (or fragment) as it is internalized by the cell (14,15). In contrast, antibodies directed to antigenic sites in the B or binding domain may prevent attachment of holotoxin to cell receptors, and this effectively blocks *in vitro* and *in vivo* toxicity (14,15). Unlike opsonic antibodies, toxin-neutralizing antibodies do not require complement for their action. In fact, the effector portion of the antibody molecule (located in the constant region) may be enzymatically removed, and, in some cases, the resulting Fab or $(Fab)_2$ fragments retain their toxin-neutralizing capacity.

The therapeutic potential of toxin-neutralizing antibodies has not been fully evaluated in many infectious diseases because the identity and/or pathogenic role of candidate toxins has not been established. This is particularly true in the case of invasive infections whose pathogenesis is clearly complex and may involve extracellular toxins as well as other virulence factors (16).

Viral Neutralization

Antibodies to viral surface antigens, most commonly glycoproteins, can neutralize the infectivity of virions by mechanisms which are as yet incompletely understood (17). Bound antibody likely blocks attachment of viruses to host cells, but may also inhibit virus engulfment and uncoating. Antibodies to internal viral structures or enzymes usually do not have neutralizing activity. Moreover, viral neutralization is not synonymous with inactivation, as illustrated by the fact that fully infectious virus can be recovered from virus-antibody complexes following their dissociation. In addition, some types of virus-antibody complexes are infectious; the antibodies in this case are said to be "non-neutralizing". Intracellular viruses are protected from antibodies in serum and other extracellular fluids. Some viruses, however, such as polio and influenza viruses, are released from cells in which they multiply. During this cell-free phase, the viruses are exposed to neutralizing antibodies in the extracellular millieu, and this prevents the virions from infecting other cells by blocking viral attachment or subsequent steps in the internalization process. Other viruses, like vaccinia and herpes simplex, pass directly between contiguous cells, without an extracellular phase, and thus avoid the neutralizing effects of virus-specific antibodies. Cellular immune mechanisms may be of greater relative importance in such cases.

Antibody-Dependent Cell-Mediated Cytotoxicity (ADCC)

An alternative protective mechanism mediated by anti-viral antibodies produces lysis of virus-infected cells (18,19). Antibodies recognize viral antigens displayed on the surface of infected cells. After binding to these antigens by the Fab region, the antibodies fix complement (usually via the alternative pathway) by means of the Fc domain. In some cases, this leads directly to formation of membrane attack complexes which produce cell lysis. More often, C3 deposited on the Fc portion of the antibody molecule attaches to C3 receptors on killer lymphocytes (K cells). This permits binding of the effector cells to virus-infected target cells, which are subsequently lysed by a process that likely involves discharge by the K cells of cytotoxic cytoplasmic granules (19). This mechanism of antibody-dependent cell-mediated cytotoxicity (ADCC) appears to be operative in a wide variety of viral diseases caused by both DNA and RNA viruses. Although it is apparent that virus-specific antibodies drive both complement- and killer lymphocyte-mediated lysis of virus-infected cells, the relative therapeutic significance of these two antibody activities, and of viral neutralization, is unclear.

Antibody Class, Isotype, and Species of Origin

As investigators set out to develop monoclonal antibodies for therapeutic applications in infectious disease, their primary focus is, understandably, on antibody specificity in relation to antibody function. Yet immunoglobulin class, subclass, and species of origin are critical determinants of antibody function, pharmacology, and potential toxicity.

It is clear that different kinds of antigenic stimulation or types of infection induce different classes and subclasses of antibody. Viral infections, for example, induce brisk IgG1 responses in humans (20) and IgG2a in mice (21). Encapsulated bacteria elicit human IgM and IgG2, as compared with murine IgM and IgG3 (23). Mucosal infections result in the production of secretory IgA, while parasitic infections elicit IgE responses. Recent research is revealing additional, and more specific, examples of the isotype restriction of antibody responses to diverse antigenic stimuli and the critical role of lymphokines secreted by particular lymphocyte subsets in promoting the production of antibodies representing specific isotypes (21).

Effector functions of antibodies vary with class and isotype (4,24). For example, human IgG1, IgG3, and IgM fix complement by the classical and alternative pathways and represent efficient opsonins. IgA, on the other hand, fixes complement only by the alternative pathway and is not opsonic. Moreover, this immunoglobulin class may play a unique role in blocking bacterial adherence to mucosal surfaces (25). Polymorphonuclear leukocytes and monocytes have Fc receptors for IgG1 and IgG3 but not for IgM, rendering the opsonic activity of this latter class of antibodies absolutely complement-dependent. IgG1 represents a major component of the human antibody response to protein antigens in general, and is the most common isotype identified with bacterial toxin-neutralizing activity (4,22).

It may be inferred from the above discussion that there is at least some relationship between the structure and specificity of the variable region of antibodies, on the one hand, and the structure and function of the constant region, on the other. Teleologically, it makes sense to assume that antibody responses to particular infectious agents have evolved in ways which assure optimal antibody specificity (variable region structure) and function (constant region structure), and that the two are somehow linked in genetic terms.

Pharmacology

If antibodies are to be used as therapeutic agents in man, much has to be known about their clinical pharmacology (4). Class, subclass, and species of origin are clearly major determinants of many of the pharmacologic properties of antibodies. IgM, for example, in its pentameric form the largest of the immunoglobulin classes in terms of molecular weight, is found primarily in the intravascular space. IgG, on the other hand, is more widely distributed in extravascular and intravascular spaces. IgM crosses neither the placenta nor the blood-brain barrier; IgG crosses both. Secretory IgA is found predominantly at mucosal surfaces and IgG levels increase at these locations in the face of inflammation, while IgM levels remain low.

The rates of synthesis and metabolism of different immunoglobulin classes and subclasses vary markedly, resulting in very different serum half-lives (e.g., T1/2 of 23 days for human IgG1, 16 days for IgG3, 6 days for IgA, and 5 days for IgM) (4). Murine immunoglobulins have much shorter half-lives, ranging from five days for IgG2a to one day for IgA and IgM (4). The serum half-lives of mouse antibodies administered to man are still shorter. Thus, class-, subclass-, and species-related differences in the clinical pharmacology of immunoglobulins will clearly effect the bioavailability of therapeutically administered antibodies at various sites of infection.

Potential Harmful Effects

Antibody therapy has the potential for harmful as well as beneficial effects. Heterologous antibodies in particular may induce allergic reactions which limit both their efficacy and their long term use (5). Immune complex disease is another potential problem, particularly in infections caused by encapsulated bacteria which release large amounts of capsular polysaccharide into the circulation. If anticapsular polysaccharide antibodies are introduced into this setting of potential antigen excess, precipitation of insoluble immune complexes is a realistic possibility. In fact, immune complex disease is sometimes observed to occur spontaneously in infections caused by encapsulated bacteria in the face of a vigorous antibody response by the infected patient.

So-called blocking antibodies are another source of potential concern in the context of antibody therapy. Antibodies with poor complement-fixing, opsonic, or other effector functions may preempt critical antigenic sites on a pathogen, and thus prevent other more effective antibodies from initiating a functionally useful interaction (26). Finally, there is the theoretical possibility of passively administered antibodies impeding the development of active immunity to a particular organism by occupying critical antigenic sites and thus preventing their recognition by the infected host's immune system.

Conclusion

The development of monoclonal antibodies for therapeutic applications in infectious diseases accentuates the critical relationship between antibody specificity and function. It also highlights the potentially profound impact of immunoglobulin class, subclass, and species of origin on antibody effector function and clinical pharmacology. In addition, it underscores the potential for antibody therapy to produce harm as well as good. The complexities of antibody structure and function, combined with our very rudimentary knowledge regarding the immunopathogenesis of many infectious diseases, make the development of therapeutically useful monoclonal antibodies a significant challenge in many ways (27). In response to this challenge, investigators will certainly pursue strategies which exploit innovative technologies associated with isotype switching, somatic mutation, and recombinant methodology, to custom-design optimal monoclonal antibodies for the treatment of infectious diseases.

References

1. Horwitz MA (1982). Phagocytosis of microorganisms. Rev Infect Dis 4:104.
2. Pier GB, Elcock ME (1984). Nonspecific immunoglobulin synthesis and elevated IgG levels in rabbits immunized with mucoid expolysaccharide from cystic fibrosis isolates of *Pseudomonas aeruginosa*. J Immunol 133:734.
3. Young LS, Meyer RD, Armstrong D (1973). *Pseudomonas aeruginosa* vaccine in cancer patients. Ann Intern Med 79:518.
4. Spiegelberg HL (1974). Biological activities of immunoglublins of different classes and subclasses. Adv Immunol 19:259.
5. Miller RA, Oseroff AR, Stratte PT, Levy R (1983). Monoclonal antibody therapeutic trials in seven patients with T-cell lymphoma. Blood 62:989.
6. Morell A, Nydegger UE (1986). "Clinical Use of Intravenous Immunoglobulins." London: Academic Press.
7. Pollack M (1983). Antibody activity against *Pseudomonas aeruginosa* in immune globulins prepared for intravenous use in humans. J Infect Dis 147:1090.
8. Collins MS, Roby RE (1984). Protective activity of intravenous immune globulin (human) enriched in antibody against lipopolysaccharide antigens of *Pseudomonas aeruginosa*. Am J Med 76(3A):168.
9. Joiner KA, Hammer, CH, Brown EJ, Cole RJ, Frank MM (1982). Studies on the mechanism of bacterial resistance to complement-mediated killing. I. Terminal complement components are deposited and released from *Salmonella minnesota* S218 without causing bacterial death. J Exp Med 155:797.
10. Fischetti VA, Gotschlich EC, Siviglia G, Zabriskie JB (1977). Streptococcal M protein: an antiphagocytic molecule assembled on the cell wall. J Infect Dis 136(Suppl):S222.
11. Robbins JB (1978). Vaccines for the prevention of

encapsulated bacterial diseases: current status, problems and prospects for the future. Immunochemistry 15:839.

12. Densen P, Mandell GL (1978). Gonococcal interactions with polymorphonuclear neutrophils. Importance of the phagosome for bactericidal activity. J Clin Invest 62:1161.
13. Woods DE, Straus DC, Johanson WG (1980). Role of pili in adherence of Pseudomonas aeruginosa to mammalian buccal epithelial cells. Infect Immun 29:1146.
14. Chia JK, Pollack M (1986). Functionally distinct monoclonal antibodies reactive with enzymatically active and binding domains of Pseudomonas aeruginosa toxin A. Infect Immun 52:756.
15. Pappenheimer AM, Uchida T, Harper AA (1972). An immunological study of the diphtheria toxin molecule. Immunochemistry 9:897.
16. Pollack M, Young LS (1979). Protective activity of antibodies to exotoxin A and lipopolysaccharide at the onset of Pseudomonas aeruginosa septicemia in man. J Clin Invest 63:276.
17. Beebe DP, Schrieber RD, Cooper NR (1983). Neutralization of influenza virus by normal human sera: mechanisms involving antibody and complement. J Immunol 130:1317.
18. Sissons JGP, Oldstone MBA (1980). Killing of virus-infected cells: the role of anti-viral antibody and complement in limiting virus infection. J Infect Dis 142:442.
19. Sissons JGP, Oldstone MBA (1980). Killing of virus-infected cells by cytotoxic lymphocytes. J Infect Dis 142:114.
20. Skvaril F (1986). Clinical relevance of IgG subclasses. In Morell A, Nydegger UE (eds): "Clinical Use of Intravenous Immunoglobulins." London: Academic Press, p 37.
21. Coutelier J-P, Vander Logt JTM, Heessen FWA, Warnier G, Van Snick J (1987). IgG_{2a} restriction of murine antibodies elicited by viral infections. J Exp Med 165:64.
22. Siber GR, Ambrosino DM (1986). Heavy and light chain restriction of human antibodies to bacterial polysaccharide antigens. In Morell A, Nydegger UE (eds): "Clinical Use of Intravenous Immunoglobulins." London: Academic Press, p 47.

23. Moreno C, Esdaile J (1983). Immunoglobulin isotype in the murine response to polysaccharide antigens. Eur J Immunol 13:262.
24. Yount WJ, Dorner MM, Kunkel HG, Kabat EA (1968). Studies on human antibodies VI. Selective variations in subgroup composition and genetic markers. J Exp Med 127:633.
25. Williams RC, Gibbons RJ (1972). Inhibition of bacterial adherence by secretory immunoglobulin A: mechanism of antigen exposure. Science 177:697.
26. Griffiss JM (1975). Bactericidal activity of meningococcal antisera: blocking by IgA of lytic antibody in human convalescent sera. J Immunol 114:1779.
27. Pollack M, Raubitschek AA, Larrick JW (1987). Human monoclonal antibodies that recognize conserved epitopes in the core-lipid A region of lipopolysaccharides. J Clin Invest, in press.

The Pharmacology and Toxicology of Proteins, pages 255–272

MURINE MONOCLONAL ANTIBODIES NEUTRALIZING THE EFFECTS OF PORCINE C5a

Tracy Deinhart[1], Jeff Wang[1], Karen Toy[1], Akitoshi Ishizaka[2], Kenton E. Stephens[2], Elaine W. Hall[2], Thomas A. Raffin[2], Dennis E. Chenoweth[3], Steven L. Kunkel[4], and James W. Larrick[1]

[1]Cetus Corporation Department of Immunology, 3400 W. Bayshore Rd., Palo Alto, CA 94303

[2]Division of Respiratory Medicine, Stanford University, Stanford, CA 94305

[3]RLT-02, P.O. Box 490, Route 120 and Wilson Road, Round Lake, IL 60073

[4]Department of Pathology, University of Michigan, School of Medicine, Ann Arbor, MI 48109

ABSTRACT We have generated a panel of four murine monoclonal antibodies (MAbs) recognizing porcine complement fragment C5a. These MAbs were screened for their ability to immunoprecipitate ^{125}I-labeled C5a and to bind C5a in solid phase enzyme immunoassay. These monoclonals have Ka's for C5a in the 5.2×10^8 to 4.3×10^9 M^{-1} range. All of the monoclonals block C5a-induced neutrophil chemotaxis, chemiluminescence, and polarization. They block the ability of passively administered porcine C5a to cause neutropenia in rabbits. These anti-porcine C5a neutralizing monoclonals will have potential uses in therapeutic models of complement activation.

INTRODUCTION

Human C5a is a 74-amino acid glycoprotein with a molecular weight of 11,300, which is generated as a cleavage product of complement protein C5 in both the classical and alternative pathways of complement activation (10,18). C5a possesses anaphylotoxic, chemotactic, contractile, and permeability-enhancing activities. A specific C5a receptor has been identified on human neutrophils, and a chemotactic response of these granulocytes to purified human C5a has been observed at concentrations of 0.4 to 17 nm (2,3,5,34). Binding of C5a to this receptor is thought to be the causal event mediating complement-induced granulocyte aggregation (15) and superoxide production (7,8). Such aggregation in vivo has been postulated as a mechanism of tissue damage in such clinical conditions as pulmonary dysfunction and leukostasis in hemodialysis patients, sudden blindness with retinal infarction after trauma or pancreatitis, myocardial infarction, post-pump syndrome after cardiopulmonary bypass surgery, systemic lupus erythematosis, burn injury, lung injury, and the adult respiratory distress syndrome (ARDS)[1] (4,28,30). In vivo studies of mice which are genetically deficient in C5 have established that C5a is required for pulmonary edema formation in response to pneumococcal sepsis, scald wounds, and hypoxia (12,17,27). Detection of elevated levels of C5a has been used to predict the onset of ARDS in patients at high risk who have been followed prospectively (16). Recent work from Stevens et al. (31), suggests that rabbit anti-human C5a antibodies can attenuate ARDS and sepsis in nonhuman primates.

We recently generated a panel of murine anti-human C5a MAbs to investigate the role of C5a in clinical syndromes of complement activation (20). Because these antibodies demonstrated little cross-reactivity with C5a

[1] Abbreviations used: ARDS, adult respiratory distress syndrome; BPBS, 1% bovine serum albumin in PBS; HBSS, Hanks balanced salt solution; i.p., intraperitoneally; i.v., intravenously; MAbs, monoclonal antibodies; PBS, phosphate-buffered saline; PBST, phosphate-buffered saline + 0.1% Tween 20; PCMA, PBS + Ca^{++}, + Mg^{++} + albumin; PMN, polymorphonuclear leukocyte; SDS, sodium dodecyl sulfate; ZPS, zymosan-activated pig serum.

derived from animals commonly used for preclinical studies (e.g., pig, guinea pig, dog, sheep), we have generated a series of specific anti-pig C5a monoclonals. These antibodies bind to porcine des Arg C5a with high affinity, immunoprecipitate ^{125}I-labeled porcine C5a, block porcine C5a-induced neutrophil polarization, chemotaxis, chemiluminescence, and protect rabbits from neutropenia induced by systemic infusion of porcine C5a. The potential therapeutic use of these reagents in animal models of complement activation is discussed.

METHODS

Monoclonal Antibody Production.

BALB/c mice were immunized intraperitoneally (i.p.) with 20 μg of purified porcine C5a in complete Freund's adjuvant. The porcine C5a antigen was affinity-purified as previously described (21). Mice were boosted with immunogen in incomplete Freund adjuvant then in PBS at 2-week intervals. Mice were boosted intravenously (i.v.) with 1 μg of C5a 3 days prior to fusion. Monoclonals were generated by standard methods (26). Spleen cells were fused to mouse myeloma parent SP2/0Ag14 (29) at a ratio of 5:1, and hybridomas were selected in hypoxanthine/azaserine selection media (11). Positive wells were cloned by limiting dilution.

C5a Enzyme-linked Immunosorbent Assay (ELISA).

Immulon 2 96-well flat bottom plates were coated with 100 μℓ porcine C5a (2.5 μg/mℓ) in PBS for 1 h at 37°C. Antigen-coated plates were washed three times with wash buffer PBST. Test supernatant (50 μℓ) or purified antibody was added to each well and incubated for 30 min at room temperature. Wells were washed three times as above and then 50 μℓ of peroxidase-conjugated rabbit antimouse immunoglobulin G (IgG) or IgM (Zymed, South San Francisco, Calif.) diluted 1:1000 in dilution buffer was added to each well. Wells were incubated 60 min at room temperature and washed three times. One hundred microliters of the peroxidase substrate, o-phenylenediamine (Sigma, St. Louis, MO) 3 mg/mℓ in 0.1 M sodium citrate buffer containing 0.015% H_2O_2 was then added to wells.

The reaction was stopped with an addition of 100 μℓ of 1N HCl after 5 min. Plates were read on a multiscan plate reader at 492 nm (Flow Laboratories, Inc., McLean, Va).

Multiwell Immunoprecipitation.

Wells of a 96-well polyvinylchloride microtiter plate (Dynatech Laboratories, Inc., Alexandria, Va) were blocked with 1% BSA in phosphate-buffered saline (BPBS), pH 7.2, for 60 min at 37°C. The plates were then flicked, and 50 μℓ of MAb (supernatant or purified) was added to the wells. A 100,000-cpm sample of ^{125}I-labeled C5a (50 μℓ) was then added to each well, and the mixture was incubated for 60 min at room temperature. After this incubation, 50 μℓ of a 1:5 dilution of rabbit antimouse immunoglobulin immunobeads (Bio-Rad Laboratories, Richmond, Calif.) was added to each well and incubated for 60 min at room temperature with agitation. The immunobeads were then washed three times with PBS + 0.1% Tween 20 (PBST), and the plates were cut. The individual wells were counted on a gamma counter (LKB, Wallac, Finland).

After counting, the immunobeads were resuspended in 30 μℓ of sample buffer (0.06 M Tris, pH 7.0, 5%-2-mercaptoethanol, 2% sodium dodecyl sulfate [SDS], 1.5% glycerol). This suspension was boiled for 3 min, and then the supernatant was analyzed by SDS + polyacrylamide gel electrophoresis (SDS-PAGE) according to the stacking gel procedure of Laemmli (19). The gels were dried and exposed to Cronex X-ray film (Du Pont Co., Wilmington, Del.) at -70°C with intensifying screens (1).

Antibody Purification.

Pristane (Tetramethylpentadecane, Aldrich Chemical Co., Inc. Milwaukee, Wis.) primed BALB/c mice were injected i.p. with 5 x 10^6 cloned hybridoma cells. Ascites fluid was collected 14 to 28 days later, centrifuged, and stored at -76°C until needed. Ascites fluid was filtered and applied to an AffiGel Protein A Monoclonal Antibody Purification System (MAPS, Bio-Rad) according to the manufacturer's instructions. Active fractions were collected and dialyzed against PBS. Fractions were then filtered through a 0.2-μm Acrodisc filter (Gelman Sciences, Inc., Ann Arbor, Mich.) and stored in the same buffer at 4°C.

Determination of Monoclonal Binding Affinities for C5a.

Dilutions of the antibodies giving 75% of maximum C5a immunoprecipitation were determined. Various concentrations of cold C5a (8 ng to 10 μg/mℓ) diluted in 200 μℓ PBS and 5% bovine serum albumin were incubated with 200 μℓ of antibodies for 2 h. All procedures were carried out at room temperature. An aliquot of ^{125}I-labeled C5a (100 μℓ) was added for 1 h followed by 100 μℓ of Bio-Rad rabbit antimouse IgG immunobeads for an additional hour. The beads were washed with PBS and transferred to an LKB gamma counter. Fifty percent inhibition of binding was determined, and the affinity constant was calculated by the method of Müller (22).

Isotype Determination.

A MonoAb-ID EIA Kit (Zymed) was used according to the manufacturer's instructions for isotype determination. Briefly, a 96-well polyvinylchloride plate was coated with C5a and then blocked with 1% BPBS, pH 7.2. Then 50 μℓ of culture supernatant was added to wells and incubated 60 min at room temperature. Plates were washed and 50 μℓ of each subclass-specific rabbit antimouse Ig antibody was added to separate wells. After a 60-min incubation at room temperature and a 3x wash, 50 μℓ of peroxidase-labeled goat antirabbit IgG was added to each well and incubated 60 min at room temperature. Plates were washed three times, 0.1 mℓ of substrate solution was added to each well, and the reaction was stopped by adding 0.1 mℓ stopping reagent. Results were read with a multiscan spectrophotometer.

Neutrophil Polarization.

Polarization of human polymorphonuclear leukocytes (PMNs) was performed essentially as described (6) with minor modifications. PMNs were isolated from fresh heparinized blood by dextran sedimentation followed by

Ficoll-Hypaque gradient centrifugation. The PMNs were adjusted to 1.25 x 10^6 cells/mℓ of Hanks balanced salt solution (HBSS), and 0.9 mℓ was added to each of duplicate tubes (12 x 75 mm polypropylene) containing 0.1 mℓ of stimulant or HBSS. The samples were incubated 10 min in a 37°C shaking water bath. To end the reaction, 1 mℓ of ice-cold 0.1 M phosphate-buffered formaldehyde (10% vol/vol [pH 7.2]) was added to each tube. Cells were kept cold (4°C) until they were examined by phase microscopy. Tubes were coded and read blindly. The percentage of cells polarized (i.e., elongated morphology) was determined by counting 200 cells per tube.

To block C5a induced polarization, 75 ng of purified porcine C5a in 100 μℓ of HBSS was incubated with 100μg of the monoclonals in 100 μℓ of HBSS for one hour at RT. The mixture was added to 0.8 mℓ of cells adjusted to 1.4 x 10^6 cells/ mℓ as described above.

Measurement of Neutrophil Chemiluminescence.

PMNs were isolated from heparinized blood on monopoly resolving medium (Flow Labs). Cells were washed twice in HBSS after hypotonic lysis of red cells. Cells were incubated in 5 mℓ scintillation vials at 1 x 10^6/mℓ in PBS containing 0.1% bovine serum albumin and 25 μM lucigenin (Sigma). Various stimuli or inhibitors were added, and counts per minute were recorded on an LKB scintillation counter equipped for chemiluminescence.

Procedure for Human PMN Chemotaxis

Human PMN were isolated using a modification of the procedure used by Goldstein et al., for isolation of PMN for lysosomal degranulation (14). Eight mℓ of venous blood were drawn into a syringe containing 2 mℓ of acid citrate dextrose (Sigma) solution. Five mℓ of a cold solution of 6% w/v dextran (531,000 average molecular weight. [Sigma]) in normal saline were then added and the syringe was inverted 50 times to mix. The mixture was allowed to sediment in the upright syringe at room temperature for 45 min. The serous supernatant, containing the PMN, was carefully expelled through a catheter into a sterile 15 cc centrifuge tube and then diluted up to a final volume of 15 mℓ with PBS (Sigma). The cells were centrifuged at 400 g for 10 min, the supernatant was discarded, and the cell pellet was gently resuspended in 2 mℓ of cold PBS. Residual erythrocytes were lysed isotonically by adding 6 mℓ of cold, deionized water, inverting for 25 seconds and then isotonicity was restored with 2 mℓ of 3.5% NaCl. The cells were again diluted to 15 mℓ with cold PBS, and the centrifugation and wash described above was repeated twice more. The final cell pellet was resuspended in 2 mℓ of PBS containing 6 x 10^{-4} M Ca^{++}, 1 x 10^{-3} M Mg^{++} and 1% albumin (PCMA for PBS with Ca, Mg and albumin). To 0.1 mℓ of this cell suspension, 0.9 mℓ of Trypan blue dye and 9 mℓ of PCMA were added. The viable cells (those excluding Trypan blue dye) were counted in a hemacytometer and the final volume was adjusted with PCMA to bring the cell concentration to 2 x 10^6 cells/mℓ. All cell and chemoattractant solutions were prepared using PCMA as the buffer.

Chemotaxis was assayed in a 48-well micro chemotaxis chamber (Neuro Probe, Inc., Cabin John, MD), and the cells were separated from the chemoattractant solutions by cellulose nitrate filters, 13 mm diameter with a 3 micron mean pore size (Sartorius Co., Hayward, CA). Four wells were covered by one filter and duplicate filters were run for each test solution. Each assay (n=1) was begun using a different blood sample. The bottom wells were filled with 35 μℓ of the test solution, the filters were placed on top and the chemotaxis chamber was assembled. In the upper wells, 50 μℓ of the cell suspension was added and the chemotaxis chamber was incubated at 37°C for 40 min at 100% humidity and 5% CO_2. The chamber was removed, disassembled quickly, and the filters were carefully transferred to a staining chamber fashioned of steel screen. The filters were prepared for mounting using the method of Zigmond (36). The cells were fixed briefly in 90% methanol, stained with Mayer's hematoxylin, dehydrated in successive washes increasing from 70%-100% ethanol and cleared in xylene.

The filters were then mounted on slides, cell side down, using Permount (Harleco, Gibbstown, NJ) and after drying, were read at 400 X magnification. The leading front technique (36) was used to determine the furthest extent of chemotaxis. Migration distances were expressed as a percentage of the leading front for the positive control to correct for inter-assay variations.

Zymosan-activated pig serum (ZPS) was prepared using the method of Goldstein (14) and was the positive control for C5a-induced chemotaxis. The chemoattractant and control solutions consisted of 2% of ZPS, either alone or with antibody, or 2% normal pig serum (NPS). The anti-C5a MAb was antibody 288-26F7 described below, and the control antibody was an antitransferrin MAb prepared in a similar manner. The parent antibody solution was adjusted to 1 mg/mℓ and dilutions of 1:100, 1:1000 and 1:10,000 were tested for the anti-C5a MAb and 1:100 for the anti-transferrin MAb. All data is reported as mean and SEM.

In Vivo Studies.

Purified anti-C5a monoclonals were injected i.v. into an ear vein of 2-kg rabbits. A blood sample was taken from each rabbit at Time 0 (before C5a or monoclonal was injected into each rabbit). The monoclonals (4 mg) were then injected i.v. A second blood sample was drawn after an hour, and 10 μg of C5a was injected i.v. in the contralateral ear. Blood samples were taken at 1, 5 and 8 min post-C5a injection. Each blood sample was analyzed for PMNs and total white blood cells. The differential between the total white blood cells and the PMNs (% PMN) was calculated.

RESULTS

1. Characterization of Anti-C5a Monoclonals.

We have generated a panel of four murine anti-porcine C5a MAbs (Table 1). The C5a-specific ELISA binding titers of the ascites range from $1:10^3$ to $1:10^6$. All of these antibodies are of the IgG_1,κ isotype. The panel of monoclonals was also tested for the ability to immunoprecipitate soluble radioiodinated des-Arg C5a. The affinities of the MAbs for porcine C5a ranged from 5.2×10^8 M^{-1} to 4.3×10^9 M^{-1} (Table 1).

Table 1
ANTI-C5a MAbs: ISOTYPES, ASCITES TITER AND AFFINITIES

Antibody	Isotype	Ascites Titer	Affinity (M^{-1})
302-21G6	IgG_1,κ	$1:10^4$	4.3×10^9
302-16G7	IgG_1,κ	$1:10^4$	1.7×10^9
288-26F7	IgG_1,κ	$1:10^6$	8.9×10^8
302-11E9	IgG_1,κ	$1:10^3$	5.2×10^8

Table 2
CROSS-REACTIVITY OF ANTI-PORCINE C5a MONOCLONALS: IMMUNOPRECIPITATION OF ^{125}I-PORCINE C5a AND ^{125}I-HUMAN C5a

Mab[1]	Specificity	^{125}I-porcine C5a	^{125}I-human C5a
288-26F7	porcine C5a	9.4[2]	0.6
302-16G7	porcine C5a	7.6	0.6
302-11E9	porcine C5a	6.1	0.5
302-21G6	porcine C5a	4.2	0.6
260-11G5	human C5a	0.1	8.1
L113	<u>Pseudomonas</u> endotoxin	0.4	0.6

[1] Purified antibody concentration = 2 μg/mℓ
[2] % of input cpm bound =
$$\frac{\text{(cpm bound by MAb)}}{\text{cpm input - background cpm}} \times 100$$

2. Specificity of Monoclonals for Porcine C5a.

The specificity of binding of the monoclonals for porcine C5a versus human C5a was determined by immunoprecipitation (Table 2).

The porcine C5a-specific MAbs immunoprecipitated only ^{125}I-porcine C5a. Anti-human C5a antibody 260-11G5 bound human C5a but not porcine C5a. A negative control MAb, L113, did not bind to porcine or human C5a.

3. Neutralization of C5a-Induced Chemiluminescence.

Porcine C5a produces a rapid transient increase in neutrophil chemiluminescence. Chemiluminescence of PMNs stimulated by C5a preincubated with various monoclonals is shown in Figure 1. A nonbinding isotype control monoclonal did not diminish the C5a-induced signal (not shown). Each of the four monoclonals tested blocked the ability of C5a to stimulate neutrophil chemiluminescence.

Figure 1:

Monoclonal antibodies block porcine C5a-induced granulocyte chemiluminescence. C5a was used at a final concentration of 100 ng/mℓ. Monoclonals were mixed with C5a 45 min prior to assay at a final concentration of 100 µg/mℓ. The mixture was added to cells at time 0 (arrow). Experiment was performed 4 times with essentially the same results as shown. Anti-C5a monoclonals block the C5a-induced chemiluminescence signal, whereas a nonbinding monoclonal has no effect on the assay.

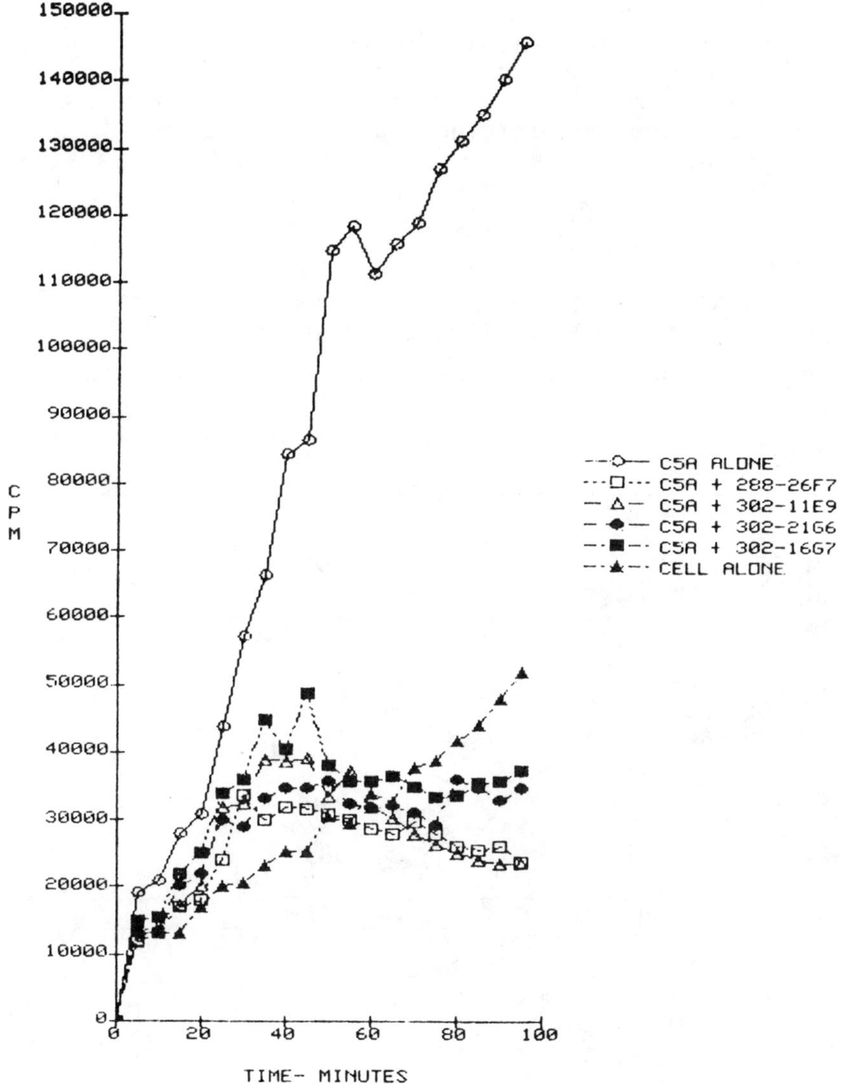
150000
140000
130000
120000
110000
100000
90000
80000
70000
60000
50000
40000
30000
20000
10000
0
C P M
0
20
40
60
80
100
TIME- MINUTES
C5A ALONE
C5A + 288-26F7
C5A + 302-11E9
C5A + 302-21G6
C5A + 302-16G7
CELL ALONE

4. Neutralization of C5a-Induced Neutrophil Polarization.

Membrane binding of C5a causes PMNs to undergo a dramatic change in shape called polarization. This phenomenon is dose-dependent. Fifty percent of the cells became polarized at a concentration of porcine C5a of 30 ng/mℓ. The anti-porcine C5a MAbs demonstrated 60 to 90% inhibition of C5a-induced neutrophil polarization (Table 3).

Table 3
ANTI-PORCINE C5a MAbs BLOCK POLARIZATION OF HUMAN GRANULOCYTES INDUCED BY PORCINE C5a

Sample	% Polarized[1]	
	Test 1	Test 2
1. Media only[2]	3	2
2. C5a only[3]	98	98
3. 288-26F7 only[4]	7	3
4. 288-26F7 + C5a[5]	12	11
5. 302-11E9 only[4]	5	3
6. 302-11E9 + C5a	18	10
7. 302-16G7 only[4]	4	3
8. 302-16F7 + C5a	10	15
9. 302-21G6 only[4]	10	8
10. 302-21G6 + C5a	52	39
11. L113 only[4,6]	2	2
12. L113 + C5a	98	97

[1]% polarized = $\frac{\text{\# of cells polarized}}{\text{total \# of cells counted (=200)}} \times 100$

[2]Hanks' Balanced Salt Solution (HBSS).

[3]50 ng of porcine C5a in HBSS.

[4]90 μg MAb in HBSS.

[5]Blocking step = 1 hr RT.

[6]Nonspecific MAb, IgG_1 (α*Pseudomonas aeruginosa* LPS)

5. Inhibition of C5a-Induced PMN Chemotaxis.

A microchemotaxis assay was used to measure migration of human neutrophils along a concentration gradient of zymosan-activated pig serum (ZPS). ZPS stimulated significantly more migration than normal pig serum (Table 4). This migration is thought to be due to C5a generated by the yeast cell walls. A nonspecific isotype control MAb did not block chemotaxis, whereas anti-porcine C5a MAb 288-26F7 showed significant inhibition of chemotaxis at a final concentration of 100 ng/mℓ.

Table 4
HUMAN PMN CHEMOTAXIS TOWARDS ZYMOSAN-ACTIVATED AND NORMAL PORCINE SERUM: INHIBITION WITH ANTI-PORCINE C5a MAb 288-26F7

Test*	n=	Mean Migration (microns)	SEM
ZPS	6	136.90	±4.97
NPS	6	74.30	5.50
C100	6	61.71	2.76
C1,000	6	61.00	3.49
C10,000	6	116.68	11.58
T100	6	137.42	4.27

*ZPS = Zymosan-activated pig serum
NPS = Normal pig serum
C100 = α-C5a diluted 1:100 from 1 mg/mℓ
C1000 = α-C5a diluted 1:1000 from 1 mg/mℓ
C10,000 = α-C5a diluted 1:10000 from 1 mg/mℓ
T100 = α-transferrin MAb diluted 1:100 from 1 mg/mℓ

6. Neutralization of C5a Activity *In Vivo*.

C5a administered to rabbits produces a rapid, though transient, fall in circulating PMNS (23,35). We have used this method to demonstrate *in vivo* neutralization of C5a with the panel of C5a monoclonals. Table 5 presents the results of these experiments. All of the anti-porcine MAbs, when administered prior to porcine C5a, blocked the C5a-induced neutropenia when compared to injected saline.

Table 5
PORCINE C5a INDUCED NEUTROPENIA IN RABBITS:
ANTI-C5a MAbs PROVIDE PROTECTION

	Pre-MAb T = 60 min	Post-MAb T = 0'	C5a	Post-C5a T = 1'	Post-C5a T = 5'	Post-C5a T = 8'
	Total PMNs (per mm^3 blood)					
288-26F7	4836	5150		4860	5320	4416
302-11E9	1872	4100		3692	3886	4698
302-16G7	3840	2720		2296	2340	3610
302-21G6	4000	3729		1932	2366	2464
Neg Control (saline)	3276	2774		56	58	1568

DISCUSSION

This report describes the generation of a panel of murine MAbs that bind to and neutralize the *in vitro* and *in vivo* effects of porcine complement fragment C5a. They were generated by a screening procedure called plate immunoprecipitation. This technique using ^{125}I-labeled antigen selects monoclonals that recognize the C5a in a native, soluble form. In previous studies we have found that antigen binding to plastic could alter its antigenicity, i.e., some monoclonals that bound solid phase antigen did not bind soluble antigen (J. W. Larrick, unpublished data). The panel of monoclonals we have generated have a high affinity for porcine C5a.

Several pieces of information suggest these high affinity monoclonals block the receptor binding of porcine C5a. They block C5a-induced neutrophil polarization, C5a-induced neutrophil chemiluminescence and C5a-induced neutrophil chemotaxis.

In vivo passively administered C5a is known to provoke *a rapid*, though transient, neutropenia (13,24). It has been postulated that the continued production of this molecule during sepsis and other states of complement activation leads to neutrophil aggregation and activation (7,8). Rabbits passively pre-immunized with the monoclonals were protected from subsequent C5a-induced neutropenia.

A number of workers have generated data supporting the role of activated complement in the pathogenesis of ARDS (7,9,17,32,33). The major mediator of damage appears to be the neutrophil, activated by C5a, although studies of ARDS developing in neutropenic patients suggest that other factors may also contribute to the syndrome (25). In the future we plan to directly test these monoclonals in porcine models of sepsis and reperfusion injury. If C5a contributes to the pathogenesis of these syndromes, we expect that one or more of the MAbs will give protection because they neutralize passively administered porcine C5a in rabbits. A rabbit anti-human C5a serum used by Stevens and co-workers (31) shared enough cross-reactivity with C5a from *Cygnomolous* spp. monkeys to show protection in this primate *Escherichia coli* sepsis-ARDS model. This antiserum also neutralized C5a-induced neutrophil aggregation *in vitro*.

We have recently generated high-affinity monoclonals to human C5a (20). It is anticipated that these anti-human C5a monoclonals will have therapeutic use based upon their capacity to neutralize the biological effects of human C5a both *in vitro* and *in vivo*. The anti-porcine C5a monoclonals described above will permit us to directly test the role of C5a in sepsis and reperfusion injury.

ACKNOWLEDGMENTS

The authors acknowledge the expert word processor help provided by Joan Murphy. The technical assistance of George Senyk, Jon Silver, and Pam Cato is gratefully acknowledged.

REFERENCES

1. Bonner WM, Laskey RA (1974). A film detection method for tritium-labeled proteins and nucleic acids in polyacrylamide gels. Eur J Biochem. 46:83.
2. Chenoweth DE, Hugli TE (1978). Demonstration of specific C5a receptor on intact human polymorphonuclear leukocytes. Proc Natl Acad Sc. USA 75:3943.
3. Chenoweth DE, Hugli TE (1980). Human C5a and C5a analogs as probes of the neutrophil receptor. Molecular Immunol 17:151.
4. Chenoweth DE, Cooper SW, Hugli TE, Stewart RW, Blackstone EH, J. W. Kirklin JW (1981). Complement activation during cardiopulmonary bypass. New Engl J Med 304:497.
5. Chenoweth DE, Coodman HG (1983). The C5a receptor of neutrophils and macrophages. Agents Actions 12:252.
6. Cianciolo GF, Snyderman R (1981). Monocyte responsiveness to chemotactic stimuli is a property of subpopulation of cells that can respond to multiple chemoattractants. J Clin Invest 67:60.
7. Craddock PR, Hammerschmidt A, White JG, Dalmasso AP, Jacob HS (1977). Complement (C5a)-induced granulocyte aggregation *in vitro*: A possible mechanism of complement-induced leukostasis and leukopenia. J Clin Invest 60:260.
8. Dahinden CA, Fehr J, Hugli TE (1983). Role of cell surface contact in the kinetics of superoxide production by granulocytes. J Clin Invest 72:113.
9. Fein AM, Lippman M, Holtzman H, Eliraz A, Goldberg SK (1983). The risk factors, incidence, and prognosis of ARDS following septicemia. Chest 83:40.
10. Fernandez HN, Hugli TE (1978). Primary structural analysis of the polypeptide portion of human C5a anaphylatoxin. J Biol Chem 253:6955.
11. Foung SKH, Sasaki DT, Grumet FC, Engleman EE (1983). Production of functional human T-T hybridomas in selection medium lacking aminopterin and thymidine. Proc Natl Acad Sci USA 79:7484.
12. Gelfand JA, Donelan M. Hawiger A, Burke JF (1982). Alternative complement pathway activation increases mortality in a model of burn injury in mice. J Clin Invest 70:1170.

13. Gilbertsen RB, Carter GW, Quinn DJ (1980). Effects of F-Met-Leu-Phe and Zymosan-activated serum on rat neutrophils in vivo. J Reticuloendothelial Soc 27:485.
14. Goldstein IM, Brai M, Osler AG, Weisman G (1973). Lysosomal enzyme release from human leukocytes: Mediation by the alternative pathway of complement activation. J Immuno 111:33.
15. Hammerschmidt DE, Bowers TK, Lammi-Keefe CJ, Jacobs HS, Craddock PR (1980). Granulocytes aggregometry: A sensitive technique for the detection of C5a and complement activation. Blood 55:898.
16. Hammerschmidt DE, Weaver LJ, Hudson LD, Craddock PR, Jacobs HS (1980). Association of complement activation and elevated plasma-C5a with adult respiratory distress syndrome: Pathophysiological relevance and possible prognostic value. Lancet 1:947.
17. Hosea SF, Brown E, Hammer C, Frank M (1980). Role of complement activation in a model of adult respiratory distress syndrome. J Clin Invest 66:375.
18. Hugli TE (1981). The structural basis for anaphylatoxin and chemotactic functions of C3a, C4a, and C5a. CRC Critical Reviews in Immunology Vol 1:321.
19. Laemmli UK (1970). Cleavage of structural proteins during the assembly of the head of bacteriophage T4. Nature (London) 227:680.
20. Larrick JW, Deinhart T, Wang J, Fendly BM, Chenoweth DE, Kunkel SL (1987). Murine monoclonal antibodies recognizing neutralizing epitopes on human C5a. Infect and Immun (in press).
21. Manderino GL, Suarez AF, Kunkel SL, Ward PA, Hirata, AA (1982) Purification of human C5a by immunoabsorbant and molecular sieve chromatography. J Immunol Methods 53:41.
22. Müller R (1980). Calculation of average antibody affinity in antihapten sera from data obtained by competitive radioimmunoassay. J Immunol Methods 34:345.
23. O'Flaherty JT, Kreutzer DL, Showell HJ, Ward PA (1977) Influence of inhibitors of cellular function on chemotactic factor-induced neutrophil aggregation. J Immunol 119:1751.
24. O'Flaherty JT, Showell HJ, Ward PA (1977). Neutropenia induced by systemic infusion of chemotactic factors. J Immunol 118:1586.

25. Ognibene SE, Martin SE, Parker MM, Schlesinger T, Roach P, Burch C, Shelhamer JH, Parrillo JE (1986). Adult respiratory distress syndrome in patients with severe neutropenia. New Engl J Med 315:547.
26. Oi V, Herzenberg L (1980). Immunoglobulin-producing hybrid cell lines, p.351-370. In Mishell B, Schiigi S (eds): "Selected Methods In Cellular Immunology." San Francisco, CA. W. J. Freeman, Co.
27. Parrish DA, Mitchell BC, Henson PM, Larsen GL (1984). Pulmonary response of fifth component of complement-sufficient and deficient mice to hyperoxia. J Clin Invest 74:956.
28. Sacks T, Moldow CF, Craddock PR, Bowers TK, Jacobs HS (1978). Oxygen radicals mediate endothelial cell damage by complement stimulated granulocytes. An in vitro model of immune vascular damage. J Clin Invest 61:1161.
29. Shulman M, Wilde CD, Kohler G (1978). A better cell line for making hybridomas secreting specific antibodies. Nature (London) 276:269.
30. Stevens JH, Raffin TA (1984) Adult respiratory distress syndrome - I. Etiology and mechanisms. Postgrad Med J 60:505.
31. Stevens JH, O'Hanley P, Shapiro JM, Mihm FG, Satoh PS, Collins JA, Raffin TA (1986). Effects of anti-C5a antibodies on the adult respiratory distress syndrome in septic primates. J Clin Invest 77:1812.
32. Tate RM, Repine JE (1983). Neutrophils and the adult respiratory distress syndrome: state of the art. Am Rev Resp Dis 128:802.
33. Tonnesen MG, Smedly LA, Henson PM (1984). Neutrophil endothelial cell interactions: Modulation of neutrophil adhesiveness induced by complement fragments C5a and C5a des-Arg and formyl-methionyl-leucyl-phenylalanine in vitro. J Clin Invest 74:1581.
34. Webster RO, Hong SR, Johnston Jr RB, Henson PM (1980). Biological effects of the human complement fragments C5a and C6a des-Arg on neutrophil function. Immunopharm 2:201.
35. Webster RO, Larsen GL, Henson PM (1982). In vivo clearance and tissue distribution of C5a and C5a desarginine complement fragments in rabbits. J Clin Invest 70:1177.
36. Zigmond SH, Hirsch JG (1972). Effects of cytochalasin B on polymorphonuclear leukocyte locomotion, phagocytosis and glycolysis. Exp Cell Res 73:383.

The Pharmacology and Toxicology of Proteins, pages 273–285

POTENTIALS FOR THERAPEUTIC APPLICATIONS OF INTERFERONS WITH OTHER CELL MODULATING PROTEINS[1]

Joan H. Schiller[2] and Ernest C. Borden[3]

Department of Human oncology
University of Wisconsin Clinical Cancer Center
Madison, Wisconsin 53792

ABSTRACT Recent investigations have revealed a complex world of lymphokine and cytokine interactions. In this review, we have briefly outlined some of the preclinical investigations involving interferon in combination with other biologicals which are currently ongoing in both laboratory and animal model systems, and we discuss their potential application in clinical trials.

INTRODUCTION

Interferons (IFNs), naturally produced proteins with antiviral and anticellular properties, are potent modulators of cellular function. They regulate gene expression, modulate expression of proteins on the cell surface, and induce synthesis of new enzymes. These alterations result in modulation of levels of receptors for other cytokines, regulatory proteins on the surface of immune effector cells, and enzymes which modulate cell growth. On a cellular basis, these effects translate into an

[1]Research in our laboratories in this area was supported by the American Cancer Society, the National Cancer Institute Biological Response Modifiers Program, XOMA Corporation, and Triton Biosciences, Inc.
[2]Joan H. Schiller is a recipient of a Stetler Research Award for Women Physicians
[3]Ernest C. Borden is an American Cancer Society Professor of Clinical Oncology

alteration in the state of differentiation, rate of proliferation, and functional activity of various cell types. IFNs may be thus used to augment the effects of other molecules on endothelial, immune, hematopoeitic or endocrine cells. The therapeutic use for combinations of IFNs with other cell modulating agents has yet to be defined, but will depend on judicious preclinical studies *in vitro* and *in vivo*.

COMBINATIONS OF INTERFERON

The study of cytokine interactions began in the early part of this decade with the study of the interactions between Type I and Type II IFNs. As prototype lymphokines, differences in physicochemical, antigenic, and biological properties between IFN types prompted investigation into their combined *in vivo* and *in vitro* effects. Studies with impure preparations of murine Type I and Type II IFNs demonstrated synergistic antitumor activity (1-3). Subsequent studies with purified preparations of human IFN have observed a similar potentiation in human tumors or transformed cell lines, including transformed human amnion WISH cells (4), melanoma (5,6), histiocytic lymphoma (7), fibrosarcoma, colon, transitional cell, renal, and bronchogenic carcinoma cells (8) Synergy has been observed in tumors resistant to the antiproliferative effects of one or both IFNs administered as single agents (8), although not all tumor types display this phenomena (7).

Although the mechanism of this synergistic interaction is unknown and the subject of current investigations, these observations have prompted clinical investigations to determine whether the combination of Type I and Type II IFNs also results in enhanced antitumor efficacy. Initial Phase I trials designed to determine maximally tolerated doses (MTD) of combinations of Type I and Type II IFNs and to identify qualitatively or quantitatively unexpected toxicities are currently at or nearing completion. Kurzrock *et al.*, have completed a Phase I protocol in which patients with metastatic solid tumors received daily IM injections of IFN-αA and IFN-γ (9). The MTD was 1-2 x 10^6 U/m^2/day of each IFN, with dose limiting side effects consisting of persistent fever and worsening fatigue.

At the University of Wisconsin Clinical Cancer Center, we have initiated a series of trials to: 1) identify the MTD and the side effects of combinations of IFN-γ and IFN-β_{ser}; 2) determine the immumomodulatory effects and biological response modification of this combination; 3) determine the optimal biological dose; and 4) determine the clinical antitumor effects of this regimen in malignant melanoma. Our Phase I trial evaluated dose, safety, and pharmacokinetics in 24 patients treated with a 2 hour infusion of IFN-γ followed by a 10 minute IV injection of IFN-β_{ser}, 3 times a week (10). Patients were entered on fixed dose levels of 1×10^6, 3×10^6, 10×10^6, 30×10^6, and 100×10^6 units of each IFN. The maximally tolerated dose when administered by this schedule for 4 or more weeks was 30×10^6 units of each IFN. Dose limiting side effects at the 100×10^6 unit dose level consisted of fatigue, nausea, vomiting, anorexia, paralytic ileus, and neutropenia. The most common side effects at the 3 highest dose levels were fever, rigors often requiring parenteral meperdine, and constitutional symptoms. Serum IFN levels were dose related, with peak titers occurring immediately after IFN administration. One patient with a poorly differentiated nodular lymphoma had a partial response.

We have subsequently completed a second Phase I trial which was designed to determine the doses of combined IFN-β_{ser}/IFN-γ which produced the greatest effect on NK cell cytotoxicity, monocyte activation, HLA-DR expression, 2-5A synthetase activity, and $\beta_2$microglobulin expression. The effects of the combined administration of IFN-β_{ser} and IFN-γ on these parameters as compared to either IFN alone will also be determined. Phase II trials evaluating the therapeutic effects of this concept are underway in patients with unresectable melanoma and bronchogenic carcinoma.

INTERLEUKIN-2 AND INTERFERONS

Interleukin-2 (IL-2) is a lymphokine produced by stimulated T lymphocytes that has the ability to initiate and support the proliferation of various types of T cells and large granular lymphocytes. It has generated considerable interest clinically because of its effects on natural killer (NK) and lymphokine-activated killer (LAK) cell activity (For review, see 11-13). The ability of IFNs to enhance the expression of such cell surface proteins as HLA determinants and tumor specific antigens prompted investigation into their effects on the expression of the interleukin-2 receptor, the Tac antigen. Partially purified preparations of IFN-γ induced the expression of IL-2 receptors on peripheral T cells (14), and recombinant IFN-γ increased both the number of Tac positive monocytes and the number of anti-Tac binding sites, whereas IL-2 had no influence on the IL-2 receptor expression on monocytes (15). Combinations of partially purified preparations of mouse fibroblast IFN and IL-2 resulted in additive effects on NK cell-mediated cytotoxicity against mouse spleen cells (16). In addition, IL-2 can induce IFN-γ mRNA synthesis (17), and IFNs-α,-β, and -γ can stimulate IL-2 production (18), further suggesting an interaction between the IFN and IL-2 systems.

Recombinant IFN-αA_2 has resulted in a marked inhibition of IL-2 production in melanoma patients treated intramuscularly three times weekly, which was followed by a gradual return to pretreatment levels or above. This delayed increase correlated with a delayed increase in NK activity against melanoma target cells (19). A recent Phase I trial of recombinant IL-2 administered by either bolus or constant infusion showed an increase in IFN-γ levels (20). Another recently compiled Phase I investigation of recombinant IL-2 and IFN-β observed *in vivo* activation of LAK cells with the combination of IL-2 and IFN-β, but not IL-2 alone (21). These results have formed the basis for an upcoming Eastern Cooperative Oncology Group Phase II trial in non-small cell lung cancer, in which patients would be randomized to the combination of IL-2 and IFN-β or IL-2 alone, and response and LAK cell activation *in vivo* would be correlated.

TUMOR NECROSIS FACTOR AND INTERFERONS

Tumor necrosis factor (TNF) and lymphotoxin are proteins with antitumor activity which were originally identified in the serum of bacillus Calmette Guerin-sensitized mice injected with endotoxin (22-24). These agents have cytostatic and cytotoxic activity against a wide range of human and murine tumor cells but little antiproliferative activity for normal tissues (25-26). However, not all tumor cells are sensitive; in one study, TNF had a cytostatic or cytotoxic effect on 12 of 23 human tumor cell lines (25), and in another, on 7 of 22 (26).

Administration of IFNs may be one method of enhancing the sensitivity of cancer cells to growth inhibition of TNF. IFN-γ has been shown to enhance the expression of cellular receptors for TNF (27-29). Synergistic antiproliferative effects using combinations of IFN-γ and TNF have been described in human breast and cervical carcinoma and melanoma cell lines _in vitro_ (25,26,30,31), and human breast and bowel tumor xenografts _in vivo_ (32).

We assessed the antiproliferative effects of TNF-α and IFN-γ, alone and in combination, on nine human colon carcinoma cell lines (33). All were resistant (<30% inhibition) to TNF-α alone. Four cell lines were resistant to IFN-γ alone, two exhibited a minimal degree of sensitivity (30-50% inhibition), one was moderately sensitive, and two were inhibited 70% or greater. A synergistic antiproliferative effect occurred in eight of the nine cell lines treated with a combination of TNF-α and IFN-γ. In seven of these eight, the combination of cytokines resulted in 30-40% more growth inhibition than predicted had an additive interaction occurred ($p<.005$).

The clinical development of TNF as a single agent is currently underway. Phase I studies have evaluated escalating and fixed doses of TNF by various routes and schedules of administration (34-36). The most frequently observed systemic toxicities include fever, chills, and fatigue. The intensity of fever and chills tended to decrease following repetitive treatment, suggesting that tolerance to these symptoms developed. Rises in serum hepatic transaminases were also occasionally seen. In an ongoing study at the University of Wisconsin Clinical Cancer Center, 15 patients have been treated with TNF-α by a 30 minute IV infusion on a 3 times a week schedule (37). Dose levels include 5, 10, 25, 50, 75, 100, 125, and 150 mcg/m^2. Three patients have been treated with 125 mcg/m^2 without dose-limiting toxicity. One patient with locally recurrent colon carcinoma treated at the 50 mcg/m^2 dose level has had a partial response of his pelvic disease.

Based upon the extensive preclinical data demonstrating the synergistic effects of the combination of TNF and IFN-γ in laboratory and animal studies, we are currently planning a Phase I trial to determine the MTD of TNF-α that can be administered with a "low," "intermediate," or "high" dose of IFN-γ (10%, 50%, and 100%, respectively of the MTD of IFN-γ when administered intravenously 3x/wk as a single agent). The effects of combined TNF and IFN-γ at each of these dose levels will be assessed on such biological responses as Class I and Class II HLA expression, F_c receptor expression, T cell activation and serum triglyceride, cholesterol, and lipoprotein levels.

MONOCLONAL ANTIBODIES AND INTERFERONS

The advent of hybridoma technology in 1975 (38) resulted in the generation of monoclonal antibodies (MAbs) to various tumor associated antigens (TAAs). Potential various applications of these molecules to the treatment of cancer have included the use of radiolabeled MAbs for tumor targeting, bone marrow purging with MAbs, the use of MAbs to growth factors and growth factor receptors, and the generation of radiolabeled MAb and MAb conjugates for therapy (for review see 39,40). The success of these potential applications depends in part upon the expression of the TAA, which may vary between tumors of patients with a given tumor type, between tumors in different anatomical sites on the same patient, and between cells within a given tumor lesion. As IFNs are capable of augmenting the synthesis of various cellular proteins, one approach to the problem of heterogeneity of TAA expression has been the use of IFNs potentially to enhance the expression of TAA on the surface of carcinoma cells.

Initial reports utilizing partially purified IFN preparations gave somewhat conflicting results: leukocyte IFN enhanced (41) or did not affect (42,43) the expression of various melanoma associated antigens, and reduced the expression of a common acute lymphoblastic leukemia antigen on four lymphoblastoid cell lines (44). Studies involving purified preparations of IFNs demonstrated that recombinant IFN-α preparations can increase the expression of tumor antigens on human breast and colon carcinoma cells (45) and human melanoma cells (46,47). Species of IFN-α differed in their ability to alter the level of antigen expression on the surface of MCF-7 breast carcinoma cells (48). Recombinant IFN-β enhanced the expression of two melanoma associated antigens (46), while recombinant IFN-γ modulated the expression of melanoma associated antigens (47) and of some myeloid cell-associated antigens expressed on small cell carcinoma cell lines (49). However, not all human carcinoma cells demonstrate enhanced antigenic expression. Clones of MCF-7 breast carcinoma cells have been isolated which do not express a tumor antigen either before or after treatment with IFN-αA (50). Reduction of antigen expression on melanoma cells by IFN-γ was due to changes in synthesis or shedding of the antigen, and not due to a loss of the antigen by subpopulations analyzed (51).

Preliminary _in vivo_ studies have provided further evidence to support the continued investigation and possible clinical application of IFN and MAb. Combination therapy with recombinant IFN-α and an anti-idiotype MAb produced synergistic antitumor effects in murine lymphoma (52). Partially purified IFN-α enhanced the tumor uptake of ^{111}Indium labelled anti-melanoma MAb 96.5 (Hybritech, Inc.) in five patients with metastatic malignant melanoma (53).

Thus both _in vitro_ and initial _in vivo_ results suggest this approach will have clinical merit. Considering the diversity of macromolecules recognized by MAb, it is hardly surprising IFN has not resulted in augmented expression of all TAA. Further Phase I trials will be required to elucidate optimal doses and timing of IFNs and MoAb.

CONCLUSION

Phase II clinical trials have established the antitumor effects of interferons (IFNs) as single agents in more than a half dozen malignancies. Single modalities, however, are not usually optimum treatment for any but early stage cancers. IFNs will not be an exception to this clinical axiom. Optimum uses of IFN in conjunction with other modalities must thus be defined. Preclinical investigations with IFN in combination with other biological response modifiers and cell modulating proteins support further investigation in this area.

REFERENCES

1. Fleischmann WR Jr, Kleyn KM, Baron S (1980). Potentiation of antitumor effect of virus-induced interferon by mouse immune interferon preparations. J Nat Cancer Inst 65 (5):963-966.
2. De Clercq E, Zhang Z-X, Huygen K (1982). Synergism in the antitumor effects of type I and type II interferon in mice inoculated with leukemia L1210 cells. Cancer Letters 15:223-228.
3. Fleischmann WR Jr (1982). Potentiation of the direct anticellular activity of mouse interferons: mutual synergism and interferon concentration dependence. Cancer Res 42:869-875.
4. Oleszak E, Stewart WE II (1985). Potentiation of the antiviral and anticellular activities of interferons by mixtures of HuIFN-γ and HuIFN-α or HuIFN-β. J IFN Res 5:361-371.
5. Schiller JH, Willson JKV, Bittner G, Wolberg WH, Hawkins MJ, Borden EC (1987). Antiproliferative effects of interferons on human melanoma cells in the human tumor colony forming assay. J IFN Res in press.
6. Czarniecki CW, Fennie CW, Powers DB, Estell DA (1984). Synergistic antiviral and antiproliferative activities of _Escherichia coli_-derived human alpha, beta, and gamma interferons. J Virol 49:490-496.
7. Denz H, Lechleitner M, Marth C, Daxenbichler G, Gastl G, Braunsteiner H (1985). Effect of human recombinant alpha-2- and gamma-interferon on the growth of human cell lines from solid tumors and hematologic malignancies. J IFN Res 5:147-157.
8. Schiller JH, Groveman DS, Schmid SM, Willson JKV, Cummings KB, Borden EC (1986). Synergistic antiproliferative effects of human recombinant α54- or β_{ser}-interferon with γ-interferon on human cell lines of various histogenesis. Cancer Res 46 (2):483-488.
9. Kurzrock R, Rosenblum MG, Quesada JR, Sherwin SA, Itri LM, Gutterman JU (1986). Phase I study of a combination of recombinant interferon-alpha-A and recombinant interferon-gamma in cancer patients. J Clin Oncol 4 (11):1677-1683.

10. Schiller JH, Storer B, Willson JKV, Borden EC (1986). Phase I trial of combinations of recombinant interferons β_{ser} and γ in patients with advanced malignancy. Cancer Res submitted.
11. Gillis S, Conlon PJ, Cosman D, Hopp TP, Dower SK, Price V, Mochizuki DY, Urdal DL (1986). Lymphokines: From conjecture to the clinic. Sem Oncol 13:218-227.
12. Rosenberg S (1985). Lymphokine-activated killer cells: a new approach to immunotherapy of cancer. J Natl Cancer Inst 75 (4):595-603.
13. Rosenberg SA (1986). The adoptive immunotherapy of cancer using the transfer of activated lymphoid cells and interleukin-2. Sem Oncol 13 (2):200-206.
14. Johnson HM, Farrar WL (1983). The role of gamma interferon-like lymphokine in the activation of T cells for expression of interleukin 2 receptors. Cell Immunol 75 (1):154-159.
15. Holter W, Grunow R, Stockinger H, Knapp W (1986). Recombinant interferon-γ induces interleukin 2 receptors on human peripheral blood monocytes. J Immunol 136:2171-2175.
16. Kuribayashi K, Gillis S, Kern DE, Henney CS (1981). Murine NK cell cultures: Effects of interleukin-2 and interferon on cell growth and cytotoxic reactivity. J Immunol 126 (6):2321-2327.
17. Farrar WL, Birchenall-Sparks MC, Young HB (1986). Interleukin 2 induction of interferon-γ mRNA synthesis. J Immunol 137 (12):3836-3840.
18. Rosztoczy I, Siroki O, Beladi I (1986). Effects of interferons-α, -β, and -γ on human interleukin-2 production. J Interferon Res 6:581-589.
19. Hersey P, MacDonald M, Hall C, Spurling A, Edwards A, Coates A, McCarthy W (1986). Immunological effects of recombinant interferon alfa-2a in patients with disseminated melanoma. Cancer 57:1666-1674.
20. Lotze MT, Matory YL, Ettinghausen SE, Rayner AA, Sharrow SO, Seipp CAY, Custer MC, Rosenberg SA (1985). In vivo administration of purified human interleukin-2 II. Half life, immunologic effects, and expansion of peripheral lymphoid cells in vivo with recombinant IL-2. J Immunol 135 (4):2865-2875.
21. Kriegel R. Personal communication.

22. Carswell EA, Old LJ, Kassel RL, Green S, Fiore N, Williamson B (1975). An endotoxin-induced serum factor that causes necrosis of tumors. Proc Nat Acad Sci USA 72 (9):3666-3670.
23. Wang AM, Creasey AA, Ladner MB, Lin LS, Strickler J, Van Arsdell JN, Yamamoto R, Mark DF (1985). Molecular cloning of the complementary DNA for human tumor necrosis factor. Science 228:149-154.
24. Shirai T, Yamaguchi H, Ito H, Todd CW, Wallace RB (1985). Cloning and expression in Escherichia coli of the gene for human tumour necrosis factor. Nature 313:803-806.
25. Williamson BD, Carswell EA, Rubin BY, Prendergast JS, Old LJ (1983). Human tumor necrosis factor produced by human B-cell lines: Synergistic cytotoxic interaction with human interferon. Proc Natl Aca Sci USA 80 (17):5397-5401.
26. Sugarman BJ, Aggarwal BB, Hass PE, Figari IS, Palladino MA, Shepard HM (1985). Recombinant human tumor necrosis factor-α: Effects on proliferation of normal and transformed cells in vitro. Science 230 (4728):943-945.
27. Aggarwal BB, Eessalu TE, Hass PE (1985). Characterization of receptors for human tumour necrosis factor and their regulation by γ-interferon. Nature 318:665-667.
28. Ruggiero V, Tavernier J, Fiers W, Baglioni C (1986). Induction of the synthesis of tumor necrosis factor receptors by interferon-γ. J Immunol 136:2445-2450.
29. Tsujimoto M, Feinman R, Vilcek J (1986). Differential effects of type I IFN and IFN-γ on the binding of tumor necrosis factor to receptors in two human cell lines. J Immunol 137 (7):2272-2276.
30. Stone-Wolff DS, Yip YK, Kelker HC, Le J, Henriksen-Destafano D, Rubin BY, Rinderknecht E, Aggarwal BB, Vilcek J (1984). Interrelationships of human interferon-gamma with lymphotoxin and monocyte cytotoxin. J Exp Med 159:828-843.
31. Lee SH, Aggarwal BB, Rinderknecht E, Assisi F, Chiu H (1984). The synergistic anti-proliferative effect of γ-interferon and human lymphotoxin. J Immunol 133 (3):1083-1086.

32. Balkwill FR, Lee A, Aldam G, Moodie E, Thomas JA, Tavernier J, Fiers W (1986). Human tumor xenografts treated with recombinant human tumor necrosis factor alone or in combination with interferons. Cancer Res 46:3990-3993.
33. Schiller JH, Bittner G, Storer B, Willson JKV (1987). Synergistic antitumor effects of tumor necrosis factor and gamma interferon on human colon carcinoma cell lines. Cancer Res in press.
34. Blick MB, Sherwin SA, Rosenblum MG, Gutterman JU (1986). A phase I trial of recombinant tumor necrosis factor (rTNF) in cancer patients. Proc Amer Soc Clin Oncol 5:14 (abstract).
35. Khan A, Pardue A, Aleman C, Dickson J, Pichyangkul S, Hill JM, Hilario R, Hill NO (1986). Phase I clinical trial with recombinant tumor necrosis factor (TNF). Proc Amer Soc Clin Oncol 5:226 (abstract).
36. Chapman PB, Lester TJ, Casper ES, Gabrilove JL, Kempin S, Welt S, Sherwin S, Old LJ, Oettgen HF (1986). Phase I study of recombinant tumor necrosis factor (rTNF). Proc Amer Soc Clin Oncol 5:231 (abstract).
37. Trump DL (1986). Personal communication.
38. Kohler G, Milstein C (1975). Continuous culture of fused cells secreting antibody of predefined specificity. Nature 256:495-496.
39. Houghton AN, Scheinberg DA (1986). Monoclonal antibodies: Potential applications to the treatment of cancer. Sem Oncol 13 (2):165-179.
40. Schlom J (1986). Basic principles and applications of monoclonal antibodies in the management of carcinomas: The Richard and Hinda Rosenthal Foundation Award Lecture. Cancer Res 46:2335-2339.
41. Liao S-K, Kwong PC, Khosravl M, Dent PB (1982). Enhanced expression of melanoma-associated antigens and β_2-Microglobulin on cultured human melanoma cells by interferon. JNCI 68:19-25.
42. Basham TY, Bourgeade MF, Creasey AA, Merigan TC (1982). Interferon increases HLA synthesis in melanoma cells: Interferon-resistant and -sensitive cell lines. Proc Natl Acad Sci USA 79 (10):3265-3269.

43. Imai K, NG A-K, Glassy MC, Ferrone S (1981). Differential effect of interferon on the expression of tumor-associated antigens and histocompatibility antigens on human melanoma cells: Relationship to susceptibility to immune lysis mediated by monoclonal antibodies. J Immunol 127 (2):505-509.
44. Hokland M, Ritz J, Hokland P (1983). Interferon-induced changes in expression of antigens defined by monoclonal antibodies on malignant and nonmalignant mononuclear hematopoietic cells. J Interferon Res 3 (2):199-210.
45. Greiner JW, Hand PH, Noguchi P, Fisher PB, Pestka S, Schlom J (1984). Enhanced expression of surface tumor-associated antigens on human breast and colon tumor cells after recombinant human leukocyte α-interferon treatment. Cancer Res 44:3208-3214.
46. Giacomini P, Aguzzi A, Pestka S, Fisher PB, Ferrone S (1984). Modulation by recombinant DNA leukocyte (α) and fibroblast (β) interferons of the expression and shedding of HLA- and tumor-associated antigens by human melanoma cells. J Immunol 133 (3):1649-1655.
47. Murray JL, Pillow JK, Rosenblum MG, Gutterman JU, Hersh EM, Carlo DJ (1986). Differential _in vitro_ effects of alpha recombinant interferon (rIFNαA) and gamma recombinant interferon (rIFNγ) on the expression of melanoma-associated antigens (MAA) P97 and 240 Kd on melanoma cell line Ts294. Proc Amer Assoc Cancer Res 27:313.
48. Greiner JW, Fisher PB, Pestka S, Schlom J (1986). Differential effects of recombinant human leukocyte interferons on cell surface antigen expression. Cancer Res 46:4984-4990.
49. Ball ED, Sorenson GD, Pettengill OS (1986). Expression of myeloid and major histocompatibility antigens on small cell carcinoma of the lung cell lines analyzed by cytofluorogaphy: Modulation by γ-interferon. Cancer Res 46:2335-2339.
50. Greiner JW, Tobi M, Fisher PB, Langer JA, Pestka S (1985). Differential responsiveness of cloned mammary carcinoma cell populations to the human recombinant leukocyte interferon enhancement of tumor antigen expression. Int J Cancer 36:159-166.

51. Giacomini P, Imberti L, Aguzzi A, Fisher PB, Trinchieri G, Ferrone S (1985). Immunochemical analysis of the modulation of human melanoma-associated antigens by DNA recombinant immune interferon. J Immunol 135 (4):2887-2894.
52. Basham TY, Kaminski MS, Kitamura K, Levy R, Merigan TC (1986). Synergistic antitumor effect of interferon and anti-idiotype monoclonal antibody in murine lymphoma. J Immunol 137 (9):3019-3024.
53. Murray JL, Rosenblum MG, Lamki L, Talpaz M, Hersh EM, Gutterman JU, Carlo DJ (1986). Enhancement of tumor uptake of ^{111}indium (^{111}In)-labeled anti-melanoma monoclonal antibody (MOAB) 96.5 in melanoma patients receiving partially purified alpha interferon (αIFN). Proc Amer Soc Clin Oncol 5:226.

The Pharmacology and Toxicology of Proteins, pages 287–298

DIFFERENTIAL INCREASES OF TUMOR ANTIGEN EXPRESSION BY HUMAN LEUKOCYTE INTERFERONS

F. Guadagni, J. Schlom, and J.W. Greiner

Laboratory of Tumor Immunology and Biology, National Cancer Institute, NIH, Bethesda, MD 20892

ABSTRACT Human rHu-IFN-α is a multigene family. We compared rHu-IFN- αA,B,C,D,F,I,J and K for their abilities to growth inhibit human breast carcinoma cells as well as to augment the expression of surface antigens. A high degree of heterogeneity with respect to these biological parameters was found among the different rHu-IFN-α species. The A and B species were most potent in cell growth inhibition and augmenting mAb binding to surface antigens such as HLA, carcinoembryonic antigen (CEA) and a high-molecular weight tumor-associated glycoprotein, termed TAG-72. In contrast, the D and J species were virtually inactive in altering the expression of these or any other surface antigen both _in vitro_ and _in vivo_. rHu-IFN-αD and -αJ did interact with the interferon receptor by their ability to interfere with the rHu-IFN- αA induced antigen augmentation on the MCF-7 cell surface. In conclusion, the rHu-IFN-α family exerts a wide range of biological actions. These findings are important for identifying those interferon species that are capable of augmenting tumor antigen expression, and thus, may be earmarked for consideration in further study of the enhancement of mAb binding to human carcinoma cell populations.

INTRODUCTION

Recombinant human leukocyte (alpha) interferon (rHu-IFN-α) consists of a family of individual species of which at least eight different species have been expressed in _Escherichia coli_, purified and their biological activities compared (1-3). Besides the inherent differences in amino acid composition, several studies have found differences in the antiviral, antiproliferative and immunomodulatory activities of these species of leukocyte interferon (4-7). For example, a single leukocyte (alpha) interferon species (rHu-IFN-αJ) has been shown to be an effective antiviral agent, yet completely inactive in eliciting human natural killer activity (7). Therefore, as originally observed with the various natural leukocyte interferon species (4) there seems to exist quantitative and qualitative differences with respect to their abilities to regulate a variety of biological properties of human target cells.

Our laboratory has reported that the clone A of recombinant human leukocyte interferon (rHu-IFN-αA) can increase the binding of monoclonal antibodies (mAbs) to the surface of human breast and colon carcinoma cells (8-10). We have shown that such an increase is a result of the enhanced expression of tumor antigens, such as the 180 kD CEA and the high molecular weight (>10^6d) mucin, TAG-72, which react with their respective mAbs (8,11). The present study was carried out to evaluate the effectiveness of the different species of rHu-IFN-α for their abilities to alter cell proliferation and modulate the level of cell surface antigen expression. The antigens monitored were the CEA, the TAG-72 antigen and HLA, all of which are constitutively expressed by the human breast carcinoma cell line. Furthermore, studies were also conducted to compare the activities of selected rHu-IFN-α species for their ability to enhance tumor antigen expression in human breast carcinoma cells grown as tumors in athymic mice.

MATERIALS AND METHODS

Recombinant Human Leukocyte Interferons: The isolation, expression, and purification of eight different clones (i.e., clones A,B,C,D,F,I,J, and K) of human leukocyte interferon have been described (2,3,12,13).

Hybridoma Methodology: The details of the generation and characterization of the mAbs B1.1, B72.3, and B6.2 have been reported (14,15). The W6/32 monoclonal antibody recognizes the A,B,C loci of the HLA major histocompatibility complex (15).

Determination of Cell Surface Antigen Expression: A complete description of this radioimmunoassay to detect antigen expression on live cells has been published (16).

Effect of rHu-IFN-αA on the Reactivity of ^{125}I-B6.2 and ^{125}I-MOPC-21 to MCF-7 Tumor Extracts: The B6.2 IgG was purified from ascites obtained from pristane-primed Balb/c mice inoculated i.p. with 10^7 hybridoma cells. MOPC-21, a mouse myeloma IgG, was purchased from Litton Bionetics, Inc. (Rockville, MD). Both immunoglobulins were labeled with Na ^{125}I with iodogen (17). MCF-7 tumors were grown in female, athymic mice supplemented with 17 β-estradiol. The mice were treated with either rHu-IFN-αA or -αD by twice daily intramuscular injections of 100,000 units/injection. After the fifth injection the MCF-7 tumors were removed, extracts prepared (15) and the level of ^{125} I-labeled antibody measured. The mice were also bled by eye puncture and the plasma levels of rHu-IFN-αA and -αD were measured as previously described (18).

RESULTS

MCF-7 cells were treated with 10-2000 antiviral units of each clone of rHu-IFN-α. Table 1 summarizes the antiviral titers of each rHu-IFN-α species required to achieve a 25% reduction in MCF-7 cell growth. This level of inhibition was chosen because neither rHu-IFN-αD or -αJ caused more than 40% inhibition of MCF-7 cell growth at antiviral levels of >5000 units/ml (15). There was a greater than 100-fold range in the antigrowth activities of these eight rHu-IFN-α species. As shown, rHu-IFN-αB, the most potent species, inhibited MCF-7 cell growth by 25% with the addition of only 2 units/ml. For comparison, it required 125-fold more units of rHu-IFN-αD (250 units) to evoke the same degree of growth inhibition. The order of potency of these eight rHu-IFN-α species for the inhibition of MCF-7 growth was B>A=F>K=C>J>I >>D. The MCF-7 cells express two distinct surface tumor antigens, CEA and TAG-72, as well as the class I major histocompatibility HLA complex. The eight different species of rHu-IFN-α were evaluated for their ability to increase the binding of mAbs B1.1, B72.3 and W6/32 to their respective cell surface tumor and normal (HLA) antigens. Table 2 summarizes the relative potencies of each rHu-IFN-α cloned species for modulating the expression of these antigen on the MCF-7 cell surface. Interestingly, of the eight Hu-IFN-α species tested, the -αD and -αJ species which were the least effective for inhibiting MCF-7 cell growth, were also relatively ineffective in enhancing HLA expression and completely inactive for altering the level of CEA or TAG-72 expression. Whereas, other IFN -α species enhanced cell surface HLA ex-

Table 1: Comparison of the antiproliferative properties of the different recombinant leukocyte interferons on a human breast carcinoma cell.

rHu-IFN-α	MCF-7 cell growth inhibition (I_{25})[a] (antiviral units/ml)
A	38 (25-60)[b]
B	2 (0.5-12)
C	60 (40-110)
D	250 (180-410)
F	33 (25-80)
I	92 (80-125)
J	75 (60-95)
K	58 (45-70)

[a]After 5 days in the interferon-containing growth medium, the cells were removed by trypinization and counted with a hemocytometer. I_{25} represents the approximate antiviral titer for each leukocyte (alpha) interferon which resulted in a 25% inhibition of cell growth.

[b]The values are shown as the mean of 3-5 experiments for each leukocyte interferon species with the appropriate range in parentheses. The values for rHu-IFN-αB were estimated by using 1-1000 units for testing.

Table 2: Comparison of the effects of leukocyte interferons on antigen expression on the surface of breast carcinoma cells.

Interferon species	Relative Enhancement in MAb Binding to Respective Tumor Antigens		
	B1.1 (CEA)	B72.3 (TAG-72)	W6/32 (HLA-ABC)
A	++	+++	+++
B	++	++	+++
C	++	+	++
D	0	0	+
F	+	+	+++
I	++	+++	+++
J	0	0	+
K	++	++	+++

0 < 25% increase in mAb binding
+ - 25-49% " " " "
++ - 50-100% " " " "
+++ - >100% " " " "

pression at least 50-100%, IFN-αD and -αJ at antiviral titers of greater than 1000 units/ml could only boost HLA expression by 30% (Table 2). Similarly, analysis of whole cell extracts from the same rHu-IFN-αD or -αJ treated cells also revealed no appreciable change in the level of B1.1 or B72.3 binding following incubation with 2,000-5,000 antiviral units of either rHu-IFN-αD or IFN-αJ. The rHu-IFN-αI species that was a relatively weak inhibitor of the growth of MCF-7 cells *in vitro* (Table 1), was found to be a potent inducer of the expression of all three antigens. The ranking of the eight clones of rHu-IFN-α for antigen enhancement was thus different than that for growth inhibition of the MCF-7 cells. The order of potency for enhancement of CEA, TAG-72 and HLA expression was A=I≥B≥K>C>F>>J,D.

The rHu-IFN-αA and -αD were chosen for further evaluation of their abilities to alter tumor antigen expression *in vivo*. For these experiments, MCF-7 cells were grown as subcutaneous tumors in female, athymic mice and administered either rHu-IFN-αA and -αD as intramuscular injections. Previous studies have shown that the MCF-7 tumor extracts express a 90 kD antigen that is recognized by mAb B6.2

Table 3: Comparison of the abilities of rHu-IFN-αA and D to enhance the level of binding of ^{125}I-B6.2 to extracts of MCF-7 tumors.[a]

Treatment	Serum Interferon (antiviral units/ml)	^{125}I-B6.2 (cpm/20ug protein)	^{125}I-MOPC-21 (cpm/20ug protein)
None	<40	6250	189
	<40	6700	193
	<40	8270	352
rHu-IFN- α A	140	13850 (1.96)	74
	160	17400 (2.46)	185
	<40	8340 (1.18)	198
rHu-IFN- α D	<40	7210 (1.02)	248
	220	6329 (0.90)	357
	180	8610 (1.22)	186

[a]Athymic mice were implanted with a pellet containing 17β-estradiol and four days later received a subcutaneous injection of 2 x 10^7 MCF-7 cells. After 45-50 days the mice had tumors with an average diameter of 1.0-2.0 cm. At that time the mice were divided into three groups: control mice received i.m. injections of 0.1 ml saline and the other two groups received 105 units/injection of either rHu-IFN-αA or D. The mice were given two injections/day and after the fifth injection were sacrificed and extracts of the MCF-7 tumors were prepared as previously described (15).

$$(\ ) = \frac{\text{cpm/20 ug extract protein - rHu-IFN treated}}{\text{cpm/20 ug extract protein - control } (\bar{X} = 7073)}$$

(8,17). Therefore, studies were conducted to determine whether the rHu-IFN- αs can alter the level of expression of the B6.2-reactive 90 kD antigen in the MCF-7 tumors. As shown in Table 3, the constitutive level of ^{125}I-B6.2 binding to the MCF-7 tumor extract ranged from 6000-8000 cpm/20 ug of tumor protein. Treatment of the MCF-7 tumor-bearing athymic mice with rHu-IFNαA resulted in significant (approx. 150 antiviral units/ml) serum levels of the interferon in 2 of 3 animals. Furthermore, extracts made from the MCF-7 tumors of these mice bound approximately 2-fold more of the ^{125}I-labeled mAb B6.2. The mice that received rHu-IFN-αD also had high serum levels of the interferon, yet, did not have increased levels of binding of the ^{125}I-B6.2 mAb. Thus, the in vitro data which demonstrate the inability of the D clone of rHu-IFN-α to enhance tumor antigen expression is corraborated. Finally, treatment with either rHu-IFN-α did not alter the binding of MOPC-21, a mouse myeloma IgG.

The combination of weak antiproliferative and relatively ineffective tumor antigen enhancing activities by IFN-αD, in particular, might be explained by inefficient binding of this IFN-α species to the MCF-7 surface interferon receptor. This hypothesis was tested by preincubating MCF-7 cells in the presence of 1000 units/ml of rHu-IFN-αD followed by the addition of 1000 units/ml rHu-IFN-αA. As shown previously, 1000 units rHu-IFN-αA/ml resulted in a significant increase in B1.1 binding to CEA (approximately 2-fold) on the surface of MCF-7 cells. A 2 hr reincubation with IFN-αD or -αJ completely blocked the rHu-IFN-αA induced increase in B1.1 binding. These findings indicate that rHu-IFN- αD can interact with the MCF-7 cell membrane and block the enhancement of tumor antigen expression induced by rHu-IFN-αA. It was also observed that the inhibition of the rHu-IFN-αA mediated increase in B1.1 binding by rHu-IFN-αD could be abrogated by substantially increasing the titer of the rHu-IFN-αA to >25,000 units/ml which suggests the competitive nature of the rHu-IFN-αD inhibition (15).

DISCUSSION

Treatment of the breast carcinoma cell line, MCF-7, with eight different species of rHu-IFN- α, which display a high degree of amino acid sequence homology with complete identity at 85 of their 165 amino acids results in both quantitative and qualitative differences in antiproliferative activity and enhancement in the cell surface expression of tumor and HLA antigens. All eight rHu-IFN-α species showed some activity for boosting the level of HLA expression as measured by the binding of mAb W6/32. In contrast, there seems to be a broader range of activities among the different rHu-IFN-α species with respect to their abilities to enhance tumor antigen expression. rHu-IFN-αB, -αA, -αI, and -αK induced the most dramatic increases in CEA and TAG-72 expression on the surdace of the MCF-7 cells. Treatment with rHu-IFN-αC and -αF evoked an intermediate elevation in B1.1 and B72.3 binding to their respective tumor antigens. An interesting observation was that although at high antiviral titers IFN-αD and -αJ could induce a small increases in HLA expression, these species failed to evoke any change in the expression of either tumor antigen. As stated previously, other investigators have reported a wide diversity among the rHu-IFN-α species for eliciting a particular biological function (i.e., antiviral, antigrowth, augmentation of the natural killer activity, augmenting tumor antigen expression and inducing alterations in cellular differentiation). In particular, Ortaldo et al. (7) reported that the J species of rHu-IFN-α lacked any ability to boost natural killer activity. It would be attractive to postulate that since rHu-IFN-αJ fails to alter NK activity and cell surface tumor antigen expression, yet, exhibits effective antiviral and antigrowth activity that it represents an interferon species that is functionally inactive as an immunomodulatory compound.

These and previous results suggest that the mechanisms which mediate augmentation of HLA and tumor antigen expression by rHu-IFN-α are different. Cell sorting experiments have shown that HLA-A,B,C is expressed in approximately 100% of the human carcinoma cells and, therefore, the rHu-IFN-α induced HLA expression is a result of more antigen being expressed per cell or an increased antigen production by a subset of responsive cells. These observations are in accordance with those of other investigators who showed that the interferon-mediated increase in HLA seem to be a result of an increased synthesis of the surface antigen. The augmentation of tumor antigen expression by rHu-IFN-αA is different from that of HLA in several aspects. First, mAb B1.1 reacts with 40-60% of the MCF-7 cells, but following rHu-IFN-αA treatment more than 90% of the cell population is CEA positive (8), indicating a recruitment of antigen negative cells to become antigen positive. Second, analysis of single cell subclones of the MCF-7 cell line indicate that three different classes of cells exist: cells expressing either a high or low levels of constitutive tumor antigen expression which are enhanced following IFN-αA treatment; cells expressing a constitutive level of antigen but do not respond with increased expression by rHu-IFN-αA treatment; and cells that are negative for a particular tumor antigen either before or after interferon treatment. These results suggest that the parental MCF-7 cell line consists of a heterogeneous population of cells which vary in their capacity to respond to rHu-IFN-αA with increased tumor antigen expression. Third, the data presented indicates that rHu-IFN-αD and -αJ can induce a modest increase in HLA expression, whereas, both species are unable to alter tumor antigen expression.

The lack of antigen enhancing ability of the rHu-IFN-α D and -αJ species may be a result of: a) preferential inactivation of these IFN-α species by the MCF-7 target cell; b) the inability of the cell surface interferon receptor to recognize these compounds leading to ineffective binding to the receptor; and/or c) their binding to a component of the interferon receptor which cannot elicit the required transmembrane signals to evoke alterations in tumor antigen expression. Present and previous data demonstrate that both rHu-IFN-αD and - αJ inhibit the growth of MCF-7 cells as well as reduce the antigen enhancement induced by rHu-IFN-αA or -αB suggests that the -αD and -αJ species can competitively interact with the interferon receptor. The third possibility, which requires further experimentation, is suggested by the observations that various rHu-IFN-α preparations differ in their ability to induce a number of interferon related properties, including antiviral, antiproliferative, differentiation modulating and antigen enhancing effects. Because of the differences in the structures of rHu-IFN-αD and -αJ versus active rHu-IFN- α species, it is possible that the primary amino acid sequence or the tertiary structures of these molecules which allow for binding and/or activation of only a portion of the cell membrane IFN-α/β receptor, i.e., the region which regulates antiviral (rHu-IFN-αD and -αJ), natural killer cell (rHu-IFN-αD) and HLA expression (rHu-IFN-αD and -αJ), but not the receptor domain required for modulating tumor antigen expression may be responsible for the observed varied effects. Different rHu-IFN-α species, as a result of structural dissimilarities could then exhibit a high, intermediate, or low affinity for specific binding domains within the interferon receptor which are required to elicit the appropriate biological and biochemical responses after interferon binding. These observations provide additional support for the hypothesis that the interferon receptor contains multiple trigger sites that regulate specific subsets of biochemical pathways distal to receptor occupancy (4,15).

Some select mAbs are currently being evaluated in experimental models as well as available clinical material to determine their usefulness in the diagnosis and management of human carcinomas. These mAbs have been tagged with different radionuclides and used to localize occult tumor lesions *in situ*. In addition to the above immunodiagnostic uses, the potential of mAbs to delivery therapeutic effectors to the tumor site is being investigated. The use of recombinant interferon to augment tumor antigen expression in each of these situations merits consideration. The results suggest that rHu-IFN-αA can act as a potent inducer of antigen expression *in vivo*. Moreover, the *in vitro* data showing the lack of tumor antigen enhancement by rHu-IFN-αD was further corraborated by the *in vivo* data establishing the heterogeneity of the rHu-IFN-α species for antigen modulation. While the mechanism responsible for the *in vivo* increase in B6.2 reactivity to MCF-7 tumor extracts following rHu-IFN-αA treatment is unclear, one can speculate that increased synthesis of the 90 kD tumor antigen is a component. Likewise, rHu-IFN-αA treatment may increase the percentage of MCF-7 cell expressing the B6.2-reactive 90 kD antigen thus resulting in a more homogeneous tumor cell population for antigen expression. In a previous study, similar findings resulted in a high localization index for ^{125}I-B6.2 (Fab')$_2$ to a human colon carcinoma grown in athymic mice (18). Furthermore, our data suggest that normal cells which are negative for a distinct mAb-defined tumor antigen (i.e., TAG-72) remain so even after treatment with high levels of recombinant interferon. These findings suggest that interferon treatment *in vivo* could enhance the signal (i.e., surface antigen expression) and thus be considered an adjuvant for the detection and treatment of occult tumor cells by therapeutically-conjugated mAbs. Testing these hypotheses is integral to the development of combination therapies which include a biological response modifier and mAb.

REFERENCES

1. Nagata S, Mantei N, Weissman C (1980). The structure of one of the eight or more distinct chromosomal genes human interferon -α. Nature 287: 401.

2. Pestka S (1983). The human interferons - From protein purification and sequence to cloning and expression in bacteria: before, between and beyond. Arch. Biochem. Biophys. 221:1.

3. Goeddel DV, Leung DW, Dull TJ, Gross M, Lawn RM, McCordliss R, Seeburg PH, Ullrich, A, Yelverton E, Gray

PH (1981). The stucture of eight distinct cloned human leukocyte interferon cDNAs. Nature 290:20.

4. Evinger M, Rubinstein M, and Pestka S (1981). Antiproliferative and antiviral activities of human leukocyte interferons. Arch Biochem Biophys 210:319.

5. Kramer MJ, Dennin R, Kramer C, Jones G, Connell E, Rolon N, Gruarin A, Kale R, Trown PW (1983). Cell and virus sensitivities studies with recombinant alpha interferons. J Interferon Res 3: 425.

6. Lee SH, Kelley S, Chiu H, Stebbing N (1982). Stimulation of natural killer cell activity and inhibition of various leukemic cells by purified human leukocyte interferon subtypes. Cancer Res 42:1312.

7. Ortaldo JR, Herberman RB, Harvey C, Osheroff P, Pan YE, Kelder B, Pestka S (1984). A species of human α interferon that lacks the ability to boost human natural killer activity. Proc Natl Acad Sci USA 81:4926.

8. Greiner JW, Horan Hand P, Noguchi P, Fisher PB, Pestka S, Schlom J (1984). Enhanced expression of surface tumor-associated antigens on human breast and colon tumor cells after recombinant human leukocyte α-interferon treatment. Cancer Res 44:3208.

9. Giacomini P, Aguzzi A, Pestka S, Fisher PB, Ferrone S (1984). Modulation by recombinant DNA leukocyte (α) and fibroblast (β) interferons of the expression and shedding of HLA and tumor associated antigens by human melanoma cells. J Immunol 133:1649.

10. Greiner JW, Horan Hand P, Pestka S, Noguchi P, Fisher PB, Colcher D, Schlom J (1984). Detection and enhancement (by recombinant interferon) of carcinoma cell surface antigens, using monoclonal antibodies. In Levine AT, Vande Woude GF, Topp WC, Watson JD (eds): "Cancer Cells/The Transformed Phenotypes 1," New York:Cold Spring Harbor Press, Vol. II, p 265.

11. Greiner JW, Tobi M, Fisher PB, Langer JA, Pestka S (1985). Differential responsiveness of cloned mammary carcinoma cell populations to the human recombinant leukocyte interferon enhancement of tumor antigen expression. Int J

Cancer 26:159.

12.Allen G, Fantes KH (1980). A family of structural genes for human lymphoblastoid (leukocyte type) interferon. Nature 287:408.

13.Maeda S, McCordliss R, Gross M, Sloma A, Familletti PC, Tabor JM, Evinger M, Levy WP, Pestka S (1980). Construction and identification of bacterial plasmids containing nucleotide sequence for human leukocyte interferon. Proc Natl Acad Sci USA 78:4648.

14.Colcher D, Horan Hand P, Nuti M, Schlom J (1981). A spectrum of monoclonal antibodies reactive with human mammary tumor cells. Proc Natl Acad Sci USA 78:3199.

15.Greiner JW, Fisher PB, Pestka S, Schlom J (1986). Differential effects of recombinant leukocyte interferons on cell surface antigen expression. Cancer Res 46:4984.

16.Greiner JW, Horan Hand P, Wunderlich D, Colcher D (1986). Radioimmunoassay for detection of changes in cell surface tumor antigen expression induced by interferon. Meth Enzymol 119:682.

17.Colcher D, Zalutsky M, Kaplan W, Kufe D, Austin F, Schlom J (1983). Radiolocalization of human mammary tumors in athymic mice by a monoclonal antibody. Cancer Res 43:736.

18.Greiner JW, Guadagni F, Noguchi P, Pestka S, Colcher D, Fisher PB, Schlom J (1987). The use of recombinant interferon to enhance monoclonal antibody-targeting of carcinoma lesions in mice. In press, Science.

The Pharmacology and Toxicology of Proteins, pages 299–306

THROMBOLYTIC THERAPY WITH ACTIVASE (rt-PA)

Elliott B. Grossbard, M.D.

Department of Clinical Research, Genentech, Inc.
South San Francisco, California 94080

Thrombolytic therapy has been demonstrated to reduce mortality following myocardial infarction. Recombinant DNA produced human tissue-type plasminogen activator (Activase[R]) has been shown to be a superior thrombolytic agent. This paper reviews the clinical development of Activase against the background of major coronary thrombolysis trials.

Thrombolytic therapy is based on the concept that accelerated resolution of a thrombus or embolism can have salutary clinical effects. The Food and Drug Administration acknowledged this concept when Streptokinase and Urokinase were approved for use in pulmonary embolism almost ten years ago.

Recently, however, thrombolytic therapy has been applied to the early treatment of coronary thrombosis, which is the precipitating event in over 80 percent of acute (Q wave) myocardial infarction(1). The results from several large randomized trials utilizing Streptokinase suggest that thrombolytic therapy reduces mortality in myocardial infarction(2-5). These studies truly vindicate the notion that, at least in myocardial infarction and probably in other clinical conditions as well, rapid restoration of blood flow through vessels pathologically occluded by thromboembolism is clinically beneficial.

Clinical benefit must be balanced against risk and the use of the "first-generation" thrombolytics (Streptokinase and Urokinase) was obligatorily accompanied by systemic fibrinogenolysis, a phrase denoting the degradation of several important coagulation factors, which may increase the risk of clinically significant bleeding(6). The development of thrombolytic agents with fibrin - or clot-specificity raised the possibility that the risk-benefit ratio for thrombolytic therapy could be significantly enhanced. Tissue plasminogen activator is a natural protein whose affinity for fibrin coupled with its marked activation by fibrin made it a likely candidate to be a superior thrombolytic agent(7). Recombinant DNA technology was necessary to produce sufficient quantities for adequate clinical development and subsequent commercialization(8). The final product of this effort is called recombinant tissue plasminogen activator or Activase[R]. This review will summarize salient points of the clinical trials with Activase that have been undertaken over the past three years.

Biology of Fibrinolysis

The conversion of the inactive proenzyme plasminogen to plasmin, (i.e., plasminogen activation) is the key step in fibrinolysis(9). Plasmin is a nonspecific protease that in addition to digesting fibrin (which is the main constituent of a thrombus) also degrades fibrinogen, factor V, factor VIII and platelet glycoproteins, all of which are important hemostatic factors. Streptokinase and Urokinase have no particular affinity for fibrin and their thrombolytic capability is inseparable from their fibrinogenolytic effect. Activase binds to fibrin and its plasminogen activator property is markedly enhanced by fibrin(7); thus its thrombolytic capacity is enhanced and its fibrinogenolytic effect is reduced compared to the "non-specific" activators. The fibrinogen sparing effect of Activase is relative, not absolute, and high doses and long durations of therapy can produce fibrinogen breakdown(10-12), though less than that observed with "non-specific" activators. Furthermore, a hemostatic plug (a "good" clot) is also composed of fibrin and bleeding might be anticipated in such cases even if fibrinogen were well maintained. Such bleeding, however, would be easier to manage clinically than bleeding associated with extensive fibrinogenolysis, unless it was in a particularly critical and unforgiving location (intracerebral).

Clinical Trials with rt-PA (G11021) in Myocardial Infarction

The first trials with rt-PA in myocardial infarction used a product (code G11021) made on a small scale that was predominantly in the two-chain (linked by a disulfide bond) form. These studies demonstrated that 40-80 mg of Activase over one to three hours could produce recanalization of angiographically proven coronary thrombosis in approximately 70-75 percent of cases(13). These early trials also demonstrated two important safety features. One was that fibrinogen was indeed largely spared. The other was that when aggressive anticoagulation with heparin accompanied thrombolytic therapy in trials with multiple invasive diagnostic procedures, bleeding was frequent, but usually confined to puncture sites.

The two critical trials during this early period were studies comparing Activase to Streptokinase(14,15). Both trials were randomized, both used core radiography laboratories, and both showed Activase to be significantly more effective than Streptokinase in promoting coronary reperfusion. The larger trial (290 patients) was conducted by the NIH and was double blinded. Both studies showed Activase to be more fibrinogen sparing than Streptokinase, although only the European trial showed a difference in bleeding incidence (lower in Activase patients). Since Streptokinase was the drug that had been used in the trials reporting that thrombolytic therapy reduced mortality, and since coronary reperfusion is uniformly felt to be the main effect of thrombolytic therapy in myocardial infarction, a drug that was significantly superior to Streptokinase in promoting this effect and which was arguably safer was recognized as a significant breakthrough.

Trials with rt-PA (G11044) in Myocardial Infarction

As development of Activase continued, a new predominantly single chain form (code G11044) was introduced into clinical trials. Pharmacokinetic studies in primates and rabbits suggested that G11044 was cleared approximately 30-40 percent more rapidly than G11021(16). The clinical significance of this observation was demonstrated by trials conducted by the NIH and at Massachusetts General Hospital which confirmed that the more rapid clearance of G11044 necessitated an increase in dose(17,18). A series of trials conducted at doses of 80, 100 and 150 mg over three to six hours established several key points. The first was that 100 mg of G11044 was clinically equivalent to 80 mg of G11021, and that 80 mg of G11044 was less effective than 80 mg of G11021. The second was that 150 mg of G11044 promoted more rapid recanalization than 100 mg, but that the final incidence of recanalization observed with the two regimens was similar. Third, fibrinogen was depressed most by the highest dose, and bleeding, including clinically significant bleeding, was more frequent at the highest dose. Finally, G11044 at 100 mg consumed less fibrinogen than G11021 at 80 mg.

Adjuncts to Coronary Thrombolysis

If, as it now appears, thrombolytic therapy reduces mortality in myocardial infarction, and Activase is the most effective agent, the next critical question centers around the timing of and need for other pharmacologic measures (Beta blockade, nitrates) and coronary angioplasty (PTCA). The recently concluded TAMI trial(19), (Thrombolysis and Angioplasty in Myocardial Infarction) found in a study of almost 400 patients that PTCA need not be done immediately following Activase thrombolysis in that the ventricular function and reocclusion rates did not differ between patients randomized to immediate PTCA and those who underwent PTCA electively. The more ambitious (4,000 patient) NIH TIMI trial will study the need for and timing of PTCA and ß-blocker therapy. In both these trials, all patients receive Activase thrombolysis. A pilot study for this trial (with over 300 patients) had a remarkably low (4.8 percent) in-hospital mortality(20).

Other Trials with Activase

Studies with Activase are underway in a variety of other indications. Preliminary results in pulmonary embolism(21) shows significant angiographic improvement within six hours in 90 percent of treated patients. Placebo controlled trials in deep venous thrombosis(22) and unstable angina(23) have demonstrated statistically significant improvement in patients receiving Activase. Studies in peripheral arterial occlusion(24) and central retinal vein occlusion are encouraging as well. Perhaps the most challenging clinical indication is stroke and protocols are currently under consideration for the use rt-PA in stroke.

Complications of Therapy

The clinical experience with Activase in the U.S., Europe, and worldwide now exceeds 3,000 patients. The principal complication of therapy is bleeding, usually at puncture sites. The most significant complication is intra-cranial bleeding. With doses on the order of 100 mg the incidence of intracranial bleeding is 0.3 to 0.5 percent. When the dose is increased to 150 mg, some studies suggest the incidence may increase to 1.0 percent or greater. These studies usually include therapeutic doses of heparin, so the precise relationship of thromoblytic therapy to stroke is unclear. Furthermore, controlled trials suggest a 0.5 percent incidence of stroke in the immediate post-infarction period in nonthrombolytic-treated groups. Thus, it is not immediately clear that all intracranial hemorrhage in the immediate post-infarct period is related to thrombolytic therapy.

Conclusions

Recombinant DNA technology has enabled scientists to produce on a commercial scale a scarce protein with properties that suggest a significant role for it in the treatment of the most lethal disease in the Western World. It is reasonable to speculate that additional and nonobvious applications of fibrinolytic agents like Activase and its third-generation successors will be forthcoming shortly and that some of these applications will have a significant impact on quality of life as well as duration.

REFERENCES

1. DeWood, MA, Spores, J, Notske R, Mouser LT, Burroughs R, Golden MS, Lang HT (1980). Prevalence of total coronary occlusion during early hours of transmural myocardial infarction. New Engl J Med 303:897.
2. Kennedy JW, Ritchie JL, Davis KB, Fritz JK (1983). Western Washington randomized trial of intracoronary streptokinase in acute myocardial infarction. New Engl J Med 309:1477.
3. Yusuf S, Collins, Peto R, Furberg C, Stampfers MJ, Goldhaber SZ and Hennekens CH (1985). Intravenous and intracoronary fibrinolytic therapy in acute myocardial infarction: Overview of results on mortality, reinfarction and side effects from 33 randomized controlled trials. Eur. Heart J. 6:556.
4. Simoons, ML, Serruys PW, V/D Brand M, Bar F, DeZwaans C, Res J, Verheught FWA, Krauss XH, Remme WJ, Vermeer F, Lubsen J (1985). Improved survival after early thrombolysis in acute myocardial infarction. Lancet 1:578.
5. GISSI (1986). Effectiveness of intravenous thrombolytic treatment in acute myocardial infarction. Lancet 1:397.
6. Sharma GVRK, Cella G, Parisi AF, Sasahara AA (1982). Thrombolytic therapy. New Engl J Med. 306:1268.
7. Collen D (1980). On the regulation and control of fibrinolysis. Thrombos. Haemostas. 43:77.
8. Pennica D, Holmes WE, Kohr WJ, Harkins RN, Vehar GA, Ward CA, Bennett WF, Yelverton E, Seeburg PH, Heyneker HL, Goeddel DV, Collen D (1983). Nature 301:214.
9. Robbins KC (1982). The plasminogen-plasmin enzyme system. In Colman RW, Hush J, Marder VJ, Salzman EW (eds): "Haemostasis and Thrombosis", Philadelphia: J. B. Lippincott.
10. Sobel BE, Gross RW, Robison AK (1984). Thrombolysis, clot selectivity and kinetics. Circulation 70:161.
11. Topol, EJ, Bell WR, Weisfeldt ML (1985). Coronary thrombolysis with recombinant tissue-type plasminogen activator. Annal of Int. Med 103:837.
12. Collen D, Bounameaux H, De Cock F, Lijnen HR, Verstraete M (1986). Analysis of coagulation and fibrinolysis during intravenous infusion of recombinant human tissue-type plasminogen activator in patients with acute myocardial infarction. Circulation 73:511.

13. Collen D, Topol EJ, Tiefenbrunn J, Gold HK, Weisfeldt ML, Sobel BE, Leinbach RC, Brinker JA, Ludbrook PA, Yasuda T, Bulkley BH, Robison AK, Hutter AM, Bell WR, Spadaro JJ, Khaw BA, Grossbard EB (1984). Coronary thrombolysis with recombinant human tissue-type plasminogen activator: a prospective randomized placebo controlled trial. Circulation 70:1012.
14. The TIMI Investigators (1985). Special report from the NHLBI thrombolysis in myocardial infarction (TIMI) trial. New Engl J Med 312:932.
15. Verstraete M, Bernard R, Bory M, Brower RW, Collen D, de Bono DP, Erbel R, Huhmann W, Lennane RJ, Lubsen J, Mathey D, Meyer J, Michels HR, Rutsch W, Schartl M, Schmidt W, Uebis R, von Essen R (1985). Randomized trial of intravenous recombinant tissue-type plasminogen activator versus intravenous streptokinase in acute myocardial infarction. Lancet 1:842.
16. Baughmann RL (1987). The pharmacokinetics of tissue plasminogen activator. In Sobel BE, Collen D, Grossbard EB (eds): "Tissue plasminogen activator in thrombolytic therapy", New York: Marcel Dekker.
17. Mueller HS, for TIMI investigators (1986). Different fibrinolytic potencies of two forms of tissue-type plasminogen activator. Clin Res 34:631A.
18. Garabedian HD, Gold HK, Leinbach RC, Johns JA, Yasuda T, Kanke M, Collen D (1987). Comparative properties of two clinical preparations of recombinant human tissue-type plasminogen activator in patients with acute myocardial infarction. J An Coll Card. in press.
19. Topol EJ, Califf RM, George BS, Kereiakes DJ, Abbotsmith CW, Candela RJ, Lee KL, Pitt B, Stack RS, O'Neill WW (1987). A multicenter randomized trial of intravenous recombinant tissue plasminogen activator and emergency angioplasty in acute myocardial infarction. Submitted for publication.
20. Braunwald E (1986). Presented at George Washington University Symposium on thrombolytic therapy. Dallas, Texas.
21. Goldhaber SZ, Vaughn DE, Markis JE, Selwyn AP, Meyerovitz MF, Loscalzo J, Kim DS, Kessler CM, Dawley DL, Sharma GVRK, Sasahara A, Grossbard EB, Braunwald E (1986). Acute pulmonary embolism treated with tissue plasminogen activator. Lancet 1:886.

22. Turpie AGG, Jay RM, Carter CJ, Hirsh J (1985). A randomized trial of recombinant tissue plasminogen activator for the treatment of proximal deep vein thrombosis. Circulation 72:III-193.
23. Gold HK, Johns JA, Leinbach RC, Yasuda T, Grossbard E, Zusman R, Collen D (1987). A randomized blinded placebo-controlled trial of recombinant human tissue-type plasminogen activator in unstable angina pectoris. Submitted for publication.
24. Groor RA, Risius B, Young JR, Denny K, Beven EG, Geisinger MA, Hertzer NR, Krajewski LP, Lucas FV, O'Hara PJ, Ruschhaupt WF, Winton S, Zelch MG, Grossbard EB (1986). J Vasc Surg 3:115.

The Pharmacology and Toxicology of Proteins, pages 307–323

PROTEASE NEXIN I. STRUCTURE AND POTENTIAL FUNCTIONS

Joffre B. Baker, Michael McGrogan,[2]
Christian C. Simonsen,[2] Randy W. Scott,[2]
Robert S. Gronke and Allen Honeyman[3]

Department of Biochemistry, University of Kansas
Lawrence, Kansas 66045

ABSTRACT Protease nexin I (PNI) is a potent trypsin, thrombin and urokinase inhibitor that is secreted by foreskin fibroblasts and several other types of extravascular cells. The recent sequencing of PNI cDNA indicates that this inhibitor is a member of the serpin family of proteins, which includes several of the major serine protease inhibitors of blood plasma. The primary structure of PNI is virtually identical to a glial-derived neurite promoting factor which stimulates extension of neurite-like projections from neuroblastoma cells, and inhibits the migration of granule cell neurons. In fibroblast cultures secreted PNI inhibits the mitogenic response to thrombin and regulates secreted urokinase. Purified PNI protects extracellular matrix proteins from destruction by human fibrosarcoma cells, and inhibits the growth of these tumor cells. Immunofluorescence studies indicate that PNI becomes deposited on cell surfaces and extracellular matrices. These interactions may be related to the high affinity

[1]This work was supported by National Institutes of Health Research Grants CA-29307 (J.B.B.), CA-00886 (J.B.B.), and GM-37891 (M.M.) and a grant-in-aid from the American Heart Association, Kansas Affiliate.

[2]Invitron Corporation, Redwood City, CA 94063.

[3]Department of Microbiology, University of Kansas, Lawrence, KS 66045.

of PNI for heparin and other sulfated polysaccharides. A factor that is similar to PNI is also present on platelets and regulates thrombin binding to platelet receptors.

INTRODUCTION

It has long been suspected that cellular proteases play important roles in tissue matrix destruction and remodeling, cell movement, invasion and metastasis, and cell growth. In the last decade the natures of the proteases and other regulators that participate in these processes have begun to be elucidated. It is now clear that the same kinds of regulatory mechanisms, and to some extent even the same proteins, that function in blood coagulation and fibrinolysis also function in these processes. Thus, thrombin, the ultimate protease in the coagulation cascade, is also a fibroblast mitogen (1) and, possibly, a neuroregulator (2). Urokinase, a plasminogen activator, has been implicated in tissue matrix destruction in normal tissue remodeling (3), and cancer (4).

In the coagulation and fibrinolysis pathways regulatory serine proteases become activated and activate other serine proteases, culminating in the cleavage of prominent structural proteins (fibrinogen and fibrin, respectively). These processes can be initiated (a) rapidly, because serine proteases are continually present as circulating zymogens or latent proteases, and (b) at precisely the sites where they are needed, by the binding of the proteases to prelocalized specific acceptor sites that can amplify or modify their enzymatic activities (5,6). Deactivation of these processes is also extremely rapid, because plasma contains more than half a dozen specific circulating protease inhibitors, which belong to the evolutionarily related family of proteins called the serpins (7,8). All of the protease inhibitors in the serpin family are suicide inhibitors. Their reactive centers, which are always located 30-40 residues from their carboxyl ends, form stable tetrahedryl or acyl complexes with the catalytic site seryl residues of their protease targets (8). Most of the known serpins are present at ~μM levels in plasma. Several pathological conditions have been traced to genetic or environmentally-induced deficiencies in them (9-11).

The present manuscript reviews recent findings regarding protease nexin I, (PNI) a serpin which is secreted by several types of anchorage-dependent extravascular cells in culture but which is not present at more than trace levels in plasma. Several lines of evidence indicate that this inhibitor may regulate the extravascular actions of thrombin and urokinase. In addition, despite its low concentration in plasma, PNI could also control the activity of intravascular thrombin, because a PNI-like protein is bound to the surfaces of platelets where it inhibits thrombin binding to platelet receptors.

Protease Nexins.

In 1979 Baker *et al*. reported that a large fraction of the ^{125}I-thrombin that bound to human foreskin fibroblasts was taken into 77 kilodalton SDS-resistant complexes with a cellular component (12). At the time, this component was assumed to be a cell surface receptor for thrombin, but subsequently it was shown (13) that this "receptor" accumulated in the medium in soluble form, and bore several similarities to antithrombin III (AT3), the major plasma inhibitor of thrombin. Like AT3 (14), the fibroblast factor (i) did not complex thrombin that was blocked at its catalytic site with diisopropyl phosphofluoridate, (ii) had a high affinity for heparin, and (iii) complexed thrombin at a vastly accelerated rate in the presence of heparin (12). However, the fibroblast factor was distinct from AT3 immunologically and electrophoretically, and in its ability to rapidly bind urokinase.

Low *et al*. (15) demonstrated why these 77 kilodalton complexes were found associated with the cells. Foreskin fibroblasts possess cell surface receptors which bind the thrombin-factor complexes that form in the culture medium. The receptor-bound complexes are rapidly internalized, and the protease moiety hydrolyzed in lysosomes (15). The fate of the thrombin-complexing factor is not known because to date only the ^{125}I-thrombin has been labelled in uptake experiments.

Because the only known function of the thrombin and urokinase-binding factor was to mediate the binding of these proteases to the cells, it was named protease nexin, or PN ("nexin" is Latin for to bind or link). In 1982 it was shown that human foreskin fibroblasts produce two other protease nexins which are distinguished from the thrombin and urokinase-binding protease nexin in terms of their electrophoretic mobilities and protease specificities, but which also are cleared from culture medium by receptor-mediated endocytosis (16,17). PNI (~43 kilodaltons) is now the term used for the thrombin and urokinase-binding protease nexin. PNII (~95 kilodaltons) binds the epidermal growth factor-binding protein (16). PNIII (~40 kilodaltons) binds the gamma subunit of nerve growth factor (17).

Little is known about the receptors for any of the protease nexins. Howard and Knauer have recently shown that the PNI receptors have a remarkably low and broad pH optimum for thrombin-PNI binding, extending from pH 4-5.5 (18). In contrast, the pH optimum for binding of EGF-binding protein-PNII complexes is pH 7.5, suggesting that the receptors for PNI-protease complexes and PNII-protease complexes are distinct.

To date PNI has been the most intensively studied of the PNs. W. E. Van Nostrand and D. D. Cunningham have recently developed a procedure for purifying PNII (personal communication).

STRUCTURE AND BIOCHEMICAL PROPERTIES OF PNI

PNI was first purified by Scott and Baker (19), who used as starting material serum-free medium conditioned by human foreskin fibroblasts in microcarrier cultures. The medium was concentrated by ultrafiltration, passed down a heparin-agarose column at ~0.3 M NaCl, and eluted with 1.0 M NaCl. The eluted PNI was $\geq$ 75% pure after this step. A second step involving chromatography on octyl agarose was later found to give variable recoveries, and was replaced by a gel exclusion step (20). Recently, Farrell *et al*. used dextran sulphate-Sepharose and anion exchange chromatography to purify PNI from medium conditioned by foreskin fibroblasts in roller bottles (21).

SDS-polyacrylamide gel electrophoresis and sedimentation equilibrium analysis indicate PNI molecular weights of ~50 and 43 kilodaltons, respectively (20). The presence of 6% carbohydrate (20,22) in PNI may explain this discrepancy. Nonequilibrium pH gradient electrophoresis of PNI resolves half a dozen isoforms (19), which may be generated by differential sialylation. In addition, use of a linear NaCl gradient to elute PNI from heparin-agarose reveals two different heparin affinity forms of PNI (20). It is possible that these forms of PNI correspond to the α and β forms which have been detected by cDNA sequencing (vide infra).

Reactivity with Serine Proteases.

Analysis of the rates of which PNI inactivates purified serine proteases (20) indicates that PNI is a remarkably fast-acting inhibitor of several serine proteases that cleave proteins after Lys or Arg residues, in particular trypsin ($k_{assoc.}$ ~4×10^6 M^{-1} S^{-1}), thrombin ($k_{assoc.}$ ~6×10^5 $M^{-1}S^{-1}$) and urokinase ($k_{assoc.}$ ~2×10^5 $M^{-1}S^{-1}$). PNI is a substantially faster inhibitor of thrombin than is either AT3 or heparin cofactor II (20,21). In the presence of heparin, PNI inhibits thrombin at nearly the diffusion-controlled rate. Heparan sulfate also accelerates the rate of thrombin inhibitor, but by a smaller increment than does heparin (25).

PNI is a relatively rapid inhibitor of urokinase, although two other recently described inhibitors, plasminogen activator inhibitor-1 (PAI-1) (26), and plasminogen activator inhibitor-2 (PAI-2) (27), inhibit urokinase more rapidly. Heparin does not accelerate the rate at which PNI inhibits urokinase (28).

Homology to Other Serpins and to Glial-derived Neurite Promoting Factor.

Recently, cDNA encoding PNI has been cloned, sequenced and expressed (M. McGrogan, J. Gohari, M. P. Li, C. Hsu, R. W. Scott, C. C. Simonsen and J. B. Baker; manuscript submitted). A mixed 14-mer oligonucleotide probe corresponding to amino acids 20-24 at the N-terminus of PNI was used to screen a λGT10 foreskin fibroblast cDNA library. Several positive clones were shown to contain nucleotide sequence corresponding to the N-terminal sequence of PNI. That these cDNAs code for PNI has been verified by comparing the deduced amino acid sequences to the amino acid sequences of a number of tryptic cleavage fragments of PNI protein. The deduced Mr of the PNI polypeptide is 41,919.

Sequencing several PNI cDNAs has revealed that fibroblasts contain two forms of PNI mRNA which differ only by a three base insertion and yield polypeptides containing Thr_{310}-Arg-Ser (αPNI) or Thr_{310}-Thr-Gly-Ser (βPNI).

```
                                                         MNW    PNI

                                ↓
MYSNVIGTVTSGKRKVYLLSLLLIGFWDCVTCHGSPVDICTAKPRDIPMNPMCIYRSPEK    AT3

                ↓        10        20        30        40
HLPLFLLASVTLPSICSHFNPLSLEELGSNTGIQVFNQIVKSRPHDNIVISPHGIASVLG    PNI
                                             ***  **  *
KATEDEGSEQKIPEATNRRVWELSKANSRFATTFYQHLADSKNDNDNIFLSPLSISTAFA    AT3

     50        60              70        80         90
MLQLGADGRTKKQLAMVMRY------GVNGVGKILKKINKAIVSKKNKDI-VTVANAVFV    PNI
*  ***   *  **  *                   * *     * **      **  *
MTKLGACNDTLQQLMEVFKFDTISEKTSDQIHFFFAKLNCRLYRKANKSSKLVSANRLFG    AT3

 100       110       120       130        140       150
KNASEIEVPFVTRNKDVFQCEVRNVNFEDPASAC-DSINAWVKNETRDMIDNLLSPDLID    PNI
                *         *   *      ** ** * *   *        *
DKSLTFNETYQDISELVYGAKLQPLDFKENAEQSRAAINKWVSNKTEGRITDVIPSEAIN    AT3

   160       170       180       190       200       210
GVLTRLVLVNAVYFKGLWKSRFQPENTKKRTFVAADGKSYQVAMLAQLSVFRCGSTSAPN    PNI
  ** *****  ******** * **** *  *  *** *    *  *   **
E-LTVLVLVNTIYFKGLWKSKFSPENTRKELFYKADGESCSASMMYQEGKFRYRRVAEGT    AT3

  220       230       240       250       260       270
DLWYNFIELPYHGESISMLIALPTESSTPLSAIIPHISTKTIDSWMSIMVPKRVQVILPK    PNI
       ***  *  * *   **      *              *          *  *
QV----LELPFKGDDITMVLILP-KPEKSLAKVEKELTPEVLQEWLDELEEMMLVVHMPR    AT3

  280       290       300        310       320       330
FTAVAQTDLKEPLKVLGITDMFDSSKANFAKI-TTGSENLHVSHILQKAKIEVSEDGTKA    PNI
*       *** *   *  * *   *     *   *   * **    **  ** * *  *
FRIEDGFSLKEQLQDMGLVDLFSPEKSKLPGIVAEGRDDLYVSDAFHKAFLEVNEEGSEA    AT3

   340        350         360       370
SAATTAILIARS-SPPW--FIVDRPFLFFIRHNPTGAVLFMGQINKP                 PNI
 * *      **  *    *   **** ***  *     ***    *
AASTAVVIAGRSLNPNRVTFKANRPFLVFIREVPLNTIIFMGRVANPCVK              AT3
          / \
         P1  P1'
```

FIGURE 1. Alignment of the primary sequence of PNI with that of AT3 (29). The arrows indicate the N-terminal residues of the secreted proteins. Asterisks denote shared amino acids.

The α form is present in a 2:1 excess. Both forms of PNI cDNA have been expressed in CHO cells and yield PNI that complexes thrombin and urokinase. Southern blotting experiments indicate that the PNI gene is present in a single copy. The inserted triplet of βPNI is a splice acceptor site junction. Moreover, genomic sequencing demonstrates that an AG dinucleotide preceeds the inserted triplet in the PNI gene. Thus it seems likely that the α and β forms of PNI are generated by alternative splicing.

Fig. 1 shows the deduced amino acid sequence of βPNI aligned with the amino acid sequence of AT3. PNI shares 31% of its sequence with AT3, confirming that PNI, like antithrombin III (8), is a member of the serpin family. Table 1 shows the optimal homology scores for PNI and 7 compared serpins. PNI is clearly more closely related to AT3 than to the other serpins, although its homology to ovalbumin is also noteworthy. The deduced P1 and P1' (reactive center) residues of PNI, Arg 346 and Ser 347, respectively, align with and are identical to those of AT3 (Fig. 1).

TABLE 1
COMPARISON OF HUMAN PNI TO OTHER SERPINS

Protein	Optimal Homology Score
PNI (self)	1881
Antithrombin III	526
Ovalbumin	431
Heparin cofactor II	314
α1 antichymotrypsin	312
α1 proteinase inhibitor	301
C1 inhibitor	255
Angiotensinogen	131
Rod C protein*	50

*not a member of the serpin family; a B. subtilis protein chosen for random comparison.

Fast P program. Sequence data was obtained from the NBRF-PIR data base except for C1 inhibitor (30), heparin cofactor II (31), and rod C (A.L. Honeyman and G.C. Stewart, unpublished data).

Intriguingly, the amino acid sequence of βPNI is virtually identical to that of a glial cell-derived neurite promoting factor (GdNPF) which was recently been cloned and sequenced by Gloor et al. (32). The sole difference in primary structures is at position 241 (Glu in PNI and Ser in GdNPF). The effects of PNI/GdNPF on neurons and other cells are reviewed below.

POTENTIAL FUNCTIONS

Neuronal Cells

In 1973 Monard et al. reported that a factor secreted by cultured glioma cells (GdNPF) promoted the outgrowth of axon-like projections from neuroblastoma cells (33). The GdNPF exhibited thrombin inhibitory activity which was apparently required for its neurite-promoting functions, since other thrombin inhibitors, namely hirudin and a synthetic thrombin inhibitor, also promoted neurite outgrowth (2). Guenther, Nick, and Monard purified GdNPF from serum-free medium conditioned by rat glioma cells in microcarrier culture using chromatography on heparin-Sepharose and blue dye-agarose (34). The purified protein inhibited thrombin, urokinase, and trypsin and stimulated neurite outgrowth in neuroblastoma cell cultures (half maximally effective dose: 1.6×10^{-10} M). Recently, Lindner et al. showed that purified GdNPF inhibited the migration of ^{3}H-thymidine-labeled granule cells in cultured explants from early postnatal mouse cerebellum (35). This action was inhibited by thrombin, suggesting that the effect required the protease inhibitory activity of GdNPF. Strikingly, granule cell migration was not affected by several other inhibitors of thrombin and plasmin, suggesting that either the critical protease is recognized by GdNPF but not the other inhibitors, or that another feature of GdNPF apart from its protease inhibitory activity is also required for inhibition of granule cell migration. The observations that GdNPF promotes neurite outgrowth in the neuroblastoma system but inhibits granule cell migration, suggests that a delicate balance between the levels of protease and protease inhibitor (GdNPF) regulates cell migration or extension. GdNPF could prevent granule cell migration by preventing release of cells from adhesive contacts with the extracellular matrix (35), and could promote neurite formation by protecting matrix components, cellular adhesion to which is necessary for projection of cell extensions (2,32).

The similarity of GdNPF to PNI is indicated by findings that rat brain astrocytes and rat G6 glioma cells produce a factor that forms SDS-resistant complexes with thrombin and urokinase which are of the appropriate molecular weights and are immunoprecipited by anti-PNI antibody (36,37).

Connective Tissue Cells.

Secreted PNI regulates the interaction of foreskin fibroblasts with thrombin, which is mitogenic, and with the plasminogen/plasminogen activator system. Unlike dose-response curves generated with other mitogens, that for thrombin is markedly sigmoidal (38,39). Consistent with findings that thrombin requires its proteolytic activity for mitogenic activity (40), the minimum dose of thrombin that stimulates a detectable mitogenic response in serum-free fibroblast cultures roughly corresponds to the steady state concentration of PNI in conditioned culture medium (~2 nM). Supplementation of fibroblast cultures with additional PNI moves the thrombin dose-dependence curve further to the right (41). TC2 Chinese hamster fibroblasts, which produce little or no PNI, proliferate in response to very low (pM) doses of thrombin 42).

Like many anchorage dependent cells, foreskin fibroblasts secrete urokinase proenzyme (28,43). This zymogen does not interact with PNI, but once it is activated (which occurs in the presence of exogenous thrombin, trypsin, or plasmin) the resulting two chain urokinase is complexed by PNI (43). Mitogens (EGF, thrombin) and phorbol myristate acetate (PMA) stimulate the secretions of both urokinase proenzyme and PNI (44). Secretions of urokinase and PNI are further correlated in that the rank order of potency of PMA, EGF, and thrombin for stimulation of PNI secretion and for stimulation of urokinase secretion are the same (PMA > EGF > thrombin).

These findings are consistent with the possibility that PNI functions as a regulator of urokinase in this system. PNI is the major anti-urokinase secreted by foreskin fibroblasts (28), but this may not be the case for fibroblasts from other tissues (42). In many cell types the major anti-activator is PAI-1, not PNI (46).

In view of the anti-urokinase activity of PNI, and evidence that secreted urokinase plays a role in tumor cell invasion (4), Bergman et al. examined the effect of PNI on the ability of a human tumor cell line (HT1080) to destroy extracellular matrix (47). In the serum-containing cultures HT1080 cells, but not normal fibroblasts, rapidly destroyed the extracellular matrix substratum that had been synthesized by rat aortic smooth muscle cells, consistent with the earlier results of Jones and DeClerk (48). PNI detectably retarded matrix destruction at 2 nM, and completely inhibited matrix destruction for the entire 10 day duration of the experiment at 0.2 μM. Secreted HT1080 urokinase was probably a target of PNI in these experiments because anti-urokinase antibody partially inhibited matrix destruction. It was also noted that PNI transiently, but dramatically, inhibited the growth of HT1080 cells when they were seeded on the matrices, but not when they were seeded on plastic.

PNI also markedly inhibited the plasminogen-dependent destruction of mouse myotube extracellular matrix that was caused by medium conditioned by mouse G8 myoblasts (J. S. Rao, C. B. Kahler, J. B. Baker and B. W. Festoff, unpublished data). The latter cells secrete urokinase in large amounts. The half maximally effective dose of PNI was ~4 nM.

Platelets.

Despite the high affinity of PNI for thrombin, PNI can not be a significant soluble plasma inhibitor of thrombin because this inhibitor is present, at most, at trace levels in plasma (49). However, small amounts of an inhibitor that is similar to PNI is associated with platelets, and accounts for most of the specific binding of ^{125}I-thrombin to platelets at ^{125}I-thrombin concentrations below ~0.3 nM (50). Complexes between ^{125}I-thrombin and the PNI-like factor (PNp) do not form when the ^{125}I-thrombin is preincubated with diisopropylphosphofluoridate (51), and are electrophoretically and immunologically similar to ^{125}I-thrombin-PNI. However, the PNp factor does not bind urokinase (50). The latter finding may indicate that the platelet factor is not identical to PNI, but it is also possible that the platelet surface alters the protease specificity of PNI.

^{125}I-thrombin-PNp complexes accumulate both on the platelets and in the binding incubation buffer. The latter complexes do not bind to platelets (51). These complexes may originally form on the platelet surface and be sloughed, or be formed as a result of platelets releasing intragranular PNp following thrombin stimulation (51; R. S. Gronke, unpublished observations).

The dose-dependence curve for the reversible binding of ^{125}I-thrombin receptors is sigmoidal (50), with little reversible binding occurring until $\geq$ 25% of PNp is saturated with ^{125}I-thrombin. This suggests that at the low thrombin doses PNp may efficiently capture thrombin that would otherwise bind to platelet receptors. Recently an anti-PNI monoclonal antibody that blocks the protease inhibitory activity of PNI has been isolated (S. Wagner, W. E. Van Nostrand and D. D. Cunningham, manuscript submitted). Strikingly, this antibody inhibits PNp binding of ^{125}I-thrombin and concomitantly increases the binding of ^{125}I-thrombin to platelet thrombin receptors (R.S.G., S.W., W.E.V.N., D.D.C. and J.B.B., unpublished data). The quantity of PNp per platelet is insufficient to significantly decrease the total number of collisions of thrombin with the platelet surface, even at subnanomolar thrombin doses (50), suggesting that PNp and thrombin receptors are colocalized.

TISSUE DISTRIBUTION

In a screening study employing 15 different types of cultured cells, 10 were found to secrete factors that resembled PNI in their heparin-accelerated formation of 77 kilodalton complexes with ^{125}I-thrombin, and in the heparin sensitive binding of these complexes to the examined cells (52). Unfortunately, the ability of the factors to complex urokinase was not examined in this study. Recently, Lijnen _et al_. have described a ~50 kilodalton urinary thrombin and urokinase inhibitor which is distinct from PNI both immunologically and in that it inhibits urokinase sluggishly in the absence of heparin (50).

The following cultured cells have been found to secrete a thrombin and urokinase inhibitor that is immunologically and electrophoretically similar to human foreskin fibroblast PNI: mouse G2 myotubes (B. W. Festoff, D. H. Hantai, J. S. Rao and J. B. Baker, manuscript submitted), human umbilical smooth muscle cells

(54; W. Laug, B. L. Bergman and J. B. Baker, unpublished data), human HT1080 fibrosarcoma cells (J.B.B., unpublished data), rat brain astrocytes (36), and rat G6 glioma cells (37). In Northern blotting experiments PNI mRNA has also been detected in human SK hepatoma cells and human colon fibroblasts (M.M. and C.C.S., unpublished data).

The cellular expression of PNI *in vitro* and *in vivo* could clearly be quite different. In the only published study on *in vivo* distribution, Gloor *et al*., using Northern analysis, detected PNI/GdNPF mRNA in rat brain and (at a lower level) heart but not in spleen or liver (32).

The high affinity of PNI for heparin and other sulfated polysaccharides could immobilize and concentrate PNI on cell surfaces and extracellular matrices. Using immunofluorescence, PNI antigen has been detected on the surfaces of low density foreskin fibroblasts (55), and on the extracellular matrix of confluent fibroblasts (D.H. Farrell, S.L. Wagner, and D. D. Cunningham, manuscript submitted), in the latter case colocalized with fibronectin. Antibody against (human) PNI also stains transverse sections of mouse stermomastoid muscle, colocalizing with the neuromuscular junctions (56; B. W. Festoff, D. H. Hantai, J. S. Rao, and J. B. Baker, manuscript submitted). This observation is provocative in view of the above mentioned effects of PNI/GdNPF on cultured neuronal cells (2,33,35), the secretion of PNI by cultured mouse skeletal muscle cells, and the deposition of heparin sulfate proteoglycan at neuromuscular junctions (57).

ACKNOWLEDGEMENTS

The authors wish to thank Dennis Cunningham, Barry Festoff, Bill Van Nostrand, J. S. Rao and Steven Wagner for sharing their unpublished data.

REFERENCES

1. Chen L-B, and Buchanan, JM (1975). Mitogenic activity of blood components. I. Thrombin and prothrombin. Proc Natl Acad Sci (USA) 72:131.
2. Monard, D, Niday, E, Limat, A, and Solomon, F (1983). Inhibition of protease activity can lead to neurite extension in neuroblastoma cells. Prog in Brain Res 58:359.

3. Ossowski, L, Biegel, D, and Reich, E (1979). Mammary plasminogen activator: Correlation with involution, hormonal modulation and comparison between normal and neoplastic tissue. Cell 16:929.
4. Ossowski, L, and Reich, L (1983). Antibodies to plasminogen activator inhibit tumor metastasis. Cell 35:611.
5. Jackson, CM, and Nemerson, Y (1980). Blood coagulation. Ann Rev Biochem 49:765.
6. Esmon, CT (1987). The regulation of natural anticoagulant pathways. Science 235:1348.
7. Travis, J, and Salvesen, GS (1983). Human plasma proteinase inhibitors. Ann Rev Biochem 52:655.
8. Carrell, R, and Boswell, DR, Serpins: The superfamily of plasma serine proteinase inhibitors. In Barrett A, Salvesen G (eds): "Proteinase Inhibitors," Amsterdam: Elsevier (in press).
9. Rosen, FS, Charache, P, Pensky, J, and Donaldson, V (1965). Hereditary angioneurotic edema: Two genetic variants. Science 148:957.
10. Eriksson, S (1964). Pulmonary emphysema and alpha-antitrypsin deficiency. Acta Med Scand 175:197.
11. Marciniak, E, Farley, CH, and DeSimone, PA (1974). Familial thrombosis due to antithrombin III deficiency. Blood 43:219.
12. Baker, JB, Simmer, RL, Glenn, KC, and Cunningham, DD (1979). Thrombin and EGF become linked to cell surface receptors during mitogenesis. Nature 278:743.
13. Baker, JB, Low, DA, Simmer, RL, and Cunningham, DD (1980). Protease-nexin: A cellular component that links thrombin and plasminogen activator and mediates their binding to cells. Cell 21:37.
14. Rosenberg, RD, and Damus, PS (1973). The purification and mechanism of action of human antithrombin-heparin cofactor. J Biol Chem 248:6490.
15. Low, DA, Baker, JB, Koonce, WC, and Cunningham, DD (1981). Released protease-nexin regulates cellular binding, internalization and degradation of serine proteases. Proc Natl Acad Sci (USA) 78:2340.
16. Knauer, DJ, and Cunningham, DD (1982). Epidermal growth factor carrier protein binds to cells via a complex with released carrier protein nexin. Proc Natl Acad Sci (USA) 79:2310.

17. Knauer, DJ, Scaparro, KM, and Cunningham, DD (1982). The gamma subunit of 7S nerve growth factor binds to cells via complexes formed with two cell-secreted nexins. J Biol Chem 257:15098.
18. Howard, EW, and Knauer, DJ (1987). Characterization of the receptor for protease nexin-1: protease complexes on human fibroblasts. J Cell Physiol (in press).
19. Scott, RW, and Baker, JB (1983). Purification of human protease nexin. J Biol Chem 258:10439.
20. Scott, RW, Bergman, BL, Bajpai, A, Hersh, RT, Rodriguez, H, Jones, BN, Watts, S, and Baker, JB (1985). Protease nexin: Properties and a modified purification procedure. J Biol Chem 260:7029.
21. Farrell, DH, Van Nostrand, WE, and Cunningham, DD (1986). A simple two-step purification of protease nexin. Biochem J 237:907.
22. Howard, EW, and Knauer, DJ (1986). Biosynthesis of protease nexin-I. J Biol Chem 261:14184.
23. Jordan, R, Beeler, D, and Rosenberg, R (1979). Fractionation of low molecular weight heparin species and their interaction with antithrombin. J Biol Chem 254:2902.
24. Tollefsen, DM, Majerus, DW, and Blank, MK (1982). Heparin cofactor II. Purification and properties of a heparin-dependent inhibitor of thrombin in human plasma. J Biol Chem 257:2162.
25. Farrell, DH, and Cunningham, DD (1986). Human fibroblasts accelerate the inhibition of thrombin by protease nexin. Proc Natl Acad Sci (USA) 83:6858.
26. Van Mourik, JA, Lawrence, DA, and Loskutoff, DJ (1984). Purification of an inhibitor of plasminogen activator (antiactivator) synthesized by endothelial cells. J Biol Chem 259:14914.
27. Kruithof, EKO, Vassalli, J-D, Schleuning, W-D, Mattaliano, RJ, and Bachmann, F (1986). Purification and characterization of a plasminogen activator inhibitor from the histiocytic lymphoma cell line U-937. J Biol Chem 261:11207.
28. Scott, RW, Eaton, DL, Duran, N, and Baker, JB (1983). Regulation of extracellular plasminogen activator in human fibroblast cultures. The role of protease nexin. J Biol Chem 258:4397.
29. Petersen, TE, Dudek-Wojciechowska, G, Sottrup-Jensen, L, and Magnusson, S (1979). Primary structure of antithrombin III (heparin cofactor): partial homology

between α_1-antitrypsin and antithrombin III. In Collen D, Wiman B, Verstraete M, (eds): "The Physiological Inhibitors of Coagulation and Fibrinolysis," Amsterdam: Elsevier-North Holland Biomedical Press, 43.

30. Bock, SC, Skriver, K, Nielsen, E, Thøgersen, H-C, Wiman, B, Donaldson, VH, Eddy, RL, Marrinen, J, Radziejewska, E, Huber, R, Shows, TB, and Magnusson, S (1986). Human C1 inhibitor: Primary structure, cDNA cloning, and chromosomal localization. Biochem 25:4292.
31. Ragg, H (1986). A new member of the plasma protease inhibitor gene family. Nucl Acids Res 14:1073.
32. Gloor, S, Odink, K, Guenther, J, Nick, H, and Monard, D (1986). A glia-derived neurite promoting factor with protease inhibitory activity belongs to the protease nexins. Cell 47:687.
33. Monard, D, Solomon, F, Rentsch, M, and Gysin, R (1973). Glia-induced morphological differentiation in neuroblastoma cells. Proc Natl Acad Sci (USA) 70:1894.
34. Guenther, J, Nick, H, and Monard, D (1985). A glia-derived neurite promoting factor with protease inhibitory activity. EMBO J 4:1963.
35. Lindner, J, Guenther, J, Nick, H, Zinser, G, Antonicek, H, Schachner, M, and Monard, D (1986). Modulation of granule cell migration by a glia-derived protein. Proc Natl Acad Sci (USA) 83:4568.
36. Rosenblatt, DE, Cotman, CW, Nieto-Sampedro, M, Rowe, JW, and Knauer, DJ (1987). Identification of a protease inhibitor produced by astrocytes that is structurally and functionally homologous to protease nexin-I. Brain Res (in press).
37. Knauer, DJ, Orlando, RA, and Rosenblatt, D (1987). The glioma cell-derived neurite promoting activity protein is functionally and immunologically related to human protease nexin I. J Cell Physiol (in press).
38. Carney, DH, Glenn, KC, and Cunningham, DD (1978). Conditions which affect initiation of animal cell division by trypsin and thrombin. J Cell Physiol 95:13.
39. Baker, JB, Low, DA, Eaton, DL, and Cunningham, DD (1982). Thrombin-mediated mitogenesis: The role of secreted protease-nexin. J Cell Physiol 112:291.

40. Glenn, KC, Carney, DH, Fenton, JW II, and Cunningham, DD (1980). Thrombin active site regions required for fibroblast receptor binding and initiation of cell division. J Biol Chem 255:6609.
41. Low, DA, Scott, RW, Baker, JB, and Cunningham, DD (1982). Cells regulate their mitogenic response to thrombin through secretion of protease nexin. Nature 298:476.
42. Low, DA, Wiley, HS, and Cunningham, DD (1985). Role of cell surface thrombin binding sites in mitogenesis. In Feramisco J, Ozanne B and Stiles C (eds): "Growth Factors and Transformation." Cancer Cells Vol. 3, Cold Spring Harbor: Cold Spring Harbor Press, 401.
43. Eaton, DL, Scott, RW, and Baker, JB (1984). Purification of human fibroblast urokinase proenzyme and analysis of its regulation by proteases and protease nexin. J Biol Chem 259:6241.
44. Eaton, DL, and Baker, JB (1983). Phorbol ester and mitogens stimulate human fibroblast secretions of plasmin-activatable plasminogen activator and protease nexin, an antiactivator/antiplasmin. J Cell Biol 97:323.
45. Saksela, O, Laiho, M, and Keski-Oja, J (1985). Regulation of plasminogen activator activity in human fibroblastic cells by fibrosarcoma cell-derived factors. Cancer Res 45:2314.
46. Blasi, F, Vassalli, J-D, and Danø, K (1987). Urokinase-type plasminogen activator: proenzyme, receptor and inhibitors. J Cell Biol 104:801.
47. Bergman, BL, Scott, RW, Bajpai, A, Watts, S, and Baker, JB (1986). Protease nexin I inhibits destruction of extracellular matrix by human tumor cells. Proc Natl Acad Sci (USA) 83:996.
48. Jones, PA, and DeClerk, YA (1980). Destruction of extracellular matrices containing glycoproteins, elastin and collagen by metastatic human tumor cells. Cancer Res 40:3222.
49. Baker, JB, and Gronke, RS (1986). Protease nexins and cellular regulation. Seminars in Thrombosis and Hemostasis 12:216.
50. Gronke, RS, Bergman, BL, and Baker, JB (1987). Thrombin interaction with platelets: Influence of a platelet protease nexin. J Biol Chem 262:3030.

51. Gronke, RS, Curry, TK and Baker, JB (1986). Formation of protease nexin-thrombin complexes on the platelet surface. J Cell Biochem 32:201.

52. Eaton, DL, and Baker, JB (1983). Evidence that a variety of cultured cells secrete protease nexin and produce a distinct cytosolic protease-binding factor with similar specificity. J Cell Physiol 117:175.

53. Stump, DC, Thienpont, M, and Colhen, D (1986). Purification and characterization of a novel inhibitor of urokinase from human urine. Quantitation and preliminary characterization in plasma. J Biol Chem 261:12759.

54. Laug, WE (1987). Protease inhibitors produced by vascular smooth muscle cells (VSMC). J Cell Biochem Suppl 11B:J020.

55. Baker, JB, Bergman, BL, Bajpai, A, and Gronke, RS (1986). Properties and actions of protease nexin I. J Cell Biochem Suppl 10A:E13.

56. Baker, JB, McGrogan, MP, and Simonsen, C (1987). Protease nexin-mediated control of proteases at cell surfaces. J Cell Biochem Suppl 11B:J018.

57. Andersen, MJ, and Fambrough, DM (1983). Aggregates of acetylcholine receptors are associated with plaques of a basal lamina heparan sulfate proteoglycan on the surface of skeletal muscle fibers. J Cell Biol 97:1396.

The Pharmacology and Toxicology of Proteins, pages 325–336

PLASMINOGEN ACTIVATOR INHIBITORS IN VASCULAR SMOOTH MUSCLE CELLS[1].

Walter E. Laug and Emil Bogenmann

Division of Hematology-Oncology, Childrens Hospital of Los Angeles, Los Angeles, California 90027

Growth of bovine aortic endothelial cells on a preformed layer of bovine aortic smooth muscle cells resulted in a structure morphologically similar to myointimal proliferation found in the atherosclerotic lesion. Large amounts of free plasminogen activator inhibitory activity was detected in the medium of these cocultures neutralizing the plasminogen activator activity normally detected in the supernatant of endothelial cell cultures. Three distinct types of plasminogen activator inhibitors were subsequently detected in serum free medium conditioned by human smooth muscle cells, namely plasminogen activator inhibitors 1 and 2, and protease nexin. Plasminogen activator inhibitor 1 was the predominant inhibitor secreted by smooth muscle cells, as determined by ELISA tests using monospecific antisera. It was present mainly in latent form.

Our *in vitro* coculture system demonstrates that vascular smooth muscle cells may display a broad spectrum of serine protease inhibition allowing them to alter the fibrinolytic activity of vascular endothelial cells. The *in vitro* reconstituted vessel wall resembles intimal smooth muscle cell proliferation, the classical atherosclerotic lesion, and enables the study of important physiological and pathological conditions involving both vascular smooth muscle cells and endothelial cells.

[1]This work was supported by grants from the NIH (HL-23500, CA-25397 and NEI-4950) and the American Heart Association, Greater Los Angeles Affiliate (781-G1).

INTRODUCTION

Plasminogen activators (PA) are serine proteases which induce local proteolysis through activation of the zymogen plasminogen to plasmin (for review see 1). Two immunologically, functionally, and genetically distinct PA have been described, urokinase (u-PA) and tissue type PA (t-PA). These enzymes play an important role in fibrinolysis, tissue remodeling, and tumor cell invasion. Their extracellular enzymatic activities are controlled by specific inhibitors (PAI) produced by various cell types. Three distinct types of PAI are presently known, PAI-1 (2), PAI-2 (3) and protease nexin (4).

Increased levels of PAI-1 were found in the plasma of patients suffering from atherosclerosis (5). The frequent occurrence of intrasvascular clot formation in this disease state is thought to be partially mediated through inhibition of fibrinolysis by PAI. The cellular origin of the PAI-1 in such patients remains speculative although endothelial cells (3) and platelets (6) were found to produce PAI-1.

Proliferation of smooth muscle cells (SMC) in the intimal layer of arteries represents the classical lesion for atherosclerosis (7). Since these lesions are often the site of intravascular clot formations we speculate that SMC may inhibit the fibrinolytic activities of endothelial cells resulting in increased thrombogenesis.

We present a novel *in vitro* coculture system of vascular SMC and endothelial cells which resembles myointimal proliferation. Large amounts of free PA inhibitory activities were found in medium conditioned by such cocultures. All three known types of PAI were found in supernatants of cultured human SMC with PAI-1 being the major constituent. We therefore speculate that the PAI-inhibitory activities displayed in cocultures of SMC and endothelial cells were secreted by the SMC.

MATERIAL AND METHODS

Cell Cultures. Bovine endothelial cells and SMC were obtained from the aorta of a 3-day old calf. Isolation, cloning, and growth conditions have been described (8). Cocultures were obtained as described (13).

Human SMC were obtained from the aorta of a cadaver kidney transplant and grown as described (9). Human endothelial cells were isolated from the same aorta and grown on gelatin (1.5%) coated dishes in RPMI medium containing hepes buffer, 20% human serum, endothelial cell growth factor, heparin, transferrin and insulin (Laug, W.E., et al, manuscript in preparation).

PA and PA Activities. Secreted and cellular PA and PAI activities were measured in aliquots of serum free medium conditioned for 16 hr and the postnuclear fraction of cell lysates, respectively. The samples were tested on fibrin plates using human urokinase as the standard (8, 10). Alternatively, the regular (11) and the reverse fibrin overlay method (2) was used. The methods were modified using 10% SDS polyacrylamide minigels (Biorad Lab., Richmond, CA).

Quantitative determination of PAI-1, PAI-2 and protease nexin were done with ELISA tests using specific antisera (3, 4, 12). Amounts of PAI-1 and PAI-2 were kindly determined by E. Kruithof, Lausanne, Switzerland and protease nexin by J. Baker, Lawrence, Kansas.

Electronmicroscopy. Fixation, embedding and morphological analysis were done as reported (16).

RESULTS

Bovine SMC were grown until a steady cell number was reached and then overlayed with bovine endothelial cells and grown for an additional 7 days. Electronmicroscopic examination of the resulting structure showed multilayers of SMC covered by a monolayer of endothelial cells (Fig. 1). The SMC were embedded in abundant extracellular matrix proteins containing collagen fibers and amorphous material. Neither an internal elastic membrane nor a continuous basement membrane normally present in arterial vessel walls could be detected. This structure, therefore, resembled an intimal SMC proliferation as observed in the atherosclerotic lesion rather than a normal vessel wall.

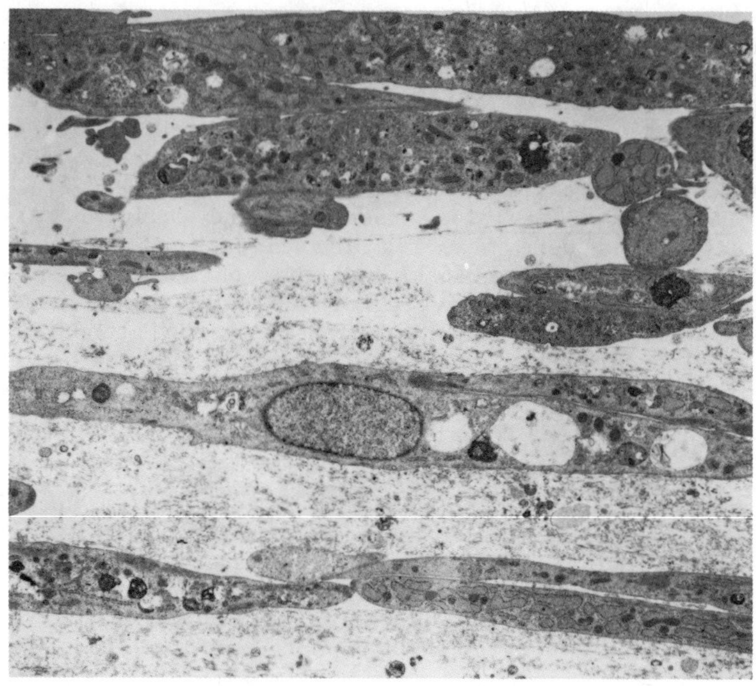

FIGURE 1: Electronmicrograph of endothelial cell-SMC coculture 2 weeks after plating. Magnification x 2200.

The fibrinolytic potential of these cocultures was assessed with the fibrin plate method (Table 1). Confluent endothelial cell cultures demonstrated large amounts of secreted PA activities as described (10). No free PA activities were found in the supernatant of either SMC or cocultures.

TABLE 1
PA ACTIVITIES IN CELL CULTURES

Cell Type	PA Activities*	
	Secreted	Cellular
Endothelial cells	21.3 ± 1.5	3.4 ± 0.2
Smooth muscle cells	0	0.08 ± 0.03
Cocultures	0	3.5 ± 0.1

*PA activities are expressed in International Units (I.U.) per dish (secreted PA) or per mg cellular protein (cellular PA).

In contrast, the cellular PA activities were similar in endothelial cells and cocultures while only small amounts were found in the SMC. These results suggested that the synthesis of PA by endothelial cells was unchanged. However, the lack of free PA activities in the supernatant of cocultures was either due to inhibition of PA secretion or neutralization of secreted PA by SMC derived PAI.

SMC have previously been shown to secrete PAI and, therefore, the medium conditioned by cocultures was tested for the presence of free PAI (Table 2).

TABLE 2
PA INHIBITORY ACTIVITIES IN CELL CULTURES

Cell Cultures	PA Inhibitory Activities* (units/culture)
Endothelial cells	0
Smooth muscle cells	5.1
Cocultures	2.4

*1 unit corresponds to amount of PAI which inhibits 50% of 1 I.U. of u-PA in 2 hr.

Free PA inhibitory activity was absent in endothelial cell conditioned medium, whereas SMC secreted large amounts of it. A lower level of free PAI activity was found in the cocultures than in the SMC cultures, suggesting complex formation between SMC derived PAI and PA secreted by the endothelial cells (8).

The findings obtained with the fibrin plate method were confirmed by regular and reverse fibrin autography of samples from the supernatants of the different cultures (Fig. 2). Free PA activity was only present in the medium conditioned by endothelial cells. PAI activities were found in all cultures with highest levels in SMC cultures and cocultures. The inhibitory activity found in endothelial cell cultures represents SDS activated PAI-1 not detectable by the fibrin plate method (2).

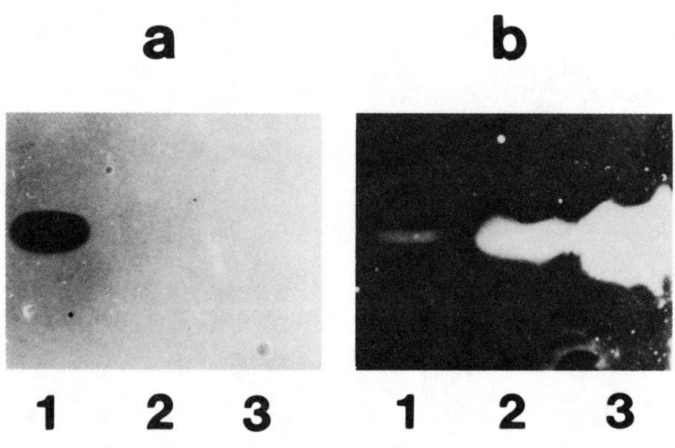

FIGURE 2. Regular (a) and reverse fibrin autography (b) of medium conditioned by the cell cultures. Lane 1 = endothelial cells, lane 2 = smooth muscle cells and lane 3 = cocultures. Dark band in (a) indicates area of fibrinolysis; white bands in (b) are zones resistant to lysis.

Subsequent experiments showed that part of the PA inhibitory activity secreted by SMC was acid sensitive (date not shown). In addition, the PAI present in SMC conditioned medium formed SDS resistant complexes with both forms of PA and thrombin (8). These data suggested the presence of at least 2 types of PAI, one of which resembled protease nexin known to bind thrombin and to be acid sensitive (4).

Immunoblotting of SMC conditioned medium using monospecific antibodies revealed the presence of the 3 known PAI (data not shown). Subsequently quantitative determination of the different PAI produced by human aortic SMC was done with ELISA tests (Table 3). Human aortic endothelial cells were included as controls.

TABLE 3
PAI CONCENTRATION IN CONDITIONED MEDIUM

Inhibitor	SMC ng/ml	Endothelial Cells ng/ml
PAI-1	1100	1750
PAI-2	82	28
Protease-nexin	35	0

All 3 known PAI could be detected in SMC conditioned medium; PAI-1 being the predominant inhibitor. The concentration was similar to that observed in human aortic endothelial cells. The PAI-1 was mainly present in the latent form as previously reported for endothelial cells (2) (data not shown). PAI-2 and protease nexin were found in much lower concentrations. Interestingly, the human aortic endothelial cells were also found to secrete small amounts of PAI-2 while protease nexin could not be detected.

These studies demonstrate that SMC produce the 3 known types of PAI which assures the SMC a broad spectrum of serine protease inhibitions. The combined action of these inhibitors covers the activities of PA (PAI-1 and PAI-2), plasmin (protease nexin and PAI-2) and thrombin (protease nexin).

DISCUSSION

A coculture system using bovine aortic endothelial and smooth muscle cells is presented. The resulting structure morphologically resembled intimal proliferation of SMC as described in the classical lesion of atherosclerosis (7). Similar structures were obtained with rat smooth muscle cells and fetal bovine endothelial cells (13). Our homologous system has the advantage of being composed of cells from the same vessel. These cocultures, therefore, may be useful to study, in vitro the physiological and pathological phenomena of vascular cells. In the present study, we investigated the influence of SMC on the fibrinolytic activity of endothelial cells.

In previous short term studies we have shown that bovine SMC inhibit the PA activities secreted by endothelial cells (8). We have also shown that SMC conditioned medium contains inhibitors of PA which form SDS resistant complexes with both types of PA. Similar results were obtained with long term cocultures in the present study. Free PA inhibitory activities were found in the supernatant of cocultures, whereas, no free PA activities could be detected. Inhibition of PA synthesis in endothelial cells by SMC seems unlikely since the cellular PA activities remained unchanged in the cocultures. We speculate that SMC either inhibit the secretion of PA by endothelial cells or that secreted PA activities are neutralized by SMC derived PAI. Experiments are under way to further explore these possibilities.

Since we were successful in growing human aortic SMC in long term tissue culture, we chose human cells to characterize and isolate PAI. The 3 known types of PAI were identified in serum free medium conditioned by human SMC using monospecific anti-PAI antisera. This finding was rather suprising since most normal cells produce one type of PAI (1). Only the human sarcoma cells HT1080 have been recently found to produce all 3 types of PAI. Contamination of the SMC with endothelial cells known to produce PAI-1 (2) and/or adventitial fibroblasts, a possible source of protease nexin (4), was avoided by restrictive culture conditions, and careful cell isolation.

SMC produced about 10 and 20 times more PAI-1 than PAI-2 and protease nexin, respectively. A similar level of PAI-1 was detected in the supernatant of human aortic endothelial cells. The majority of this inhibitor was present in the latent form (data not shown) as observed with other cell types (1). Subsequently, we were able to separate the 3 PAI by heparin-affigel chromatography using a continuous NaCl gradient (Laug, W.E., et al, manuscript in preparation). In these experiments PAI-2 had very low heparin affinity, confirming previous reports (3). In contrast to earlier publications (2), PAI-1 was found to bind to heparin-affigel although it eluted at salt concentrations considerably lower than protease nexin, from which it was easily separated.

A comparison of the characteristics of these inhibitors and their spectrum of protease inhibition is given in Table 4. These data represent a summary of results retrieved from the recent literature (3, 4, 13).

TABLE 4
COMPARISON OF PA INHIBITORS

Characteristics	PAI-1	PA1-2		Protease Nexin
Mol. Weight	54,000	47,000 & 60,000		51,000
pI	4.5-5.0	5.0	4.4	7.5-7.8
Glycosylation	+	-	+	+
Heparin Binding	+	-	-	+
Enzyme Inhibition*				
u-PA	5×10^7	9×10^5		1.5×10^5
t-PA	1×10^7	2×10^5		3×10^4
Plasmin		1×10^2		1.3×10^5
Trypsin				4×10^6
Thrombin				12×10^7

*(2nd order rate in $M^{-1}S^{-1}$)

The 3 PAI show similar molecular weights. PAI-2 exists in glycosylated and unglycosylated form and both types exhibit protease inhibitory activities. The pI of PAI-1 and PAI-2 are similar while protease nexin demonstrates a slightly alkaline isoelectric point. PAI-1 inhibits only PA whereas PAI-2 also weakly inhibits plasmin. It is a less potent PA inhibitor than PAI-1. Protease nexin, in contrast, is mainly an inhibitor of thrombin, trypsin and plasmin although it still has considerable inhibitory activities for PA. The 3 PAI are members of the serpin gene family which also includes α_1-antitrypsin, C1-inhibitor, antithrombin III, heparin cofactor II, angiotensin and ovalbumin (17).

The production of 3 types of PAI secures the SMC with broad spectrum inhibition of the serine proteases PA, plasmin, trypsin and thrombin. This assures SMC a potent control of local proteolysis. This finding may also partially explain the natural resistance of large arteries, containing multiple layers of SMC, to tumor cell invasion and destruction.

In addition, the abnormal proliferation of SMC in the intimal layer of arteries as observed in atherosclerotic lesions, may result in altered functions of the overlaying endothelial cells. Our results demonstrate that SMC derived PAI neutralize the fibrinolytic activities displayed by endothelial cells. The decreased fibrinolytic potential may enhance thrombogenecity of these lesions resulting in intravascular clot formation.

ACKNOWLEDGMENT

We thank Dr. E. Kruithof, Lausanne, Switzerland and Dr. J. Baker, Lawrence, Kansas for performing the ELISA tests. The technical assistance of Mr. William Rideout, III is gratefully acknowledged. We thank Ms. Clarita Stutters for typing the manuscript.

REFERENCES

1. Dano K, Andreasen PA, Grondahl-Hansen J, Kristensen P, Nielson LS, Shriver L (1985). Plasminogen activators, tissue degradation and cancer. Adv Cancer Res 44:139.
2. Loskutoff DJ, van Mourik JA, Erickson LA, Lawrence D (1983). Detection of an unusually stable fibrinolytic inhibitor produced by bovine endothelial cells. Proc Natl Acad Sci USA 80:2956.

3. Kruithof EKO, Vassalli J-D, Schleuning W-D, Mattaliano RJ, Bachmann F (1986). Purification and characterization of a plasminogen activator inhibitor from the histiocytic lymphoma cell line U-937. J Biol Chem 261:11207.
4. Scott RW, Bergmann BL, Bajpai A, Hersh RT, Rodrigues H, Jones BN, Barreda C, Watts S, Baker JB (1985). Protease nexin. Properties and a modified purification procedure. J Biol Chem 260:7029.
5. Paramo JA, Colucci M, Collen D (1985). Plasminogen activator inhibitor in the blood of patients with coronary artery disease. Br Med J 291:573.
6. Erickson LA, Hekman CM, Loskutoff DJ (1985). The primary plasminogen activator inhibitors in endothelial cells, platelets, serum and plasma are immunologically related. Proc Natl Acad Sci USA 82:8710.
7. Ross R (1986). The pathogenesis of atherosclerosis-an update. New Eng J Med 3M:488.
8. Laug WE (1985). Vascular smooth muscle cells inhibit plasminogen activators secreted by endothelial cells. Thrombos Haemostas 53:165.
9. Cheng CY, Laug WE (1985). Inhibitors of fibrinolysis produced by vascular smooth muscle cells. In Davidson JF, Donati JF, Coccheri S (eds) "Progress in Fibrinolysis," Vol. 7. Churchill Livingstone, Edinburgh 150.
10. Laug WE (1983). Glucocorticoids inhibit plasminogen activator production by endothelial cells. Thrombos Haemostas 50:888.
11. Grannelli-Piperno A, Reich E (1978). A study of proteases and protease-inhibitor complexes in biological fluids. J Exp Med 148:223.
12. Schleef RR, Sinha M, Loskutoff DJ (1985). Immunoradiometric assay to measure the binding of a specific inhibitor to tissue-type plasminogen activator. J Lab Clin Med 106:408.
13. Jones PA (1979). Construction of an artificial blood vessel wall from cultured blood vessel wall from cultured endothelial and smooth muscle cells. Proc Natl Acad Sci USA 76:1882.
14. Kruithof EKO, Tran-Thang C, and Bachmann F (1986). The fast acting inhibitor of tissue-type plasminogen activator in human plasma is also the primary inhibitor of urokinase. Thrombos Haemostas 55:65.

15. Levin EG (1983). Latent tissue plasminogen activator produced by human endothelial cells in culture: evidence for an enzyme-inhibitor complex. Proc Natl Acad Sci USA 80:6804.
16. Jones PA, Neustein HB, Gonzalez F, Bogenmann E (1981). Invasion of an artificial blood vessel wall by human fibrosarcoma cells. Cancer Res 41:4613.
17. Ye RA, Wun TC, Sadler JE (1987). cDNA cloning and expression in escherichia coli of a plasminogen activator inhibitor from human placenta. J Biol Chem 262:3718.

The Pharmacology and Toxicology of Proteins, pages 337–349

DECAPEPTYL: RELEASE FROM MICROCAPSULES AND TESTOSTERONE SUPPRESSION IN THE DOG

A. P. Tonelli, G. Nicolau, J. R. Lawter, B. M. Silber
R. A. Lanc, M. Lanzilotti, A. R. Dente,
W. E. McWilliams and A. Yacobi

Departments of Pharmacodynamics, Formulations Research
and Toxicology Evaluation, Medical Research Division
American Cyanamid Company
Pearl River, New York 10965

ABSTRACT Decapeptyl (CL 118,532), (D) is a superactive luteinizing hormone-releasing hormone (LH-RH) analog which has demonstrated effectiveness against the growth of steroid dependent tumors. The release of D from three microencapsulated formulations (F1, F2 and F3) intended for once-a-month intramuscular dosing, was investigated in beagle dogs. Additionally, the relative bioequivalence of the three formulations was examined as reflected by plasma concentrations of decapeptyl and by their ability to reduce plasma testosterone. Statistically significant differences in the total amount of drug released from the formulations were not present. However, the time course of D release was different, as was the profile of testosterone suppression: F1 and F3 demonstrated similar patterns of D release and testosterone suppressions. Both released approximately 50% of their total by day 1 and demonstrated peak plasma concentrations at the first (2 hour) sampling time. F2 released less decapeptyl on day 1 (23% of total released) and also produced lower peak plasma concentrations than the other formulations. Testosterone concentration profiles showed an initial elevation, lasting for 1-5 days (greatest in F2), followed by depression and total suppression lasting from 30 to 60 days after dose (greatest in F1 and

F3) A positive correlation was established between the amount of D released on day 1 and testosterone suppression. The cessation of testosterone suppression was associated with the lack of detectable decapeptyl concentrations.

INTRODUCTION

Schally and coworkers have demonstrated that intramuscular administration of microencapsulated decapeptyl in rats decreased the weight of androgen dependent prostate tumors and suppressed serum testosterone more effectively than equivalent or double doses of non-encapsulated decapeptyl administered subcutaneously (1,2). Clinical studies have shown that once-a-month intramuscular injections of microencapsulated decapeptyl, with projected release rates of 100 mcg/day, was as effective as orchiectomy in treating prostate cancer patients (3). Decapeptyl has also shown effectiveness in mass reduction for hormone dependent tumors of the ovaries and breast (4,5). Decapeptyl is considered essentially nontoxic at those doses with the most serious, but only occasionally observed, adverse effect being a temporary exacerbation of the disease due to an initial stimulation of gonadotrophins (6,7).

Regulation of plasma testosterone concentrations following administration of authentic LH-RH, or super agonists such as decapeptyl, is a pharmacodynamically complex event. Currently, the proposed mechanism for suppression is thought to involve a down regulation of the LH-RH receptors on the anterior pituitary cells. This down regulation has been demonstrated following constant infusion of these agonists and with multiple dosing regimens. It is thought to result from relatively constant exposure, or overloading of the drugs at the receptor This is in contrast to the normal physiological, pulsatile type release of LH-RH which causes LH release and elevation of plasma testosterone (8).

In this study, 3 forms of microencapsulated decapeptyl were evaluated on the basis of plasma decapeptyl concentrations and extent of testosterone suppression following intramuscular administration to beagle dogs.

METHODS

Twelve male beagle dogs (Marshall Breeding Labs, North Rose, NY), weighing between 7.3 and 10.2 kg, were used in the study. The dogs were housed in individual metabolism cages in windowless air-conditioned rooms under artificial light/dark cycles of 12 hours each. Standard dog chow (300 g) was provided daily. Water was available ad libitum. The twelve dogs were divided into three treatment groups, each group receiving a different microencapsulated formulation as described in Table 1.

Decapeptyl microcapsules were prepared by a phase separation process. The encapsulating polymer is poly(gycolide-co-dl-lactide) with a glycolide to lactide mole ratio of 47 to 53 and a weight average molecular weight of approximately 60,000 (as determined by gel permeation chromatography in methylene chloride versus polystyrene standards). Decapeptyl content of the three formulations varied from 2-5%. The dose administered to each dog was 3.0 mg (F1 and F3) or 2.85 mg (F2) of decapeptyl in approximately 1.5 mL of suspension. The expected release rate from the formulations was time-dependent with a projected duration of 30 days. Decapeptyl release from the polymeric matrix is thought to result from water uptake with subsequent generation of aqueous diffusion channels within the matrix as well as surface leaching of drug (9). A single intramuscular dose, administered in the posterior thigh region, was delivered over a time period of 1 minute.

Table I
Experimental Design

Formulation Number	Decapeptyl Content (%)	Decapeptyl per syringe[1] (mg)	Microcapsule weight per syringe (mg)
1	2.78	3.85	138.5
2	4.57	3.52	77
3	2.20	3.85	175

[1]Decapeptyl content after irradiation sterilization.

Blood samples (10 mL) were obtained from the jugular vein into EDTA-coated vacutainer tubes (Becton Dickinson, Rutherford, NJ) at the following times: 0, 2, 4, 6 and 12 hours, 1, 1.33, 2, 2.33, 3, 4, 5, 8 days, and every fourth day for 60 days postdose. Following immediate centrifugation (4°C), the resultant plasma was divided into equal aliquots for decapeptyl and testosterone determinations. All samples were stored frozen.

Decapeptyl was analyzed in plasma at Hazelton, Biotechnologies Corporation, (Vienna, Virginia) using a radioimmunoassay (RIA) procedure with an antiserum provided by Dr. F. Dray of the Institute Pasteur. Plasma testosterone concentrations were determined by a direct I-125 testosterone RIA (Immuchem Corp. Carson, CA).

RESULTS

Plasma Concentrations

Mean plasma concentrations of decapeptyl are presented in Table 2 and Figure 1. Peak concentrations were observed at the initial (2 hr.) sampling time for all dogs. Formulations 1 (F1) and 3 (F3) generated initial concentrations of decapeptyl approaching 40 ng/mL (slightly in excess of the highest standard curve value of 32 ng/mL). Formulation 2 (F2) generated significantly lower peak plasma concentrations of 19.07 ng/mL.

Similarities were evident in the mean plasma concentration versus time decay curves following dosing in the three formulations. Following an initial rapid decline in plasma concentrations, characterized by a $t_{1/2}$ of approximately 2.5 hours and a duration of 0-12 hours, a slower decline was observed ($t_{1/2} \sim$ 4-5 days), which varied for the formulations. The third phase of declining plasma concentrations consisted of relatively constant values first attained at 12 and 16 days (F1 and F2) and 3 days (F3) postdose. These concentrations declined slowly ($t_{1/2} \sim$ 50 days) for approximately 28 days with all formulations. At the end of this phase there was a relatively rapid decline in plasma concentrations to values below the assay quantitation limit of 9 pg/mL ($T_{1/2} \sim$ 5 days). No detectable concentrations of decapeptyl were evident beyond day 44 in dogs administered F3.

Table 2

Mean Plasma Concentrations of Decapeptyl Following Intramuscular Dosing in Beagle Dogs (ng/mL)

(n = 4)

Time (Days)	Formulation 1 Mean	Formulation 1 (± S.D.)	Formulation 2 Mean	Formulation 2 (± S.D.)	Formulation 3 Mean	Formulation 3 (± S.D.)
0	<0.009	(ND)	<0.009	(ND)	<0.009	(ND)
0.083	37.888[d]	(6.67)	19.075	(6.59)	36.775[e]	(2.93)
0.167	24.600	(3.22)	10.402	(4.86)	23.567[d]	(6.22)
0.25	12.167	(2.87)	5.885	(3.26)	14.519	(4.34)
0.5	2.910	(0.89)	2.973[d]	(1.98)	7.508	(2.28)
1.0	1.980	(0.86)	1.790	(0.77)	3.595	(0.87)
1.33	2.130	(0.99)	1.900	(1.01)	3.430	(0.54)
2.0	1.141	(0.60)	1.640	(1.08)	0.735	(0.27)
2.33	1.151	(0.71)	2.068	(1.0)	0.563	(0.07)
3.0	0.540	(0.29)	1.393	(0.95)	0.280	(0.07)
4.0	0.495	(0.28)	1.247	(0.67)	0.226	(0.05)
5.0	0.343	(0.22)	1.213	(0.80)	0.232	(0.06)
8.0	0.201	(0.14)	0.875	(0.86)	0.177	(0.03)
12.0	0.138	(0.13)	0.425	(0.21)	0.209	(0.05)
16.0	0.209	(0.24)	0.145	(0.06)	0.385	(0.27)
20.0	0.105	(0.10)	0.136	(0.06)	0.224	(0.10)
24.0	0.115[a]	(0.16)	0.140	(0.12)	0.221	(0.15)
28.0	0.117	(0.10)	0.079	(0.05)	0.177	(0.10)
32.0	0.114	(0.07)	0.098	(0.07)	0.137	(0.07)
36.0	0.104	(0.07)	0.102	(0.09)	0.065	(0.02)
40.0	0.110	(0.06)	0.112[d]	(0.09)	0.040	(0.01)
44.0	0.061	(0.03)	0.091[d]	(0.06)	0.017	(0.01)
48.0	0.031[a]	(0.02)	0.054[d]	(0.03)	<0.009[c]	(ND)
52.0	0.012[b]	(0.01)	0.021[ad]	(0.02)	<0.009	(ND)
56.0	<0.009[c]	(ND)	<.009[bd]	(0.01)	<0.009	(ND)
60.0	<0.009[c]	(ND)	<.009[bd]	(0.01)	<0.009	(ND)

All plasma concentrations <0.009 ng/mL were treated as 0.0 for mean data.

[a] One dog had plasma concentrations <0.009 ng/mL.
[b] Two dogs had plasma concentrations <0.009 ng/mL.
[c] Three dogs had plasma concentrations <0.009 ng/mL.
[d] N = 3;
[e] N = 2.

Figure 1

MEAN CONCENTRATION OF DECAPEPTYL IN DOG PLASMA COMPARISON OF THREE FORMULATIONS

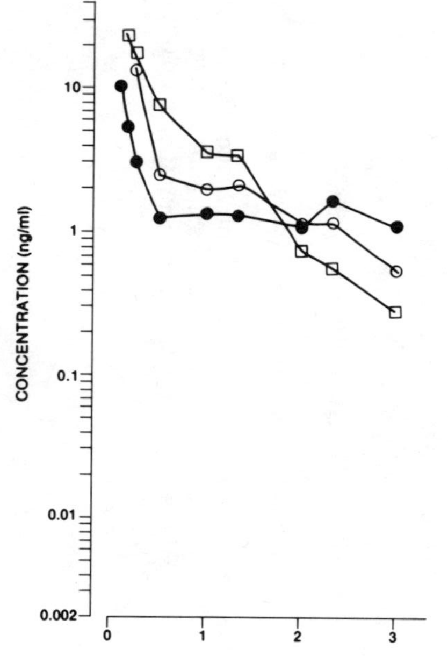

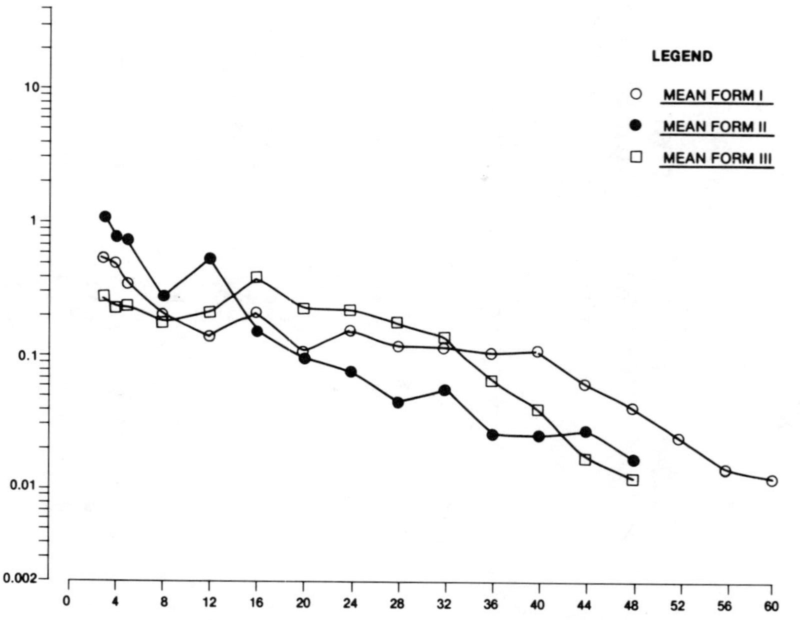

Relative Bioequivalence and Release Rates

To assess differences in the relative release rates and bioequivalence of the three microencapusalated formulations, linear clearance and distribution kinetics of decapeptyl were assumed over the range of plasma concentrations observed in this study. With this assumption, the area under the plasma concentration-time curve (AUC) should provide a direct reflection of decapeptyl release from the intramuscular dosing site.

No significant differences in the total amount of drug released from the three formulations, based on overall AUC (time 0 to time-last) comparisons were found (unbalanced ANOVA analysis, $P > 0.05$). However, significant formulation differences, in the relative amount of drug released over discrete time intervals, were present.

Specifically, the initial release of drug from the intramuscular injection site was greatest in F1 and F3 (Table 3). These formulations released enough decapeptyl in day 1 to account for approximately 50% the total dose, assuming drug delivery was complete by day 60. This rapid initial release of drug was less pronounced for F2 which released about 25% of the total dose in the same time. However, for the next 16 days of the study, this pattern was reversed. F2 generated plasma concentrations and AUCs approximately twice that of F1 and F3. Significant differences in the AUCs generated by the different formulations were not present following day 16 of the study.

Estimates for the daily release of decapeptyl were also made assuming complete mobilization of drug from the injection site by day 60 and the existence of a direct relationship between plasma AUC and released decapeptyl. As shown in Table 3, the initial, high concentrations of decapeptyl, present during day 1, corresponded to approximately 175 mcg/kg/day released from F1 and F3 and 80 mcg/kg/day for F2. From days 2-5, approximately 23 mcg/kg/day was released from each formulation. From days 8-32, which corresponded to the period of greatest testosterone suppression and relatively stable plasma concentrations of decapeptyl, approximately 5 mcg/kg/day were released from the formulations.

Table 3

Area Under the Plasma Concentration-Time Curve and Estimated Daily Release Rates of Decapeptyl in Dogs

(Mean ± S.D., N = 4)

Time Interval (Days)	Formulation 1			Formulation 2			Formulation 3		
	AUC	Percent of Total AUC[a]	ERR	AUC	Percent of Total AUC[a]	ERR	AUC	Percent of Total AUC[a]	ERR
0-1	8.83 (0.46)	48.5	171.5	4.97 (2.13)	23.3	76.0	11.15 (1.84)	52.9	188.2
1-2	1.77 (0.84)	9.7	34.4	1.79 (0.99)	8.4	27.3	2.06 (0.51)	9.8	33.2
2-5	1.88 (1.07)	10.3	12.1	4.51 (2.40)	21.2	23.0	0.98 (0.19)	4.6	5.2
5-8	0.82 (0.54)	4.5	5.3	3.13 (2.48)	14.7	16.0	0.61 (0.10)	2.9	3.3
8-16	1.37 (1.24)	7.5	3.3	3.74 (2.32)	17.6	7.1	1.96 (0.67)	9.3	3.9
16-32	2.00 (2.04)	11.0	4.8	1.91 (1.08)	9.0	3.7	3.53 (1.49)	16.8	7.1
32-48	1.40 (0.88)	7.7	1.6	1.27 (1.21)	6.0	1.2	0.77 (0.24)	3.6	0.7
48-60	0.13 (0.12)	0.7	0.2	0.18 (0.19)	0.8	0.2	0.01 (0.02)	0.05	0.01
0-Last	18.19 (6.04)	100.0	ND	21.30 (11.1)	100.0	ND	21.07 (4.23)	100.0	ND

AUC = Area under the plasma concentration-time curve in ng.day/mL.

ND = Not determined.

[a] = Percent of total AUC calculated from mean data.

ERR = Estimated daily release rate, determined from mean AUC data assuming complete release of decapeptyl from injection site (mcg/day/kg).

Plasma Testosterone Concentrations

Decapeptyl administration resulted in two alterations in basal testosterone concentrations. The initial effect, lasting from 1 to 5 days, was an elevation of testosterone. Peak values, occurring at the first (2 hour) sampling time, were approximately twice that of baseline concentrations (Table 4). Depression of plasma testosterone below baseline then followed. This was first observed on Day 5 (F1 and F3) and day 8 (F2). Complete suppression (testosterone < 0.2 ng/mL), the desired pharmacological effect, was initially observed between day 8 (F1) and day 12 (F2 and F3) postdose. This effect was of a variable duration, as shown in Table 4 and Figure 2.

Differences in the extent of plasma testosterone elevation and suppression were present among the formulations. The initial elevation of testosterone was assessed from baseline (Day 0) adjusted testosterone concentrations. Dogs administered decapeptyl in F1 had the smallest increases in plasma testosterone. Mean testosterone AUC increases for formulations 1, 2, and 3 were 2.71, 12.09 and 5.03 ng.day/mL, respectively.

Suppression of testosterone was evaluated by the number of observations and the length of time for which testosterone concentrations were below 0.2 ng/mL (Table 4). F1 generated the most desirable results. Three of the four dogs in this group had essentially total suppression of plasma testosterone from day 8 to day 60. The fourth dog had suppression for up to 32 days postdose.

Dogs administered decapeptyl in F3 showed a similar, but slightly less prolonged, profile of testosterone suppression as those dogs given F1: total suppression was evident for 40 days in three of the four dogs. Decapeptyl delivered in F2 generated the poorest testosterone suppression profile; complete suppression was present for extended times in only one of the four dogs.

In summary, the presence, and lack of positive correlations between drug and testosterone concentrations are listed below:

1. Decreases in plasma testosterone were not directly related to concomitant decapeptyl concentrations or corresponding AUC measurements. No critical threshold plasma concentration was solely responsible for a turning on, and off, of testosterone production.

Table 4
Mean Plasma Concentrations of Testosterone Following Singular Intramuscular Administration of Decapeptyl to Beagle Dogs

Mean (± S.D.)

Day	Formulation 1 Testosterone (ng/mL)	Formulation 1 No. of Observations* <0.2 ng/mL	Formulation 2 Testosterone (ng/mL)	Formulation 2 No. of Observations* <0.2 ng/mL	Formulation 3 Testosterone (ng/mL)	Formulation 3 No. of Observations* <0.2 ng/mL
0	3.99 (1.6)	0	3.64 (1.9)	0	4.22 (3.25)	0
0.083	6.67 (1.0)	0	7.23 (1.4)	0	7.05 (3.0)	0
0.167	5.49 (1.0)	0	6.20 (1.1)	0	7.43 (3.1)	0
0.25	6.16 (1.2)	0	5.61 (1.2)	0	6.83 (3.0)	0
0.5	3.91 (1.0)	0	4.79 (2.3)	0	4.86 (1.6)	0
1.0	4.43 (2.3)	0	6.26 (2.2)	0	4.46 (1.3)	0
1.33	5.09 (1.2)	0	5.65 (2.0)	0	5.93 (1.3)	0
2.0	5.56 (1.5)	0	6.83 (1.5)	0	6.01 (4.5)	0
2.33	4.62 (1.1)	0	6.07 (1.4)	0	5.12 (2.4)	0
3.0	4.75 (1.1)	0	7.90 (1.2)	0	4.67 (1.6)	0
4.0	4.04 (1.4)	0	6.06 (1.8)	0	4.76 (1.0)	0
5.0	2.63 (0.9)	0	6.55 (0.5)	0	2.81 (1.3)	0
8.0	<0.2 (0.2)	2	3.29 (1.7)	0	1.27 (2.0)	2
12	<0.2 (ND)	3	0.23 (0.5)	2	<0.2 (0.2)	3
16	<0.2 (ND)	4	0.32 (0.6)	3	<0.2 (ND)	4
20	<0.2 (ND)	4	0.67 (0.8)	1	<0.2 (ND)	4
24	<0.2 (ND)	4	0.64 (0.5)	1	<0.2 (ND)	4
28	0.5 (1.0)	3	0.76 (0.5)	1	<0.2 (ND)	4
32	<0.2 (ND)	4	0.95 (0.7)	1	<0.2 (ND)	4
36	0.66 (1.3)	3	1.41 (1.3)	1	<0.2 (0.2)	3
40	<0.2 (0.1)	3	1.22 (1.4)	2	0.23 (0.4)	3
44	<0.2 (0.2)	3	1.06 (1.3)	2	1.05 (1.4)	1
48	0.94 (1.9)	3	1.30 (1.0)	1	0.95 (0.8)	1
52	<0.2 (0.3)	3	1.85 (1.5)	1	0.55 (0.6)	2
56	0.48 (1.0)	3	2.49 (2.1)	1	2.60 (1.8)	1
60	0.30 (0.6)	3	3.05 (2.6)	1	2.06 (1.5)	1
0-60		45		18		37

* = Number of observations for which plasma testosterone was below 0.2 ng/mL detection limit.

ND = No determinations

Individual values in Table VI

For mean calculations testosterone concentrations of <0.2 ng/mL were treated as 0.0 ng/mL.

Figure 2

MEAN TESTOSTERONE CONCENTRATIONS OF DOGS DOSED WITH A SINGLE SUSTAINED RELEASE I.M. INJECTION OF DECAPEPTYL

n = 4

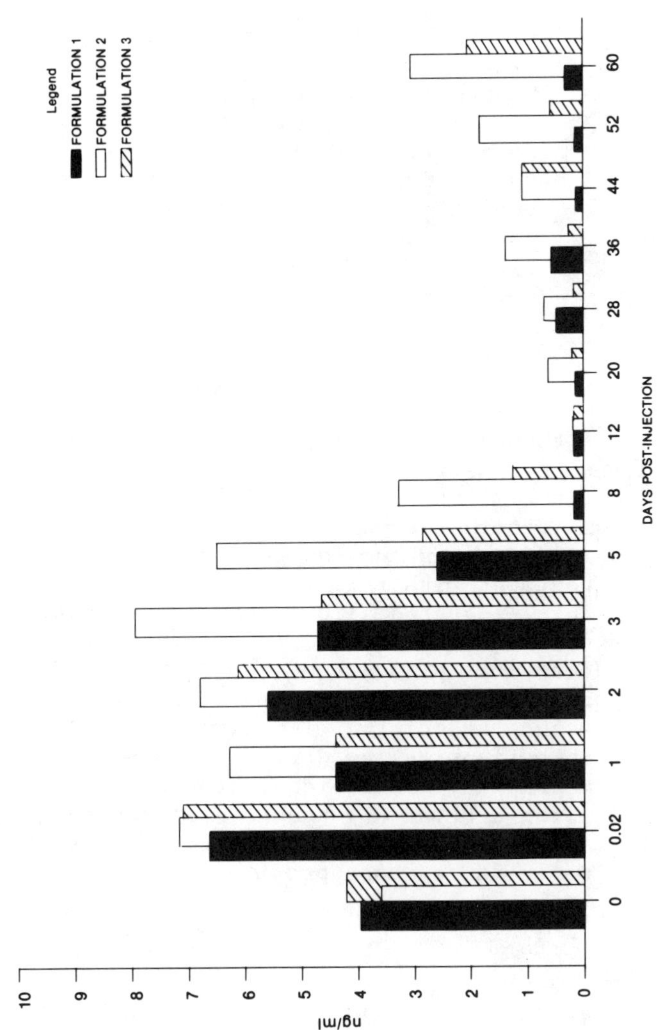

2. Testosterone suppression (the number of observations < 0.2 ng/mL), for the entire study, was positively correlated with the amount of decapeptyl released during the first two days of the study. A stronger correlation was observed with decapeptyl released during day 1.
3. Suppression of the initial testosterone elevation was also positively correlated with the amount of decapeptyl released during the first two days of the study. Again, a stronger correlation was present with decapeptyl released on day 1.
4. In most dogs, the cessation of testosterone suppression was associated with the lack of detectable decapeptyl concentrations (< 0.009 ng/mL).

DISCUSSION

Although the primary objective of this study was to determine the relative release profiles of decapeptyl from the three microencapsulated formulations, the simultaneous plasma determinations of testosterone and decapeptyl provided interesting insight towards possible optimization of treatment with this type of drug.

Data generated allowed the assessment of a decapeptyl-testosterone relationship over a specific range and profile of decapeptyl plasma concentrations. Testosterone suppression, both long term and in minimizing the initial spike, was associated with higher initial "loading" concentrations followed by sustained type release rate of 3-7 mcg/kg/day. Additional studies, with formulations utilizing a broad spectrum of release rates, are needed for better characterization and optimization of the testosterone response profile.

ACKNOWLEDGEMENT

The authors would like to express their gratitude to Mrs. Rose Van Brunt for her assistance in the preparation of this manuscript.

REFERENCES

1. Schally AV, Redding TW, Comaru-Schally AM (1984). Potential use of analogs of lutinizing hormone-releasing hormones in the treatment of hormone sensitive neoplasms. Cancer Treat Rep 68(1):281.
2. Redding TW, Chally AV, Tice TR, Meyers WE (1984). Long-acting delivery systems for peptides: Inhibition of rat prostate tumors by controlled release of [D-Trp^6] luteinizing hormone-releasing hormone from injectable microcapsules. Proc Natl Acad Sci 81:5845.
3. Purmur H, Lightman SL, Allen L, Phillips RH, Edwards L, Schally AV (1985). Randomized controlled study of orchiectomy vs long-acting D-Trp-6-LH-RH microcapsules in advanced prostatic carcinoma. Lancet 1201-1205.
4. Schally AV, Comaru-Schally AM, Redding T (1984). Anti-tumor effects of hypothalmic hormones in endocrine-dependent cancers. Proc Soc Exp Biol and Med 175:259.
5. Parmer H, Nicoll J, Stockdale A, Cassoni A, Phillip CH, Lightman SL, Shally AV (1985). Advanced ovarian carcinoma: response to the agonist D-Trp-6-LHRH. Cancer Treat Rep 69(11):1341.
6. Wakman J, Man A, Hendry WF (1985). Importance of early tumor exacerbation in patients treated with long acting analogs of gonadotrophin releasing hormone for advanced prostatic cancer. Br Med J 291:1387.
7. Deghengi R, Misset JL (1984). Disease flare induced by lutinizing hormone-releasing hormone analogs in cancer patients. Lancet 1(8389):1302.
8. Clayton RN (1984). Pharmacologic regulation of pituitary LHRH receptors. In Vickery BH, Nestor Jr. JJ, Hafez ES (eds): LHRH and Its Analogs MA. MTP Press, p 35-46.
9. Hutchinson FG, Furr BJA (1985). Biodegradable polymers for the sustained release of peptides. Biochem Soc Trans 13:520.

The Pharmacology and Toxicology of Proteins, pages 351–367

PRECLINICAL PHARMACOLOGY OF ACTIVATED PROTEIN C

S. C. Emerick[2], H. Murayama[2], S. B. Yan[1], G. L. Long[1], C. S. Harms[1], C. A. Marks[1], L. E. Mattler[1], C. A. Huss[1], P. C. Comp[3], N. L. Esmon[3] C. T. Esmon[3], and N. U. Bang[1]

[1]Lilly Research Laboratories, Eli Lilly and Company, Indianapolis, IN 46202
[2]Indiana University School of Medicine, Indianapolis, IN 46223
[3]Oklahoma Medical Research Foundation, Oklahoma City, OK 73104

ABSTRACT Although it is predicted from in vitro studies that activated human protein C (APC) may be an attractive antithrombotic agent because of its specificity of action (inactivation of factors Va and VIIIa), the antithrombotic properties in macrovascular thrombosis of APC have not been demonstrated in vivo. In this study, we investigated the ability of APC to prevent accretion of radiolabelled fibrinogen (F) to preformed jugular venous thrombi in dogs and Rhesus monkeys. Highly purified plasma-derived human protein C was activated with rabbit thrombomodulin-bovine thrombin (T). Jugular veins were isolated bilaterally in 12 dogs (6 control, 6 APC treated). Segments (5 cm) of all veins were clamped and clots formed in the clamped off segments through the injection of .5 units bovine T. Stenosing ligatures were placed downstream to prevent slippage of clots. Initially, dogs were given AMICAR 200 mg/kg followed by 200 mg/kg/h for the duration of the

[1]Present Address: Lilly Laboratory for Clinical Research, Wishard Memorial Hospital, 1001 West 10th Street, Indpls, IN 46202.

experiments to block fibrinolysis. After release of the clamps, all dogs were given approximately 25 uCi [^{125}I] dog F immediately followed by APC treatment or vehicle infusion. In treated dogs, 84 ug (2 dogs), 126 ug (2 dogs), and 168 ug APC/kg (2 dogs) was administered as an i.v. bolus followed by a constant infusion of 21, 31.5 and 42 ug/kg/h for 2 h. After 2 h infusions, all clots were removed, washed extensively in heparin saline and radiolabelled fibrin quantified. An estimate of bleeding was obtained through weighing of gauze sponges before and after placement in surgical wounds every 30 min. APC infusions resulted in a 3-6 sec prolongation in the APTT with no changes in the prothrombin time, significant decrease in fibrin accretion onto thrombi compared to controls (controls: 4.5±1.1 mg, APC-treated: 1.4±1.1 mg fibrin accreted/thrombus; P (by paired student T test): <0.01. Bleeding from surgical wounds was minimal and no greater in APC treated than control dogs. Human PC antigen by ELISA ranged from .7-3.4 ug/ml. Surprisingly, one hour after discontinuing the infusions a fall of only approximately 20% of circulating antigen was noted. In Rhesus monkey experiments, recombinant human APC was administered in a bolus of 60 60 ug/kg followed by 15 ug/kg/h for 2 h and no AMICAR was administered. The experimental design was otherwise similar to dog studies. Fibrin accreted (ug/thrombus) were controls: 55.4±33.4 (n=8), APC treated: 4.7±6.2 (n=10), p_0.001. In monkeys as in dogs, trivial prolongations in the APTT, no changes in the prothrombin time and no excessive bleeding in APC treated animals were noted. Thus, APC appears to be an attractive antithrombotic agent in dogs and monkeys and does not appear to cause increased blood loss during surgery.

INTRODUCTION

Over the last several years, the importance of the so-called protein C-protein S-thrombomodulin system as a probable major down-regulating force in blood coagulation has been established in *in vitro* experiments. Also, clinical studies screening patients with recurrent thromboembolic disease have identified families in whom heterozygous protein C or protein S deficiency appears to be the underlying cause. In a few patients with homozygous protein C deficiency, the clinical manifestations are those of purpura fulminans in early infancy (1-5). The key component in this system is protein C, a vitamin K-dependent plasma zymogen which when activated into an active serine protease profoundly influences thrombin generation through inactivation of the major cofactors of blood coagulation, clotting factors Va and VIIIa (6-8). In 1960, Mammen et al. (9) described a protein present as an impurity in purified preparations of prothrombin which when exposed to thrombin possessed anticoagulant activity in vitro. These authors named this activity autoprothrombin II-A. In 1976, Stenflo (10) upon separation of the vitamin K-dependent proteins by ion exchange chromatography noticed a peak (peak C) which contained protein immunologically nonidentical to the known vitamin K-dependent procoagulant proteins (factors II, VII, IX and X). The function of this protein now named protein C remained unknown until Seegers et al. (11) demonstrated that protein C was immunologically identical to autoprothrombin II-A previously discovered in his laboratory. The protein proved difficult to activate by physiological activators such as thrombin leading several workers to conclude that protein C was of no major physiological importance. Only the subsequent discovery of the endothelial cell surface glycoprotein, thrombomodulin (12-14), a receptor protein for thrombin and the discovery of the essential cofactor for protein C, protein S (15-16), provided conclusive evidence that protein C is readily converted into activated protein C (APC) as a regular event in human pathophysiology and that APC in consort with protein S supplies what appears to be a crucially important negative feedback loop in the development of coagulant activity in vivo. In this communication, we present preliminary preclinical pharmacology data suggesting that APC is an active antithrombotic agent.

In the discussion section, we summarize briefly theoretical reasons why activated protein C may become an attractive substitute for heparin in certain high risk patients.

MATERIALS AND METHODS

Human protein C was purified from fresh-frozen human plasma as previously described (17). The key step in this purification is immunoaffinity chromotography utilizing a Ca^{++}-dependent monoclonal anti-PC antibody immobilized on Affigel 10 beads.

Recombinant human protein C. The cloning and structure of a 461 amino acid human protein C precursor based on the nucleotide sequence of cloned human liver cDNA's has been reported previously (18).

The expression, purification, and characterization of fully biologically functional human protein C is reported elsewhere (19).

Activation of human protein C (natural sources or recombinant) involved the use of a sedimentable thrombomodulin-thrombin activator in which thrombomodulin is anchored to Affigel-10 beads via an immobilized thrombomodulin monoclonal antibody and equimolar quantities of thrombin attached to the immobilized thrombomodulin (20).

Characterization of activated protein C. Antigen was determined in an ELISA using 96-well microtiter plates coated with monovalent polyclonal PC antibody percaptil of PC or APC, an anti-human PC monoclonal antibody and horseradish peroxidase conjugated anti-murine IgG raised in a rabbit for detection and quantification of the PC or APC antigen.

Amidolytic Activity.

After activation of PC with the sedimentable thrombomodulin-thrombin activator, the amidolytic activity was determined using the tripeptide paranitroanilide synthetic substrates S2238 or S2366 essentially according to the suggestions by the manufacturers (Kabi Vitrum, Stockholm, Sweden). Biological activity utilized the one-stage activated partial thromboplastin assay (see below). Suitable calibration curves using highly purified APC of known potency allow the conversion from seconds prolongation of the APTT to nanogram APC equivalents. Alternatively, the biological, anticoagulant activity of APC was quantified by the prolongation of the factor Xa clotting time (21).

Rabbit thrombomodulin was prepared as previously described (13).

Human thrombin (specific activity appr. 2800 NIH units/mg) was a generous gift from Dr. J. W. Fenton, II (New York State Board of Health Laboratories, Albany, NY).

Bovine thrombin was purchased from Parke-Davis, Morris Plains, NJ.

Dog and monkey fibrinogens were purified from citrated dog or monkey plasma and radiolabeled with ^{125}I as previously described (22).

AMICAR was purchased from Lederle Laboratories, Wayne, NJ.

Heparin was from Eli Lilly and Company, Indianapolis, IN.

EXPERIMENTAL MODELS

Dog Experiments

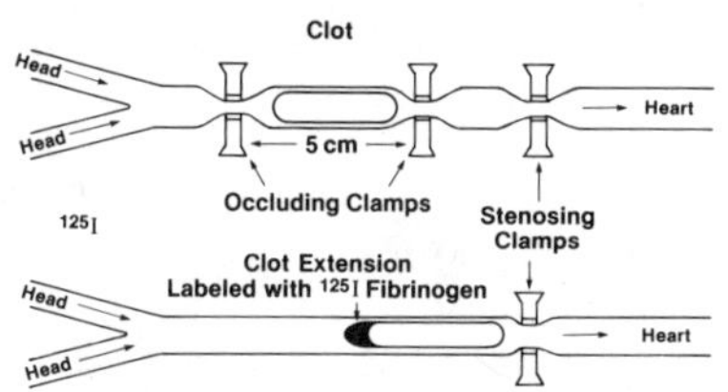

Formation and Retention of Clots in Canine External Jugular Veins

FIGURE 1. Experimental design in dog studies. For details, see text.

Figure 1 presents in schematic form the dog model utilized to assess in vivo APC antithrombotic properties. The model first described in principle by Chiu et al. (23) measures the inhibition of growth of accretion of new fibrin onto preformed thrombi. The external jugular veins were isolated bilaterally in 10-15 kg mongrel dogs anesthetized with Secobarbital Sodium. Segments exactly 5 cm in length of all jugular veins were clamped off with occluding clamps and clots formed in the clamped off segments through the injection of .5 units of bovine thrombin. Stenosing clamps were placed downstream to prevent slippage of the clots. Bilateral femoral cutdowns were established for bloodletting and infusion purposes. Immediately following the release of the occluding clamps (approximately 15 min after thrombin injection) all dogs were given approximately 25 uCi of ^{125}I-labeled dog fibrinogen into one femoral vein and immediately thereafter in treated dogs, 84, 126, or 168 ug/kg was administered as an i.v. bolus followed by a constant infusion of 21, 31.5, and 42 ug/kg/h for 2 h into one femoral vein. Two dogs were used for each of the 3 dosing schedules. Six control dogs were infused with the appropriate APC vehicle. Blood samples were withdrawn from the contralateral femoral vein before, and at 5, 30, 60, 120 and 180 minutes after the start of APC treatment.

After 2 hours infusion, all clots were removed, washed extensively in heparin saline (100 units of heparin/ml saline) following which radiolabeled fibrin was quantified. Conversion from counts/min to mg fibrin accreted per thrombus was made through quantification of plasma fibrinogen (see below) and the specific activity, that is, radiolabel/ml plasma/mg circulating fibrinogen.

An estimate of bleeding was obtained by weighing gauze sponges before and after being left in place in surgical wounds for 30 minutes. This procedure was repeated throughout the experiments to obtain an estimate of cumulative blood loss.

After initial experiments had established that no accretion of radiolabeled fibrin occurred onto control thrombi because of the dogs very active fibrinolytic enzyme system, all dogs were subsequently given E-aminocaproic acid (AMICAR) initially at 200 mg/kg followed by 200 mg/kg/h for the duration of the experiments to block fibrinolysis.

Clotting tests (activated partial thromboplastin times and prothrombin times) utilized one-stage techniques and commercial reagents (Auto APTT reagent and Simplastin automated, General Diagnostics, Morris Plains, NJ). Using fibrometers for clotting time determinations, procedures were those recommended by the manufacturers.

Clotting factors V and VIII were also one-stage procedures utilizing the Auto APTT reagent and human factor deficient plasmas from George King Biomedical, Overland Park, KS.

Fibrinogen was quantified in a chronometric assay using the Data-FI Kit from American Dade, Aguada, Puerto Rico. The procedures were performed on fibrometers in accordance with the instructions by the manufacturers.

Rhesus monkey experiments followed essentially the same protocol as dog experiments with the following modifications. Three to 4 kg Rhesus monkeys were anesthesized with Ketamine-Atropine. Unlabeled clots were formed in the internal jugular veins hanging from a surgical suture rather than being impacted at the site of a stenosing ligature which considerably improved flow conditions. Radiolabeled monkey fibrinogen rather than dog fibrinogen was used. The dose of APC was reduced in the monkeys, the monkeys being more sensitive to human APC than the dog and the standard dose used throughout was 60 ug/kg administered as a bolus followed by a 2 hour infusion of 15 ug/kg/h. Recombinant rather than plasma-derived APC was used and finally, no AMICAR was used since blocking of the fibrinolytic system proved to be unnecessary.

RESULTS

Dog Experiments

Clotting Studies (Data not shown). After the APC bolus at the 3 concentrations given, the APTT prolonged by a mean of approximately 5 seconds with no clearcut dose response. Prolongations of a mean of 1.5 to 3 seconds were maintained for the duration of the constant infusions with no clear dose response.

The prothrombin times did not change in any dog during any time of APC infusion. Clotting factors V and VIII did not change from their very high pretreatment levels of 7-800% at any time during treatment.

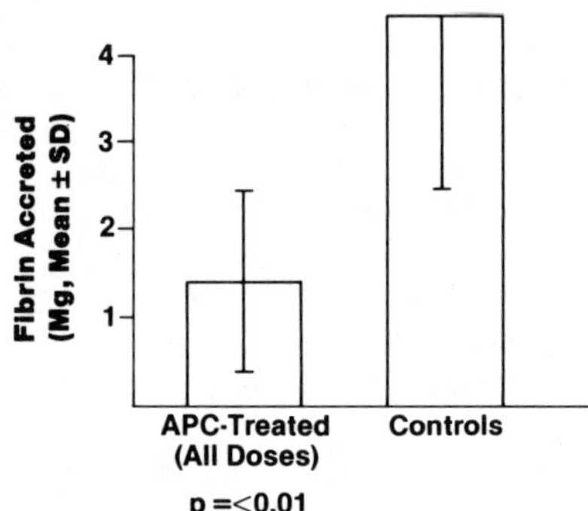

FIGURE 2. Fibrin accreted onto preformed thrombin in 6 APC-treated animals and 6 controls (mg/thrombus), (mg/thrombus, (Mean±1SD).

Fibrin accretion data for 12 veins in 6 APC treated dogs and for 12 veins in 6 control dogs are summarized in figure 2. The mean±1S.D. is 1.42±1.19 mg/thrombus in APC treated dogs and 4.54±1.30 mg/thrombus in control dogs. The difference is statistically significant with a p value of 0.01.

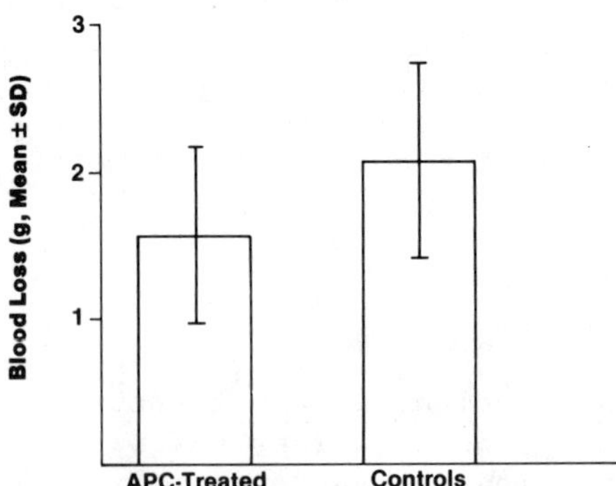

FIGURE 3. Cumulative blood loss in 6 APC-treated dogs and 6 controls (Mean±1SD).

After careful surgical hemostasis had been established initially, bleeding from surgical wounds was minimal and no greater in APC treated than in controls dogs (figure 3).

Repeated measurements of human protein C antigen by ELISA were done in all dogs (Table 1). The administration of a bolus followed by a constant infusion resulted in all cases in sustained elevations of PC antigen. The specificity of the assay, i.e., the ability of the monoclonal antibody used to see only human not dog PC is underscored by the finding of zero levels of PC antigen in control samples drawn before the administration of human APC. Surprisingly, after stopping the infusion at the 3 hour level, a fall of only approximately 20% of antigen was noted. However, we have not completely resolved at this time how much of this antigen is functionally active and how much circulates bound to the APC inhibitor.

TABLE 1
HUMAN APC LEVELS BY ELISA (ng/ml)

Dose								
Bolus (μg/kg)	Constant Infusion (μg/kg/h)	Control	5′	30′	1 h	2 h	3 h
84	21	0	1028	1112	1192	1034	784
84	21	0	1006	944	938	892	686
126	31.5	0	1988	1838	1978	1966	1496
126	31.5	0	2278	1966	2226	2082	1540
168	42	0	2588	2404	2176	1960	1644
168	42	0	3498	3564	3640	3490	3190

Rhesus Monkey Experiments.

Trivial changes in the APTT and no change in the prothrombin times, factor V or factor VIII levels observed during APC infusions into Rhesus monkeys were very similar to those observed in the dogs.

Accretion of ^{125}I fibrin to jugular venous clots expressed as mg/thrombus, mean±1 S.D. were: controls, 55.4±33.4 (N=8 veins in 4 monkeys); APC treated, 4.7±6.2 (N=10 veins in 5 monkeys). These differences are statistically significant with a p value of < 0.001.

Estimates of bleeding in the monkeys were obtained through measurements of the hematocrit which fell from 35.4±2.7% to 27.4±2.9% in the control group and from 35.3±1.7% to 28.3±2.1% in the APC-treated group, not a statistically significant difference. Hematocrit values fell substantially in all animals because of large volumes of blood removed for testing in these 3-4 kg Rhesus monkeys.

DISCUSSION

To understand the efficacy of APC as an antithrombotic agent in the experiments presented here, it is necessary to develop an understanding of two key steps in the complex coagulation reaction resulting in the generation of thrombin and fibrinogen to fibrin conversion.

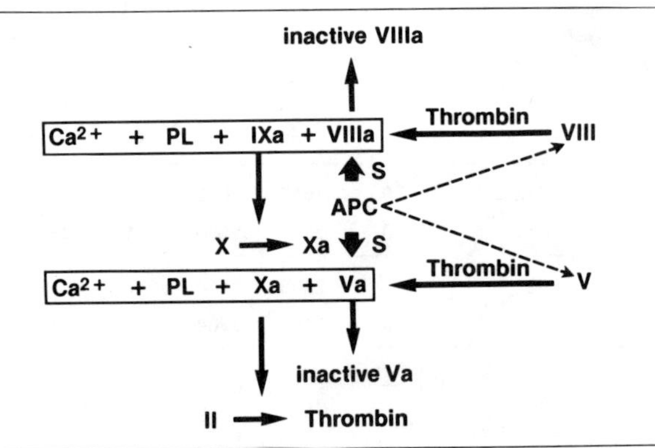

FIGURE 4. APC "on-demand" anticoagulant effect. For details, see text.

Two reactions shown in figure 4 are the conversion of clotting factor X to Xa and the conversion of prothrombin to thrombin. The two reactions are remarkably similar in several respects. They do not occur in the fluid phase, they are interphase phenomena occurring on cell surfaces be it platelets (24-26), monocytes, PMN leukocytes and lymphocytes (27-28) and under certain circumstances, even endothelial cells following stimulation with inflammatory monokines or endotoxin (29-30). Both reactions require cofactors and both reactions require ionized calcium. The key steps in coagulation involve the assembly of activated proteases, zymogen substrates, and cofactors on cell surfaces. Ionized calcium binds to clusters of specialized amino acid, gamma-carboxyglutamic or Gla residues. The binding of calcium changes the confirmation of the Gla-containing N-terminal domains of these proteins to make them readily associate with phospholipid bilayer membranes. The cofactor in the factor X to Xa conversion mediated by clotting factor IXa is clotting factor VIII and the cofactor in the prothrombin to thrombin conversion mediated by clotting factor Xa is clotting factor V. These cofactor proteins, factors VIII and V circulate as inactive precursors but as soon as small quantities of thrombin become become available, the cofactors are activated by thrombin through limited proteolysis. In the absence of activated coafactors VIIIa and Va, the rate constants for X to Xa conversion and for prothrombin to thrombin conversion decrease by about five orders of magnitude (31). It is at the level of the activated cofactors that APC so efficiently down-regulates blood coagulation. The active serine protease APC in assembly with its cofactor, protein S, attaches to cell membranes in close proximity to the clotting factor assemblies and at these sites, demonstrates remarkable specificity in that APC proteolytically degrades and destroys for all intents and purposes only the activated forms of the cofactors Va and VIIIa whereas it only very inefficiently degrades the inactive precursors, clotting factors V and VIII (6-8). Thus, when activated protein C is available on the cell surface, it effectively shuts down the generation of clotting factor Xa and the generation of thrombin. APC can be considered an on-demand anti-coagulant. Since thrombin is essential for the conversion of V to Va and VIII to VIIIa, Va and VIIIa being the vastly preferred substrates for actived protein C, APC will essentially exert its effect only if, when, and where thrombin is being generated. This is in sharp contrast to heparin which blocks all the activated serine proteases in the system and keeps the patient anticoagulated throughout the circulation and at all times whether he needs it or not and in sharp contrast to the oral anticoagulants which also maintains the patient in a constant state of anticoagulation through inhibition of the biosynthesis of the four major vitamin K-dependent clotting serine proteases in the system. Thus, we would expect APC to be at least as effective as heparin in a patient with actively developing thrombosis but would expect also that APC will be far less likely to cause serious bleeding complications that heparin or the oral anticoagulants.

To understand our thinking and strategies for developing protein C as a pharmacological agent, it is important to understand how protein C is activated physiologically. As mentioned, the endothelial cell surface receptor, thrombomodulin, is essential here. Thrombomodulin has no known biological effects in and of itself but once it combines with thrombin, it does something unusual in protein chemistry, it totally changes the spectrum of biological activities of thrombin converting it in effect from a major procoagulant enzyme into a major anticoagulant enzyme. Thrombin itself clots fibrinogen, activates platelets, converts clotting factors V and VIII to Va and VIIIa and also activates protein C at a very slow and ineffective rate. Thrombin, in complex with thrombomodulin on the other hand, does not clot fibrinogen, does not activate platelets, and does not convert V and VIII to their activated counterparts (32). The complex now turns its entire attention towards activating protein C. The rate constant for PC activation by thrombomodulin-thrombin is some 20,000-fold higher than for thrombin alone (12). Simple anatomical considerations strongly suggest that protein C, the zymogen, would be an inefficient antithrombotic agent in major vessel thrombosis. If you consider the surface area in the vascular system or rather, the ratio of endothelial cell surface to plasma volume, you'll notice that the cell surface of major vessels and therefore the availability of thrombomodulin to complex with thrombin and activate protein C is relatively small whereas the surface area in the microcirculation, the sum total of all capillaries in the body is at least 100,000-fold greater than the surface area of the macrocirculation. Thus, in microcirculatory thrombosis such as disseminated intravascular coagulation there would theoretically be ample opportunities to activate protein C and the zymogen could possibly be helpful in the treatment of microvascular thrombotic disorders. In contrast, the thrombomodulin available in major vessels at the site of a developing thrombus in all likelihood is insufficient to activate protein C. For these reasons, we have concluded that protein C preactivated in vitro is a preferred formulation in the treatment of major vessel thrombosis such as venous thromboembolism, coronary artery occlusion and thrombotic strokes.

The preliminary preclinical pharmacology experiments presented here indicate that APC indeed is an effective antithrombotic agent which causes only minor changes in conventional blood clotting tests and which carries a surprisingly low bleeding liability during surgical procedures. Crucial end-points in these studies were the reduction in fibrin accretion to preestablished thrombi and the lack of excess bleeding as measured by fairly crude criteria. We do not place much faith in the clotting time estimates, particularly, the small changes in APTT values could well be misleading. Since there is little, if any, factor Va demonstrable in the general circulation what is measured in the APTT is the inactivation of factor Va generated during the test by whatever free and active APC is present in the sample. However, it is now firmly established that a specific APC inhibitor is present in plasma of mammalian species at concentrations of about 4-5 ug/ml (32). The inhibitor, like antithrombin III, without heparin is relatively slow and inefficient. However, we have performed in vitro experiments showing that when a plasma sample is spiked with APC of known activity, this activity will decrease by about 40% in 30 minutes. Although the clotting assays in the dog experiments were all performed within 30 minutes, they undoubtedly represent an underestimate of APC anticoagulant activity. At the present time, we are uncertain as to how to monitor APC treatment with laboratory tests feeling as we do that conventional clotting tests such as the APTT or the protime useful for monitoring heparin and warfarin therapy will be inadequate, if not totally misleading in the monitoring of APC treatment. With these reservations in mind, our experiments still make us hopeful that APC in certain clinical situations in high risk patients in short-term treatment may be a suitable replacement for heparin because of its thrombus specificity as well as its apparent low liability for serious bleeding complications.

REFERENCES

1. Griffin JH, Evatt B, Zimmerman TS, Kleiss AJ, Wideman C (1981). Deficiency of protein C in congenital thrombotic disease. J Clin Invest 68:1370.
2. Broekmans AW, Veltkamp JJ, Bertina RM (1983). Congenital protein C deficiency and venous thromboembolism: a study of three Dutch families. N. Engl J Med 309:340.
3. Marciniak E, Wilson HD, Marlar RA (1985) Neonatal purpura fulminans: a genetic disorder related to the absence of protein C in blood. Blood 65:15.
4. Schwarz HP, Fischer M, Hopmeier P, Batard MA, Griffin JH (1984) Plasma protein S deficiency in familial thrombotic disease. Blood 1984:64:1297-1300.
5. Comp, P. C., Esmon, C. T. Recurrent venous thromboembolism in patients with a partial deficiency of protein S. N. Engl. J. Med. 1984:311:1525-1528.

6. Walker, F. J., Sexton, P. W., and Esmon, C. T. 1979. Inhibition of blood coagulation by activated protein C through selective inactivation of activated factor V. Biochem. Biophys. Acta. 571:333-342.
7. Suzuki, K., Stenflo, J., Dahlback, B. and Teodorsson, B. 1983. Inactivation of human coagulation factor V by activated protein C. J. Biol. Chem. 258:1914-1920.
8. Fulcher, C. A, Gardiner, J. E., Griffin, J. H. and Zimmerman, T. S. 1984. Proteolytic inactivation of human Factor VIII procoagulant protein by activated protein C and its analogy with Factor V. Blood. 63:486-489.
9. Mammen, E. F., Thomas, W. R., Seegers, W. H. Activation of purified prothrombin to autoprothrombin II (platelet cofactor II or autoprothrombin I-A) Thromb. Diath. Haemorrh 1960:5, 918-49.
10. Stenflo, J. A new vitamin K dependent protein: purification from bovine plasma and preliminary characterization. J. Biol. Chem. 1976:251:355-363.
11. Seegers, W. H., Novoa, E., Henry, R. L., Hassouna, H. I. Relationship of "new" vitamin K-dependent protein C and "old" autoprothrombin II-A. Thromb. Res. 1976:8:543-552.
12. Esmon, C. T., Owen, W. G. Identification of an endothelial cell cofactor for thrombin-catalyzed activation of protein C. Proc. Natl. Acad. Sci. USA 1981:78:2249-2252.
13. Esmon, N L., Owen, W. G., Esmon, C. T. Isolation of membrane-bound cofactor for thrombin-catalyzed activation of protein C. J. Biol. Chem. 1982:257:859-864.
14. Maryuama, I., Salem, H. H., Ishii, H., Majerus, P. W. Human thrombomodulin is not an efficient inhibitor of the procoagulant activity of thrombin. J. Clin. Invest. 1985:75:987-991.
15. Walker FJ (1980). Regulation of activated protein C by a new protein: a possible function for bovine protein S. J Biol Chem 255:5521.

16. Di Scipio RG, Hermodson MA, Yates SG, Davie EW (1977). A comparison of human prothrombin, factor IX (Christmas factor), factor X (Stuart factor), and protein S. Biochemistry 16:698.
17. Moore KL, Andreoli SP, Esmon NL, Esmon CT, Bang NU (1987). Endototoxin enhances tissue factor and suppresses thrombomodulin expression of human vascular endothelium in vitro. J Clin Invest 79:124.
18. Beckmann RJ, Schmidt RJ, Santerre RF, Plutzky J, Crabtree GR, Long GL (1985). Structure and evolution of a 461 aa human protein C precursor and its messenger. RNA based upon the DNA sequence of cloned human liver cDNAs. Nucl Acids Res 13:5233.
19. Yan SB, Grinnell BW (1987, in press). Characterization of fully functional recombinant human protein C expressed from human kidney 293 cells. Fed Proc Abstract No. 1851.
20. D'Angelo SV, Comp, PC, Esmon CT, D'Angelo A (1986). Relationship between protein C antigen and anticoagulant activity during oral anticoagulation and in selected disease states. J Clin Invest 77:416.
21. Yin ET (1974). Effect of heparin on the neutralization of factor Xa and thrombin by plasma a-2 globulin inhibitor. Thromb Diathes Haemorrhag 33:43.
22. Chang ML, Bang NU (1977). Biological behavior of higher molecular weight products of fibrinolysis. J Lab Clin Med 90(1):216.
23. Chiu HM, Hirsh J, Yung WL, et al. (1977). Relationship between the anticoagulant and antithrombotic effects of heparin in experimental venous thrombosis. Blood 49:171.
24. Miletich JP, Jackson CM, Majerus PW (1977). Interaction of coagulation factor Xa with human platelets. Proc Natl Acad Sci USA 74:4033.
25. Tracy PB, Nesheim ME, Mann KG (1981). Coordinate binding of factor Va and factor Xa to the unstimulated platelet. J Biol Chem 256:743.

26. Nesheim ME, Kettner C, Shaw E, et al. (1981). Cofactor dependence factor Xa incorporation in the prothrombinase complex. J Biol Chem 256:6537.
27. Stern DM, Drillings M, Nossel HL, Hurtlet-Jansen A, LaGamma KS, Owen J (1983). Binding of factors IX and IXa to cultured vascular endothelial cells. Proc Natl Acad Sci USA 80:4119.
28. Tracy PB, Eide LL, Mann KG (1985). Human prothrombinase complex assembly and function on isolated peripheral blood cell populations. J Biol Chem 260(4):2119.
29. Tracy PB, Rohrbach MS, Mann KG (1983). Functional prothrombinase complex assembly on isolated monocytes and lymphocytes. J Biol Chem 258(12):7264.
30. Stern DM, Nawroth PP, Kisiel W, Vehar G, Esmon CT (1985). The binding of factor IXa to cultured bovine aortic endothelial cells: Induction of a specific site in the presence of factors VIII and X. J Biol Chem 260:6717.
31. Jackson CM, Nemerson Y (1980). Blood coagulation. Annu Rev Biochem 49:765.
32. Esmon CT, Esmon NL, Harris KW (1982). Complex formation between thrombin and thrombomodulin inhibits both thrombin-catalyzed fibrin formation and factor V activation. J Biol Chem 257:7944.
33. Suzuki K (1984). Activated protein C inhibitor. Semin Thromb Hemost 10:154.

The Pharmacology and Toxicology of Proteins, pages 369–371

NEW CHALLENGES FOR PROTEIN THERAPEUTICS

John S. Holcenberg
Division of Hematology-Oncology
Department of Pediatrics,
University of Southern Califorinia,
Childrens Hospital of Los Angeles
Los Angeles, California 90054

Lymphokines and growth factors are usually released locally following specific stimuli and bind to receptors on effector cells to produce their maximal actions in the immediate environment. Thus our current therapeutic regimens with large systemic dosing are bound to produce excessive and unwanted effects. Our challenge is to target these agents to specific sites and effector cells. Possible approaches include local perfusions, chemical modifications or *in vitro* incubation of the effector cells with the agent. If these agents are administered systemically, combination therapy maybe one way to reduce the dose and still achieve the desired effects.

J. Winkelhake, Cetus Corporation, reviewed their data on combination of lymphokines for treatment of cancers. In a B_{16} melanoma model tumor necrosis factor and IL-2 showed synergistic inhibition of growth of primary tumor and of pulmonary nodules. The timing and order of the agents is critical for optimal effects. Cyclophosphamide was also shown to improve the activity of IL-2 or tumor necrosis factor. Some of this effect may be due to suppression of antigenic response to these foreign proteins or alteration in the ratio of effector to suppress cells. Clinical trials are underway to study these combinations. Similarly, M. Pallodino reported that Genentech has initiated clinical trials to study various combinations of interferon gamma and tumor necrosis factor alpha with anticancer drugs like adriamycin and cisplatin.

Another challenge for protein and peptide therapy is to expand the sites of action. Initial therapeutic uses were directed at cells and soluble factors in circulation. Examples include asparginase to convert asparagine to aspartic acid, clotting factors to replace deficiencies and antibodies to confer passive immunity. Recently, specific cell surface sites have become targets for therapy with peptide hormones, lymphokines, growth factors, monoclonal antibodies and vasoactive peptides. The new challenge is to devise methods to transport active forms of peptides and proteins across the blood-brain barrier and into cells and subcellular compartments to treat genetic diseases.

The complexity of treatment of genetic diseases was reviewed by J. Barranger, Childrens Hospital of Los Angeles. Potentially, therapy for inherited disorders of metabolism includes replacement of the missing protein, organ transplantation, bone marrow transplantation and gene therapy. He illustrated the problems using Gaucher disease, a lysosomal storage disease caused by deficiency of β-glucocerebrosidase, as an example. Solid organ transplants didn't work, because the export of glucosylcermide from storage granules into the circulation and clearance by the transplanted organ was insufficient. Bone marrow transplantation provided normal monocytes that were able to slowly clear this material from circulation and from cells in the bone marrow. Bone marrow progenetors do not enter the brain so the nervous system disorders caused by storage will probably not benefit from these transplants. Similarly, somatic gene therapy with retroviral vectors can introduce the missing enzyme activity into deficient cells in tissue culture but is limited by the distribution of the cells when infused back into the animal. Therapy with native enzyme was unsuccessful. Enzyme modified to expose mannose residues was cleared rapidly by mannose receptors in the liver and the enzyme was concentrated in lysosomes. Despite this targetting, 6 months of treatment in 6 patients did not improve their clinical or biochemical abnormalities. This failure may be due to poor association of the intracellular enzyme with old storage granules. Recent investigations have begun to define leader and stop sequences for translocation of proteins across cell membranes, receptors for translocation to specific intracellular organelles and amino acid sequences that effect the rate of intracellular degradation. This information will aid in the engineering of proteins that will concentrate in the storage granules. Thus, successful treatment of genetic diseases will require greater insites into transport and translocation of proteins and understanding of the biochemistry and molecular biology of the specific disorder.

If technology develops that will allow transport of active proteins and peptides to intracellular sites, many new treatments would be possible. These include intracellular growth factors, detoxifying enzymes, antibodies to specific proteins and nuclear regulatory proteins. For example, nuclear proteins have been shown to regulate the expression of many mammalian and viral genes by specific binding to short segments of DNA. These proteins could be a new therapy to alter gene expression in certain cells and to devise ways to target other drugs to specific segments of DNA.

Pharmacologic applications of proteins and peptides have expanded phenomenonly in the last few years. We look forward to exciting future developments.

Index